SPECTROSCOPY IN INORGANIC CHEMISTRY

Volume I

CONTRIBUTORS

C. J. BALLHAUSEN

WILLIAM L. BAUN

A. CHAKRAVORTY

K. G. DAS

D. R. EATON

DAVID W. FISCHER

J. W. HASTIE

R. H. HAUGE

J. L. MARGRAVE

H. B. MATHUR

A. S. N. MURTHY

C. N. R. RAO

HARRY D. SCHULTZ

SPECTROSCOPY IN INORGANIC CHEMISTRY

Edited by C. N. R. Rao and John R. Ferraro

Department of Chemistry
Indian Institute of Technology
Kanpur, India

Chemistry Division
Argonne National Laboratory
Argonne, Illinois

VOLUME I

 1970

ACADEMIC PRESS   New York and London

ACADEMIC PRESS, INC.
111 Fifth Avenue, New York, New York 10003

United Kingdom Edition published by
ACADEMIC PRESS, INC. (LONDON) LTD.
Berkeley Square House, London W1X 6BA

LIBRARY OF CONGRESS CATALOG CARD NUMBER: 77-117102

PRINTED IN THE UNITED STATES OF AMERICA

CONTENTS

The Measurement and Interpretation of Transition Ion Crystal Spectra

C. J. BALLHAUSEN

Inorganic Photochemistry

D. R. EATON

v

Matrix Isolation Spectroscopy

J. W. Hastie, R. H. Hauge, and J. L. Margrave

Spectroscopy of Donor–Acceptor Systems

C. N. R. Rao and A. S. N. Murthy

Mass Spectroscopy

K. G. Das

Soft X-Ray Spectroscopy as Related to Inorganic Chemistry

William L. Baun and David W. Fischer

High-Resolution Nuclear Magnetic Resonance

A. CHAKRAVORTY

Nuclear Quadrupole Resonance

HARRY D. SCHULTZ

Mössbauer Spectroscopy and Its Applications to Inorganic Chemistry

H. B. MATHUR

LIST OF CONTRIBUTORS

Numbers in parentheses indicate the pages on which the authors' contributions begin.

C. J. BALLHAUSEN (1), Department of Chemistry, Louisiana State University, Baton Rouge, Louisiana

WILLIAM L. BAUN (209), Air Force Materials Laboratory (Maya), Wright-Patterson Air Force Base, Ohio

A. CHAKRAVORTY (247), Department of Chemistry, Indian Institute of Technology, Kanpur, India

K. G. DAS (173), National Chemical Laboratory, Poona, India

D. R. EATON (29), Department of Chemistry, McMaster University, Hamilton, Ontario, Canada

DAVID W. FISCHER (209), Air Force Materials Laboratory (Maya), Wright-Patterson Air Force Base, Ohio

J. W. HASTIE* (57), Department of Chemistry, Rice University, Houston, Texas

R. H. HAUGE (57), Department of Chemistry, Rice University, Houston, Texas

J. L. MARGRAVE (57), Department of Chemistry, Rice University, Houston, Texas

H. B. MATHUR (347), National Chemical Laboratory, Poona, India

* Present address: Inorganic Materials Division, National Bureau of Standards, Washington, D.C.

A. S. N. MURTHY* (107), Department of Chemistry, Indian Institute of Technology, Kanpur, India

C. N. R. RAO (107), Department of Chemistry, Indian Institute of Technology, Kanpur, India

HARRY D. SCHULTZ (301), U. S. Department of the Interior, Bureau of Mines, Morgantown Coal Research Center, Morgantown, West Virginia

* Present address: Department of Chemistry, Indian Institute of Technology, New Delhi, India.

PREFACE

Although the applications of spectroscopy in inorganic chemistry were not as widespread as in organic chemistry until recently, the situation has changed remarkably in the last few years. Spectroscopy has become an indispensable tool to present-day inorganic chemists. This is mainly because of the advent of newer and improved techniques as well as the easy availability of sophisticated commercial instrumentation. Some of these recent innovations are the topics of several chapters in this book.

There have been several reviews and books on the well-known spectroscopic methods and their applications. Textbooks covering some of the physical methods have also appeared. We felt that it was desirable to have a collection of articles on some of the recent spectroscopic methods as well as accounts of research in some of the specialized and novel areas of inorganic chemistry in which spectroscopy has been used effectively. We have omitted discussions of the well-known aspects of spectroscopy of inorganics such as electronic and vibrational spectroscopy and their applications to coordination chemistry, since several excellent reviews are presently available. Lengthy discussions on instrumentation and experimental techniques have also been avoided. The articles are by no means complete compilations; they essentially attempt to present status reports and potentialities in the fields of study. It is hoped that this volume will be of use to chemists and spectroscopists interested in inorganic systems, and that the individual articles will assist inorganic chemists in effectively using some of the spectroscopic methods in their research. This is particularly important since one may have to employ more than one method to solve a structural problem in inorganic chemistry. The articles on newer areas could possibly encourage further research. We would feel highly rewarded if teachers and students of advanced inorganic chemistry courses find some useful material in this book.

For convenience, this treatise has been divided into two volumes. We have not adopted any specific subject classification in arranging the chapters in the two volumes, which are scheduled to appear within a short interval.

We are most thankful to the authors and to the publishers for their enthusiastic cooperation in publishing these volumes.

CONTENTS OF VOLUME II (*tentative*)

THE MEASUREMENT AND INTERPRETATION OF TRANSITION METAL ION CRYSTAL SPECTRA

C. J. Ballhausen

DEPARTMENT OF CHEMISTRY
LOUISIANA STATE UNIVERSITY
BATON ROUGE, LOUISIANA

I. Introduction

The recording and interpretation of electronic molecular spectra of transition-metal compounds started in earnest about 1950. Thanks to the easy operation of recording spectrophotometers and the relative ease with which crystal field level diagrams can be constructed, the number of papers dealing with absorption spectra and their interpretations is by now very large indeed. Most of these spectra have been obtained from solutions. However, more and more it has been realized that the spectra of crystals containing transition-metal ions are of greater value than solution spectra. In fact, a recent review paper (*32*), which dealt solely with crystal spectra and which did not aim at completeness, contained 215 pages and 460 references! The total number of references to solution spectra would now probably run to 20,000. To review such a gross number of papers is probably not possible and certainly not desirable.

Solution spectra are usually very broad and featureless, and band identifications hinge solely on the fitting of one or two absorption bands to a theoretical level diagram. All of the finer spectroscopic detail that is so essential to proper spectroscopic assignments is simply lost in the liquid phase. Indeed, one is tempted to dismiss the recording of solution spectra for all other than analytical purposes. In a serious modern discussion of excited electronic structures, these broad bands cannot provide much information.

The advantages of crystal spectra are as follows:

(1) The crystal geometry is known, or can be elicited using x-rays. One therefore knows the species with which one is dealing. The exact geometry of the species is, of course, of great importance to any interpretation.

(2) If one knows the orientation of the molecule in the crystal, one can use the powerful techniques of polarized-light spectroscopy in a meaningful way. This may often give information that greatly helps in the identification of excited states.

(3) The crystals can be cooled down, thereby often resolving vibrational structure in the absorption bands. This is of great help in the identification of the excited states.

(4) If the spectroscopic "lines" are sufficiently sharp, they can be subjected to Zeeman techniques.

(5) Polarized fluorescence and/or phosphorescence can be studied.

(6) Studies concerned with intermolecular coupling, energy transfer, and excited-state stabilizations are more readily carried out.

In the present chapter, we shall try to cover some of the general principles and experiments that have been used in the recording and interpretation of crystal spectra of molecules that contain transition-metal ions. In view of the extensive coverage up to about 1967 of d^n systems provided by the Hush–Hobbs paper (*32*), we will exemplify all discussions of the present work using, primarily, data from the literature of the last two years. No attempt will be made to discuss the electronic spectra of compounds containing rare earths.

II. Instrumentation and Technique

The handling of either pure or diluted crystals usually does not present any difficulties in absorption spectroscopy, However, the growth of sufficiently large crystals often presents difficulties; indeed, the growing of good crystals, whether pure or host crystals containing "impurities," is still somewhat of an art. To go into any details in this topic is, however, inadvisable, and the reader will have to consult the special literature dealing with crystal growth when and if he runs into trouble. It is usually feasible to grow the molecular crystals we shall deal with from aqueous solutions. However, some patience may be required!

It is usually preferable to use dilute crystals. It is quite impossible, in many instances, to grow sufficiently thin pure crystals possessing any transmittance in those regions where there are strongly allowed absorption bands. Furthermore, crystals function as effective amplifiers for impurity emissions, and it is not unusual for all the luminescence of a so-called pure crystal to originate in impurity centers present at the parts-per-million level (43). In view of this, one introduces the material one wishes to study as a "guest impurity" into some host lattice, and takes advantage of the amplification effect mentioned above. As a recent example of work using diluted crystals, we quote the investigations of $IrCl_6^{2-}$ in either a $(CH_3NH_3)_2SnCl_6$ host crystal (13, 39) or a K_2SnCl_6 host crystal (49). Many more examples, ranging from the classic example of a trivalent metal ion substituting for aluminum in Al_2O_3 to a trivalent metal ion in a $C(NH_2)_3Al(SO_4)_26H_2O$ (guanidine aluminum sulfate hexahydrate) host lattice may be found in Hush and Hobbs (32).

In order to enhance crystal transmittance as much as possible, one should take great trouble to polish the crystal faces. It is well worth the effort. In many cases a good polishing agent is a drop of water on a piece of good quality smooth pad paper. Sometimes the crystals are coated with glue or grease in order to keep them from corroding. Extreme caution must, of course, be exercised that this does not introduce undesirable absorption or emission properties.

When polished, the crystals are oriented under the polarizing microscope. The oriented small crystals are then mounted in a hole drilled through a small cylindrical copper block, across the two parallel legs of a "hairpin," or on a pinhole in a small copper plate. Rubber cement, or vacuum grease is recommended for mounting. The copper block is inserted in a cylindrical hole drilled through a cold copper finger. The finger can be cooled to the desired temperature, usually that of either liquid nitrogen (77°K), liquid hydrogen (20°K), or liquid helium (4°K). The hairpin or the copper plate can be attached to a long, thin stainless steel rod for direct immersion in the refrigerant. More than one sample can, of course, be mounted on the hairpin or copper plate. Care should be taken in mounting the crystals so as not to introduce strain by cooling, since this may lead to band shifts and splitting of degeneracies.

Care has also to be taken that the thermal isolation is good and that the sample is, in fact, at the desired temperature. Pumping on the liquid helium will bring the temperature down to 2.186°K, where liquid He(II) is formed. Helium(II) has a heat conductivity that is several thousand times greater than that of copper. Under these conditions the sample temperature is well defined; the radiant heat provided by the light source is dissipated rapidly, and the sample remains at the low temperature. In the liquid He(I) phase, however, it is not at all difficult to generate sample temperatures that are 10° above ambient.

The best Dewars seem to be glass ones. One can here use either a conventional Dewar or one where cold helium gas flows from a storage can past the mounted crystal. Metal Dewars are, by bitter experience, more likely to give trouble, especially in the vacuum part.

Usually one uses both spectrographic and spectrophotometric methods for the recording of spectra. Particularly when there are large and rapid intensity changes in a line structure, a resolving power of about 5000 (found in most recording spectrophotometers) is not sufficient to yield a reliable spectrum with all the details present (24).

If one wishes to perform polarization experiments, it is necessary to know the polarization characteristics of one's instruments. Some polarization of light occurs during every reflection event and during many absorbing events, and since such events occur in many places during the transit of light from source to detector, it follows that the dictum KNOW YOUR EQUIPMENT CHARACTERISTICS must obtain. An experimental polarization result that remains uncorrected for instrument vagary is, to put it bluntly, not a result of value.

Since light intensities may be quite low, considerable attention must be devoted to the alignment of the optical system. A good alignment may mean the difference between a spectrum and no spectrum. It is usually a good test of the alignment to shine light in " backwards " and see whether it hits the crystal.

A polarized spectrum is, as is well known, most easy to interpret in the cases where we have a uniaxial crystal. The crystals, therefore, must be either trigonal, tetragonal, hexagonal, or rather trivial cubic. One may distinguish (40, 47), in these cases, between intensity induction by either magnetic-dipole or electric-dipole mechanisms:

One measures the axial or *a*-spectrum when the unpolarized light is propagated along the unique axis **C**; in this instance, of course, both the electric vector **E** and the magnetic vector **H** are perpendicular to the optic axis. One next performs a measurement with plane-polarized light that is propagating perpendicular to **C**; two spectra may be obtained: The σ spectrum for which **E** $\perp$ **C** and **H** $\parallel$ **C**, and the π spectrum, for which **E** $\parallel$ **C** and **H** $\perp$ **C**. If the axial spectrum agrees with the σ spectrum, then we have an electric-dipole allowed transition. Correspondingly, if the axial and π spectra coincide, we have a magnetic-dipole transition.

III. The Born–Oppenheimer Approximation

The total nonrelativistic Hamiltonian for a molecular system of N nuclei and n electrons is (*13, 52*)

$$H = -\sum_{v=1}^{N} \frac{\hbar^2}{2M_v} \nabla_v^2 - \sum_{i=1}^{n} \frac{\hbar^2}{2m} \nabla_i^2 + V(\mathbf{Q}, \mathbf{r}) \tag{1}$$

The coordinates and masses of the nucleus v are designated by $(\mathbf{Q}_v, M_v)$ and the coordinates and masses for the electrons by $(\mathbf{r}_i, m)$. Let us now suppose that we have solved the "electronic" Schrödinger equation

$$\left[-\sum_{i=1}^{n} \frac{\hbar^2}{2m} \nabla_i^2 + V(\mathbf{Q}, \mathbf{r}) \right] \psi_n(\mathbf{Q}, \mathbf{r}) = W_n(\mathbf{Q}) \psi_n(\mathbf{Q}, \mathbf{r}) \tag{2}$$

for all values of **Q**. For each nuclear configuration we have, in other words, a complete set of electronic wavefunctions. In principle, therefore, we can obtain a solution to the complete Hamiltonian (1) by taking

$$\psi(\mathbf{Q}, \mathbf{r}) = \sum_n \chi_n(\mathbf{Q}) \psi_n(\mathbf{Q}, \mathbf{r}) \tag{3}$$

where the expansion coefficients $\chi_n(\mathbf{Q})$ are functions of the nuclear coordinates only. This method for the solution of the total wave equation is called the Born–Oppenheimer separation, since it leads to two separate equations, one for the electronic motion, and one for the nuclear motion.

Using the expansion (3) together with the full Hamiltonian (1) will lead to a set of coupled differential equations that determines the coefficients in the expansion, where the different coefficients $\chi_m(\mathbf{Q})$ and $\chi_n(\mathbf{Q})$ are coupled by virtue of the kinetic nuclear term in Eq. (1). The coupling operators depend, therefore, on the first and second derivatives of the electronic wave functions with respect to the nuclear coordinates.

Typical coupling terms are of the form

$$\sum_{v=1}^{N} \frac{\hbar^2}{M_v} \nabla_v \chi_n(\mathbf{Q}) \, \nabla_v \psi_n(\mathbf{Q}, \mathbf{r}) \tag{4}$$

and

$$\sum_{v=1}^{N} \frac{\hbar^2}{2M_v} \chi_n(\mathbf{Q}) \, \nabla_v^2 \psi_n(\mathbf{Q}, \mathbf{r}) \tag{5}$$

Since the electrons and the nuclei in a molecule experience the same coulombic forces, summation (5) is to good approximation given as

$$\sum_{v=1}^{N} \frac{\hbar^2}{2M_v} \chi_n(\mathbf{Q}) \, \frac{2m}{\hbar^2} \frac{\hbar^2}{2m} \nabla_i^2 \psi_n(\mathbf{Q}, \mathbf{r})$$

and we notice, therefore, that summation (5) is of the order of magnitude m/M_v smaller than the electronic kinetic energy. It may consequently safely be neglected.

The terms (4), containing the first derivative of $\psi_n(\mathbf{Q}, \mathbf{r})$ with respect to $\mathbf{Q}$, are not necessarily negligible. However, their influence will be small if the energy separation between $\psi_n(\mathbf{Q}, \mathbf{r})$ and $\psi_m(\mathbf{Q}, \mathbf{r})$ is large. To a first approximation, they may therefore also be neglected for nondegenerate wavefunctions. However, in cases of electronic degeneracies or near degeneracies the terms (4) assume importance. This is the case of the Jahn–Teller effect (*30, 56*) and we shall also see later that (4) is responsible for introducing intensity into electronically "forbidden" transitions.

Neglect of all the off-diagonal coupling operators will reduce the expansion (3) to a single term

$$\psi(\mathbf{Q}, \mathbf{r}) = \chi_n(\mathbf{Q})\psi_n(\mathbf{Q}, \mathbf{r}) \tag{6}$$

This approximation is called the adiabatic approximation.

Taking Eq. (6) together with Eq. (1) then leads to the equations

$$\left[-\sum_{v=1}^{N} \frac{\hbar^2}{2M_v} \nabla_v^2 + W_n(\mathbf{Q}) \right] \chi_n(\mathbf{Q}) = W\chi_n(\mathbf{Q}) \tag{7}$$

and

$$\left[-\sum_{i=1}^{n} \frac{\hbar^2}{2m} \nabla_i^2 + V(\mathbf{Q}, \mathbf{r}) \right] \psi_n(\mathbf{Q}, \mathbf{r}) = W_n(\mathbf{Q})\psi_n(\mathbf{Q}, \mathbf{r}) \tag{8}$$

W is here the total energy of the system.

In the simple case where we have no electronic degeneracy we can therefore consider the electronic energy $W_n(\mathbf{Q})$ to act as the potential energy for the nuclear motion, and $W_n(\mathbf{Q})$ is usually called the potential surface associated with a certain electronic state. Such a point of view, however, is meaningful only in the adiabatic approximation.

Provided that the electronic states are well separated, the Born–Oppenheimer approximation as expressed in Eqs. (7) and (8) is therefore quite good.

However, we observe that when the potential-energy surfaces belonging to functions of the same symmetry attempt to cross one another, we must expect the adiabatic approximation to break down (30).

The nuclear function $\chi_n(\mathbf{Q})$ can, as is well known (30, 63), be separated into products of rotational and vibrational wavefunctions. Since the rotational quanta are, at most, of the order of a fraction of a wave number for complexes of the transition-metal ions, we shall completely neglect these and take $\chi_n(\mathbf{Q})$ to be a product of the $(3N - 6)$ solutions to the harmonic oscillator, where we have expanded the potential $W_n(\mathbf{Q})$ in terms of the displacement coordinates ξ_l around the equilibrium configuration of the molecule (7, 30).

$$\chi_n{}^i = \prod_{l=1}^{3N-6} \chi_{v_l}^i(\xi_l) \tag{9}$$

with eigenvalues

$$\epsilon_i = \sum_{l=1}^{3N-6} (v_l + \tfrac{1}{2})h\nu_l \tag{10}$$

The indices i serve to remind us that each potential surface has a set of distinct vibrational functions associated with it.

For an electronic transition to take place between a lower state characterized by a wavefunction $\psi''(\mathbf{Q}, \mathbf{r})$ and a higher state $\psi'(\mathbf{Q}, \mathbf{r})$ the transition moment

$$\mathbf{M} = \int \psi'(\mathbf{Q}, \mathbf{r})(\mathbf{M}_Q + \mathbf{M}_n)\psi''(\mathbf{Q}, \mathbf{r}) \, d\tau_Q \, d\tau_n \tag{11}$$

must be different from zero. Here $\mathbf{M}_Q + \mathbf{M}_n$ is the dipole moment vector resolved into two parts, of which $\mathbf{M}_Q$ is due to the nuclei, $\mathbf{M}_n$ to the electrons. Using the adiabatic approximation (4) we get, because $\psi'(\mathbf{Q}, \mathbf{r})$ and $\psi''(\mathbf{Q}, \mathbf{r})$ are orthogonal to each other for the same value of $\mathbf{Q}$,

$$\mathbf{M} = \int \prod_{l=1}^{3N-6} \chi_{v_l}' (\xi_l) \prod_{l=1}^{3N-6} \chi_{v_l}''(\xi_l) \, d\xi_1 \cdots d\xi_{3N-6} \int \psi'(\mathbf{Q}, \mathbf{r})\mathbf{M}_n \psi'' (\mathbf{Q}, \mathbf{r}) \, d\tau_n \tag{12}$$

Since the electronic wavefunctions vary but slightly with $\mathbf{Q}$, one usually evaluates the electronic integral at the equilibrium nuclear configuration $\mathbf{Q}_0$. The transition probability that is proportional to $|\mathbf{M}|^2$ is therefore made up of factors that depend upon the nuclear motions alone, and one factor that depends upon the electronic motion alone. The squares of the vibrational overlap integrals are often referred to as the Franck–Condon factors.

The operator $\mathbf{M}_n$ is usually either an electric-dipole or magnetic-dipole type of term (30). The electric-dipole moment vector can be written $\sum e\mathbf{r}_i$, and it transforms as (X, Y, Z) in the point group of the molecule. The magnetic-dipole moment vector is given by $\sum (e/2mc)\mathbf{r}_i \times \mathbf{p}_i$, where $\mathbf{p}_i$ is the linear momentum of electron i. This term transforms like (L_x, L_y, L_z) in the point group of the molecule, where L_x, L_y, and L_z refer to the three components of the angular-momentum operator $\mathbf{L}$. The electronic selection rule, when both the upper and lower potential surfaces have the same conformation, is that the product $\psi'(\mathbf{r})\mathbf{M}_n\psi''(\mathbf{r})$ should contain at least one component that transforms as a totally symmetric representation in the point group of the molecule. However, if the two electronic states have different equilibrium conformations, only those symmetry elements that are common to both point groups should be considered (*30*).

IV. Electronically Allowed Transitions

In order to calculate the band shapes, we shall first deal with the case for which $\int \psi'(\mathbf{Q}_0, \mathbf{r})\mathbf{M}_n\psi''(\mathbf{Q}_0, \mathbf{r}) \, d\tau_n$ is different from zero; we say that we have an electronically allowed transition. It may be of either an electric- or magnetic-dipole type transition. Let us first treat the case where the molecule retains its shape in the excited states. If we plot the adiabatic potential curves, states with different electronic configurations will then have potential minima at different bond lengths. When the complex undergoes an electronic transition between states with different equilibrium distances, it gives rise to changes in the vibrational quantum numbers. We shall now calculate the Franck–Condon factors for such a transition, under the assumption that we are exciting only one totally symmetric vibration. In such a vibration all of the vibrational wavefunctions are totally symmetric and the vibrational quantum number may change by any number of quanta during the electronic transition.

We assume that the transition starts out from the zero vibrational level of the ground state. Using $\Delta\xi$ to denote the bond-length difference between the equilibrium conformations of the two electronic states A and B, we have in an obvious notation

$$\frac{I_v}{I_0} = \frac{I(A_0 \to B_v)}{I(A_0 \to B_0)} = \left[\frac{\int \chi_0(\xi, \beta'')\chi_v(\xi + \Delta\xi, \beta') \, d\xi}{\int \chi_0(\xi, \beta'')\chi_0(\xi + \Delta\xi, \beta') \, d\xi}\right]^2 \tag{13}$$

$$\beta'' = \frac{4\pi^2 v'' cM}{h} \qquad \beta' = \frac{4\pi^2 v' cM}{h}$$

The two values v'' and v' reflect that the vibrational frequency need not be the same for the two potential surfaces.

By expanding (*2*) $\chi_v(\xi + \Delta\xi, \beta')$ the ratio I_v/I_0 can be calculated in closed form. The additional assumption of $v'' = v'$ leads to the simple expression

$$I_v = [\tfrac{1}{2}\beta(\Delta\xi)^2]^v \frac{1}{v!} I_0 \tag{14}$$

Clearly the intensity of the lines goes through a maximum, which for reasonably large values of v can be seen to occur at $v \cong \frac{1}{2}\beta(\Delta\xi)^2$.

In the electronic ground state, the totally symmetric zero-point vibration is governed by the wavefunction $\chi_0(\xi, \beta)$. The magnitude of displacement from the equilibrium position, $\bar{\xi}$, is given by

$$\bar{\xi} = \left[\int \chi_0(\xi, \beta)\xi^2\chi_0(\xi, \beta)\, d\xi \right]^{1/2}$$

or $\bar{\xi} = (1/2\beta)^{1/2}$. In the case where $\Delta\xi$ is large compared with $\bar{\xi}$, the electronic transition will terminate on a high vibrational quantum number v in the excited state. We can then expand I_v around the maximum value of $v \cong \frac{1}{2}\beta(\Delta\xi)^2$, and get with Δv being zero in the maximum

$$I = I_{\mathrm{max}} \exp\left[-\frac{(\Delta v)^2}{\beta(\Delta\xi)^2} \right] \tag{15}$$

Evidently the absorption band will have the shape of a Gaussian curve (*36*).

The above derivation of the band shape is valid only for a band profile consisting of excitations of one totally symmetric mode in an electronically allowed transition. However, the adiabatic potential depends on all the normal modes. These also will contribute to the bandwidth; however, their influence will generally be much smaller than that of the totally symmetric modes.

Provided the molecule retains its shape during excitation, the nontotally symmetric frequencies can only be associated with vibrational quantum-number changes of 0, 2, 4, ... This follows from the fact that none of the odd Hermite functions transform as totally symmetric functions. From our assumption that the molecule is not distorted in the excited state using a nontotally symmetric vibrational coordinate, it follows that the only way a 0–2 line, for example, can get intensity is for the vibrational frequency associated with the nontotally symmetric mode to be different on the two potential surfaces. It is then easy to show (*52*) that the ratio of intensity of the $0'' \to 0'$ line to the sum of all the line intensities $0'' \to 2v'$ is given by

$$\frac{I_{0''\to 0'}}{\sum_{v=0}^{\infty} I_{0''\to 2v'}} = \frac{\sqrt{v'v''}}{\frac{1}{2}(v' + v'')} \tag{16}$$

which, with v' being not too different from v'', is seen to collect virtually all of the intensity in the 0–0 line. Consequently, all we expect to see is a progression in the totally symmetric mode(s), which may or may not produce a broad, featureless band.

A molecule that is electronically excited does, in many cases, possess an equilibrium conformation different from that of the ground state. Evidently there will be a distorting nontotally symmetric mode that will carry the molecule to the new conformation. This mode will therefore, in the *new* equilibrium conformation, transform as a totally symmetric vibration, and the "distorting" vibration will behave in the same way as a "normal" totally symmetric vibration. We expect, therefore, to see a progression in single quanta of the distorting mode.

Thus the observation of a vibrational progression in an electronic spectrum provides information on the relative geometric shapes of the two electronic states involved. It requires only that we be able to identify the occuring vibration, and this is usually done by a comparison with the vibrational frequencies found by an infrared or Raman analysis of the ground state. It is seen that in practice this means that the vibrational frequencies in the ground and excited states must not be too different, since otherwise we may not be able to make an identification. However, if possible, the measurements of isotopic ratios in the vibrational frequencies are always of great value for the correct identification of vibrations.

V. Electronically "Forbidden" Transitions

We shall now treat the case where $\int \psi'(\mathbf{Q}_0, \mathbf{r}) \mathbf{M}_n \psi''(\mathbf{Q}_0, \mathbf{r}) \, d\tau$ is equal to zero for symmetry reasons. Let us assume that we have three "electronic" states: $\psi''(\mathbf{Q}_0, \mathbf{r})$ with energy $W_0(\mathbf{Q}_0)$, $\psi'(\mathbf{Q}_0, \mathbf{r})$ having an energy of $W_1(\mathbf{Q}_0)$, and $\psi^{ex}(\mathbf{Q}_0, \mathbf{r})$ with energy $W_{ex}(\mathbf{Q}_0)$. We take $W_0 < W_1 < W_{ex}$, and with each electronic function we associate a product of two vibrational functions. In the first function $\chi(\xi_1)$, the vibrational coordinate ξ_1 is totally symmetric; ξ_2 is a nontotally symmetric coordinate in the second function $\chi(\xi_2)$.

As already mentioned, the term (4) in the full Hamiltonian can now be used to couple ψ' and ψ^{ex}. Asxsuming for simplicity of notation that we can reduce the summation over the gradients of the nuclei to one term in the nontotally symmetric vibrational coordinate ξ_2, we get, using first-order perturbation theory,

$$\Psi' = \psi' \chi'_{v_1}(\xi_1) \chi'_{v_2}(\xi_2)$$

$$+ \frac{\int \psi' \nabla_2 \psi^{ex} \, d\tau \int \chi'_{v_1}(\xi_1) \chi^{ex}_{v_1}(\xi_1) \, d\xi_1 \int \chi'_{v_2}(\xi_2) \nabla_2 \chi^{ex}_{v_2}(\xi_2) \, d\xi_2}{W_1 - W_{ex}}$$

$$\times \psi^{ex} \chi^{ex}_{v_1}(\xi_1) \chi^{ex}_{v_2}(\xi_2) \quad (17)$$

We assume now that $\int \psi''(\mathbf{Q}_0, \mathbf{r}) \mathbf{M}_n \psi^{ex}(\mathbf{Q}_0, \mathbf{r}) \, d\tau_n$ is different from zero. Because of the mixing of the electronic wavefunctions the "forbidden" electronic $\psi'' \rightarrow \psi'$ transition can then "steal" intensity from the allowed transition $\psi'' \rightarrow \psi^{ex}$.

The vibrational integrals that will occur in the expression for the "stolen" band intensity associated with the transition $\psi'' \to \psi'$ are therefore

$$\int \chi_{v_1}''(\xi_1)\chi_{v_1}^{ex}(\xi_1)\, d\xi_1 \int \chi_{v_2}''(\xi_2)\chi_{v_2}^{ex}\, d\xi_2 \int \chi_{v_1}'(\xi_1)\chi_{v_1}^{ex}(\xi_1)\, d\xi_1 \int \chi_{v_2}'(\xi_2)\nabla_2\,\chi_{v_2}^{ex}(\xi_2)\, d\xi_2$$

From the second of these integrals we get $v_2'' = v_2^{ex}$ if we are to have a value of the integral different from zero, and from the last integral $v_2' \pm 1 = v_2^{ex}$. In other words, $v_2'' = v_2' \pm 1$, and we must therefore excite one quantum of the nontotally symmetric vibration in order to get a transition probability different from zero. The totally symmetric vibration may, on the other hand, change by any number of quanta.

We will therefore not see the $(0'', 0'', \ldots, 0'') \to (0', 0', \ldots, 0')$ line; the first observed line will correspond to an excitation of a nontotally symmetric vibration on which is built a progression in the totally symmetric vibration. If more than one nontotally symmetric vibration can introduce intensity into an otherwise forbidden transition, we will have as many "false" origins, followed by progressions in the totally symmetric vibration(s), as we have perturbing vibrations.

From a group-theoretical point of view ∇_{ξ_2} transforms as the vibrational coordinate ξ_2. In order for the electronic integral occurring in Eq. (17) to be different from zero, we must therefore have

$$\Gamma(\psi_n')\Gamma(\xi_2) = \Gamma(\psi_n^{ex}) \tag{18}$$

where $\Gamma(\psi_n')$, $\Gamma(\psi_n^{ex})$, and $\Gamma(\xi_2)$ stand for the representations of ψ_n', ψ_n^{ex}, and ξ_2 in the point group of the molecule. Since we assumed that the transition $\psi_n'' \to \psi_n^{ex}$ was an allowed electric dipole transition, this leads to

$$\Gamma(\psi_m'')\Gamma(X, Y, Z) = \Gamma(\psi_n^{ex}) \tag{19}$$

The requirements (18) and (19) lead to the selection rule for the vibronic transition: the direct product of the functions $\Gamma(\psi_n'')\Gamma(X, Y, Z)\Gamma(\xi_2)\Gamma(\psi_n')$ must transform as the totally symmetric representation of the molecular point group. With $\Gamma(\text{vib})$ representing any perturbing nontotally symmetric vibration, we therefore have the general vibronic selection rule

$$\Gamma(\psi_n'')\Gamma(X, Y, Z)\Gamma(\text{vib})\Gamma(\psi_n') = \Gamma_1 \tag{20}$$

VI. Experimental Determination of the Nature of Absorbtion

Two experimental methods can be used to distinguish electronically forbidden from electronically allowed transitions. The first method uses the emission properties of a molecule. Provided we can measure both the fluorescence and absorption spectrum for the same electronic transition, it is seen that for an orbitally allowed transition the $(0, 0, \ldots, 0)'' \leftrightarrow (0, 0, \ldots, 0)'$ line will coincide in absorption and emission. The two spectra will, to good approximation, be mirror images of each other.

For vibronically allowed transitions, the $(0, 0, \ldots, 0)'' \leftrightarrow (0, 0, \ldots, 0)'$ line will be missing, and there will be a gap between the absorption and emission spectrum. The first line in the absorption spectrum will correspond to the excitation of one quantum of the "perturbing" vibration in the excited state, and the first line in the emission spectrum will have one quantum of the same vibration excited in the ground state. The gap will therefore be approximately equal to two quanta of this vibration, provided that no "hot" bands are present.

The second method uses the polarization of the absorption band to distinguish between the absorption mechanisms. Let us, for instance, consider an $A_1 \rightarrow B_2$ transition in D_{2d} symmetry. Measurements of the axial, σ, and π spectra have shown that we are dealing with an electric-dipole transition. With Z transforming like B_2 and (X, Y) transforming as E, we see that this transition is orbitally allowed and polarized along the z axis. If, for some reason or other, a vibronic mechanism based on an ϵ vibration is also active in the transition, we see that the direct product $A_1 \cdot B_2 \cdot \epsilon$ equals E, and the vibronically allowed transition will consequently be polarized in the x, y plane. A variation of the total band intensity with temperature is also sometimes used to distinguish "vibronic" from "allowed" transitions (1). This method is, however, not really conclusive.

In many molecules both "allowed" and "vibronic" types of transitions may be observed within the same band. In $Co(en)_3^{3+}$ a band is observed (19) which in O_h symmetry can be classified as an $^1A_{1g} \rightarrow {}^1T_{1g}$ transition. Since in O_h symmetry **L** transform as T_{1g}, the transition mechanism could be of the magnetic-dipole type. However, the axial and σ transverse spectra correspond both in measured intensity and in qualitative matching of the vibrational structure. Evidently the transition intensity is obtained from an electric-dipole mechanism. The symmetry of the complex is actually D_3, and in this symmetry the above transition is also orbitally allowed. Consequently we can see a progression involving a symmetric vibration with frequency 255 ± 5 cm^{-1} in the excited state, which, based on the $(0, 0)$ band, extends over five or six members before it becomes obscured. The identification of this sequence as a totally symmetric vibrational progression is confirmed by the circular dichroism measurements of Denning (14).

However, there also appear vibrations of 185 ± 5 cm^{-1}, 345 ± 5 cm^{-1} and 400 cm^{-1}, adding one quantum to the $(0, 0)$ band and followed by two or three quanta of the symmetrical 255 ± 5 cm^{-1} vibration. The intensity ratio $\mathscr{E}_\sigma/\mathscr{E}_\pi$ for each member is different from the polarizaion ratio shown by the $(0, 0)$ band. This indicates that these vibrations are not totally symmetric, and that we are consequently dealing with an intensity that is conferred by vibronic means.

A beautiful example of a magnetic-dipole-allowed transition has been observed by Ferguson et al. (25). These authors considered the electronic structures of Ni^{2+} in MgF_2 and ZnF_2. The first spin-allowed crystal field transition is $^3A_{2g} \rightarrow {}^3T_{2g}$. This transition is magnetic-dipole allowed, and one observes sharp spectral origins. That the transition mechanism really is a magnetic dipole was proven by looking at the strong polarization of the lines. On the other hand, the second spin-allowed transition is $^3A_{2g} \rightarrow {}^3T_{1g}$. This is not allowed as a magnetic-dipole transition, and the band shows a very complex structure with no sharp origins. In this case we are dealing with a "vibronic" intensity mechanism. The integrated band intensities for a magnetic-dipole transition and for a vibronic transition mechanism are expected to be very close (1); evidently the comparable band intensities in the first two Ni(II) bands may therefore be derived from different sources.

However, it is only in rare cases that vibronic coupling mechanisms lead to unambiguous assignments of broad polarized absorption bands. The low-temperature polarized spectra of trans-$K[Cr(C_2O_4)_2(H_2O)_2]3H_2O$ is a case in point (15). The spectrum of the compound shows evidence of six broad bands. The bands are strongly polarized, and the intensities are strongly temperature dependent. The molecular symmetry is D_{2h}, the factor group symmetry C_{2h}, and the site group symmetry of the Cr^{3+} ion is C_i. Parity should, in other words, be a good quantum number. This, taken together with the experimental facts, indicates that we must use a vibronic mechanism in the analysis of the spectra. The trouble is, however, that there are too many perturbing vibrations to consider; indeed, if all vibrations were equally efficient, one might expect an isotropic spectrum. Since the spectrum actually is highly anisotopic, it is necessary to suppose that some vibrations are more effective than others with regard to intensity stealing and/or to include more than one "odd" electronic state from which intensity can be stolen. Indeed there are numerous possibilities for choosing vibrations, many of which may lead to an explanation of the spectra. Evidently no unambiguous assignments can be reached.

VII. Spin-Forbidden Transitions

Since the operators for the electronic transition moments are independent of the spin coordinates, only transitions between states having the same spin quantum number should be observed. This strict spin selection rule is, however, broken down by the presence of the spin–orbit coupling term in the

Hamiltonian of the molecule. This term may be written (I) in the form

$$H^{(1)} = \sum_{i=1}^{n} A_i l_i \cdot s_i \tag{21}$$

or

$$H^{(1)} = \sum_{i=1}^{n} A_i(l_{xi}s_{xi} + l_{yi}s_{yi} + l_{zi}s_{zi}) \tag{22}$$

The summation runs over all the electrons in the molecule, and l_{xi} is the operator for the x component of the orbital angular momentum for electron i, s_{xi} is the operator for the x component of the spin angular momentum for electron i. A_i is dependent upon the radial derivative of the potential of the molecule; it is a very complicated function. In the case of a transition-metal complex one normally centers $H^{(1)}$ on the metal ion. This is usually the atom in the molecule with the highest atomic number, and consequently the electrons associated with this center "carry" most spin–orbit coupling. Thus, assuming a pseudo-central-field potential, A_i is given by (9)

$$A_i = \frac{1}{r_i^3} \frac{\partial V}{\partial r_i} \tag{23}$$

Here r_i is the distance of electron i from the metal ion, and V is the Coulomb potential. The value of A_i is now supposed to be not too different from the spin–orbit coupling parameter ζ_i for the free metal ion. This constant is known from atomic spectroscopy (23). In that case in which a summation of $H^{(1)}$ over the different atoms of the molecule is performed, the spin–orbit coupling parameter for each atom is weighted by the square of the coefficient of the molecular orbital in which the atomic orbital of the atom participate (3).

By using the identity

$$l_x s_x + l_y s_y + l_z s_z = l_z s_z + \tfrac{1}{2}l_+ s_- + \tfrac{1}{2}l_- s_+ \tag{24}$$

where $l_+ = l_x + il_y$ and similarly for l_-, s_+, and s_-, we see immediately that the quantum number $m_j = m_s + m_l$ is preserved during a "mixing" of spin states. The formal group-theoretical treatment utilizes the so-called double groups, and the spin–orbit coupling leaves the double-group state designations as the only good quantum numbers (1, 28).

Experience indicates that the intensity of a spin-forbidden band found in a complex of the first transition series is usually reduced by a factor of a hundred or so from the intensity of the band from which it is assumed to "steal" its intensity. This "stealing" is commonly calculated using perturbation techniques (1, 42).

The band profiles of spin-forbidden bands in transition-metal complexes differ widely. In the case where the electronic configurations are the same in the ground and excited states, we expect the potential surfaces to be nearly the same. Consequently we see the electronic transitions as sharp lines often associated with some vibrational structure. The best-known such example is probably the spin-forbidden $^4A_2(t_2)^3 \rightarrow {}^2E(t_2)^3$ transitions seen in Cr^{3+} complexes. They are sometimes called the ruby bands. The classic experimental work in this area is that of Spedding and Nutting (35, 50); more recent papers include the penetrating theoretical studies of Sugano and Tanabe (57, 58). Further references can be found in Hush and Hobbs (32).

Broad, structureless spin-forbidden transitions are, on the other hand, seen for instance in Co^{3+} complexes (19). In this case the observed transitions do not take place between states having the same electronic configurations, but are associated with $(t_2)^6 \rightarrow (t_2)^5(e)$ excitations. The potential surfaces for the two states are therefore not expected to be superimposable, and no fine structure is observed.

A nice example of the use of polarized spectra in the elucidation of spin-forbidden electronic transitions is provided by the work of Dingle and Palmer (18). These authors investigated the source of spectral intensity in the $Ni(en)_3$ $(NO_3)_2$ system. The second broad-band system can, in O_h symmetry, be classified as $^3A_{2g} \rightarrow {}^3T_{1g}$. In the actual D_3 symmetry of the complex $^3T_{1g}$ transform as $^3A_2 + {}^3E$, the ground state as $^3A_2$, (X, Y) as E, and Z as A_2. One therefore expects to see the $^3A_2 \rightarrow {}^3E$ transition σ polarized (X, Y), with the $^3A_2 \rightarrow {}^3A_2$ transition being orbitally forbidden. Measurements of the polarized spectra reveal, however, that the intensity of the π spectrum is about half that of the σ spectrum. If the model for $Ni(en)_3(NO_3)_2$ is right, the π intensity must therefore be vibronic in nature.

An energy-level diagram reveals that Ni(II) complexes have an $^1A_1$ state located at slightly higher wave numbers than the $^3T_1$ states. This $^1A_1$ state will couple isotropically via the spin–orbit coupling to the three orbital components of $^3T_1$, and one would therefore expect all or nearly all of the $^3A_2 \rightarrow {}^1A_1$ band intensity to be stolen from the orbitally and spin-allowed $^3A_2 \rightarrow {}^3E$ transition. Both the ground state $^3A_2$ and the $^1A_1$ state have the same electronic configuration $(t_2)^6(e)^2$. The (0–0) line in $^3A_2 \rightarrow {}^1A_1$ should therefore be sharp and be seen only in σ polarization, which is indeed exactly what is observed. The vibronic nature of the intensity in the Z polarization of the $^3A_2 \rightarrow {}^3A_2$ band is thereby likewise established.

The extreme importance of the orbital selection rules are actually best seen in the behavior of some spin-forbidden bands of V^{3+} compounds. Consider first the absorption spectrum of vanadium corundum—that is, V^{3+} dissolved in Al_2O_3. The d^2 electronic system VO_6 is exposed to a trigonal "crystal field" with the site group of V^{3+} being C_{3v}. The d^2 configuration would, in a O_h field, produce a $^3T_{1g}$ ground state. However, because of the lower field, this is split into a low-lying $^3A_2$ and a higher-lying ^3E state.

Belonging to the same electronic configuration is a $^1A_1$ state placed at about 21,000 cm^{-1}. The transition $^3A_2 \rightarrow {}^1A_1$ is therefore expected to be sharp.

$^1A_{1g}$ can couple via spin–orbit coupling to the nearby-lying $^3T_{1g}$ state (Fig. 1). Labeling the coupled components in C_{3v} symmetry after their orbital and spin quantum numbers, they are in an obvious notation $(E, \pm 1)$ and $(A_2, 0)$.

Because of second-order effects, the ground state $^3A_2$ is split into $(A_2, 0)$ and $(A_2, \pm 1)$ states, the latter being placed at ~ 8 cm^{-1}. With the electric-dipole vector transforming as $A_1(Z)$ and $E(X, Y)$ in C_{3v}, we expect a polarized spectrum as pictured in Fig. 1. This is indeed precisely what is observed (45).

We now look at the crystal spectrum (20) of [V(urea)$_6$](ClO$_4$)$_3$. The crystals are again trigonal, but the vanadium ion occupies here a site of D_3 symmetry. The mixing and the splittings in C_{3v} and D_3 are the same and carry the same symmetry designations. Yet the electric-dipole operator transforms in D_3 as $A_2(Z)$ and $E(X, Y)$. Consequently, we expect here only to see the σ-polarized transition $(A_2, \pm 1) \rightarrow {}^1A_1$. Again this is in complete accord with observation (20) (Fig. 1).

Turning now our attention toward the radiative behavior of spin-forbidden transitions, the precise emission mechanisms of even the intensely studied Cr^{3+} complexes are not too well understood (17, 27). The Cr(urea)$_6$(NO$_3$)$_2$

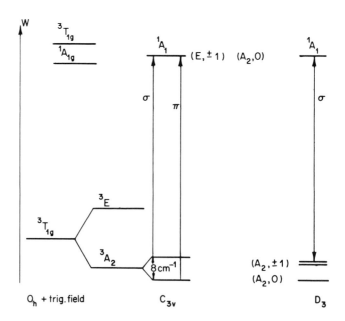

FIG. 1. Selection rules for spin-forbidden bands of V^{3+} complexes. In C_{3v} the electric-dipole vector transforms as $A_1 + E$, in D_3 as $A_2 + E$.

complex is a case in point (*17*). Both the absorption and emission bands found in the ruby band region 14160–14250 cm^{-1} was investigated using a He/Ne laser for excitation. The measured radiative lifetimes varied considerably, and no clear-cut explanations could be given. It may be possible that the (0, 0) $^4T_{2g}$ transition is placed together with the ^2E_g states. On the other hand, it may also be that some photodecomposition takes place, producing a new, emitting complex species (*27*).

The already quoted investigation of *trans*-K[Cr(C$_2$O$_4$)$_2$(H$_2$O)$_2$]3H$_2$O showed that the ruby lines found in the absorption and emission spectra at about 14000 cm^{-1} are not mirror images. On the other hand, they do have a common origin at 14432 cm^{-1}. Remembering that for this system parity was a good quantum number, a common origin is what one should expect for a magnetic-dipole process. As expected, the lines are strongly polarized; it is concluded that this system gets its intensity via a magnetic-dipole transition.

The optical absorption spectrum and the phosphorescence of MnO$_4^{3-}$ imbedded in Ca$_2$PO$_4$Cl has been studied at 2°K by Kingsley *et al.* (*33*). Both in absorption and emission two sharp lines were observed, located in the infrared at 8410 and 8703 cm^{-1}. No Stoke's shift was observed. The splitting of 293 cm^{-1} must be associated with the excited state; the radiative lifetime of 11 msec indicates a spin-forbidden transition. These lines were assigned by Orgel (*46*) as the $^3A_2 \leftrightarrow {}^1E$ transition expected in a tetrahedral d^2 system. The two intense broad bands found at higher wave numbers may be assigned, using simple crystal field theory (*46*), as $^3A_2 \rightarrow {}^3T_2$ and $^3A_2 \rightarrow {}^3T_1$.

VIII. Optical Anisotropies

In the case of a strictly octahedral molecule no optical anisotropy can, of course, occur. However, in the case where the environment of the metal ion is less than O_h or T_d symmetry, the crystals should be dichroic. Let us now look at a hexacoordinated complex where the environment of the metal ion is such that no center of symmetry exists. The low-lying "crystal field states" are predominantly made up of d-orbitals. Had the complex possessed a center of symmetry, all electric-dipole transitions between the crystal field states would be forbidden at the equilibrium configuration of the nuclei because of the parity selection rule (*30*). However, in our case, parity is no longer a good quantum number, and the unsymmetrical field ("the hemihedral field") will make an electric dipole transition slightly allowed. We may, of course, also have a "vibronic" intensity mechanism operating simultaneously with the hemihedral mechanism. Here, however, we will concentrate upon the orbitally allowed transitions.

Consider, for instance. Cr^{3+} ions dissolved in an aluminum oxide lattice where the CrO_6 complex unit has C_{3v} symmetry. The main symmetry element is a threefold axis, and we quantize our wavefunctions along that axis. The proper d^3 trigonal determinental wavefunctions have been given by McClure (*41*).

$$^4A_2: \quad |t_x t_y t_z|$$

$$^4T_2(A_1): \quad \frac{1}{\sqrt{2}}[|e_x t_z t_x| - |e_y t_y t_x|]$$

$$^4T_2(E_a): \quad \frac{1}{2}[|e_x t_y t_z| - |e_y t_z t_x|] - \frac{1}{\sqrt{2}}|e_x t_x t_y|$$

$$^4T_2(E_b): \quad \frac{1}{2}[|e_x t_z t_x| + |e_y t_y t_z|] + \frac{1}{\sqrt{2}}|e_y t_x t_y|$$

$$^4T_1(A_2): \quad \frac{1}{\sqrt{2}}[|e_x t_y t_z| + |e_y t_z t_x|]$$

$$^4T_1(E_a): \quad \frac{1}{2}[|e_x t_y t_x| + |e_y t_z t_x|] + \frac{1}{\sqrt{2}}|e_x t_x t_y|$$

$$^4T_1(E_b): \quad \frac{1}{2}[|e_x t_z t_x| + |e_y t_y t_z|] - \frac{1}{\sqrt{2}}|e_y t_x t_y|$$

Here t_x, t_y, t_z and e_x, e_y refer to the five d-orbitals quantized along the threefold axis, and each orbital is associated with an α spin.

All transitions from the ground state $^4A_2$ to the excited states $^4T_2$ and $^4T_1$ correspond to an excitation $t_2 \rightarrow e$. Consequently we can look at the electric-dipole transition moment integrals $(t_2 |\mathbf{r}| e)$ and apply the symmetry operations pertinent to the group C_{3v}. In this way it is easy to see that in the C_{3v} point group the only electric-dipole transition-moment integrals different from zero are (*15*):

$$A = (t_x |x| e_x) = -(t_x |y| e_y) = -(t_y |y| e_x) = -(t_y |x| e_y) \tag{25}$$

$$B = (t_z |x| e_x) = (t_z |y| e_y) \tag{26}$$

$$C = (t_x |z| e_x) = (t_y |z| e_y) \tag{27}$$

Using the above wavefunctions, the dipole strength for the transitions are

	σ	π
$^4A_2 \rightarrow {}^4T_2(A)$	0	0
$^4T_2(E)$	$(A - B/\sqrt{2})^2$	0
$^4T_1(A)$	0	$2C^2$
$^4T_1(E)$	$(A + B/\sqrt{2})^2$	0

The intensity ratio for the $^4A_2 \rightarrow {}^4T_1$ transition is therefore

$$\frac{I_\pi}{I_\sigma} = \frac{2C^2}{(A + B/\sqrt{2})^2} \tag{28}$$

The values of the transition-moment integrals A, B, and C, could of course in principle be calculated using the *molecular orbitals* designated t_x, t_y, t_z, e_x, and e_y. However, because of a lack of knowledge of the exact form of these orbitals, it is recommended to leave the A, B, and C integrals as parameters.

It is, however, possible to get a pure number for the intensity ratio I_π/I_σ using first-order perturbation theory (57). Again we look at a hexacoordinated Cr^{3+} complex. In octahedral symmetry the transition $^4A_{2g} \rightarrow {}^4T_{1g}$ is parity forbidden, and in order to make it " allowed " we need to mix in some " odd " wavefunctions. This can be done by expanding the hemihedral field in spherical harmonics, and picking out those which in C_{3v} symmetry transform as A_1. V_{hem} must be invariant under the trigonal rotation and change sign under an inversion. The first term in the expansion is of the form $\sum z_i$. This term is in *octahedral* symmetry a member of a T_{1u} representation. Using perturbation technique we get the ground-state wave function to be of the form

$$^4A_{2g} + \frac{\int \psi(^4A_{2g}) V_{hem} \psi(^4\Gamma_u)\, d\tau}{W(^4A_{2g}) - W(^4\Gamma_u)} \psi(^4\Gamma_u) \tag{29}$$

where we have used V_{hem} to mix in some odd state transforming like $^4\Gamma_u$.

The dipole strength of the $^4A_{2g} \rightarrow {}^4T_{1g}$ transition will evidently depend upon the value of

$$(A_{2g}|V_{hem}|\Gamma_u M') \cdot (\Gamma_u M'|\mathbf{R}|T_{1g}) \tag{30}$$

Here we have used the formalism (1, 28) that T_1 and T_2 states behave like pseudo P states. We can therefore associate a quantum $M_L = 1, 0, -1$ with the three components of T_1 and T_2. Since V_{hem} transforms like $T_{1u}(0)$ we see that the $\Gamma_u M'$ must transform like $T_{2u}(0)$ in order for the first matrix element to be different from zero.

In the final analysis the dipole strength of the transition is therefore proportional to the square of the matrix element

$$(T_{2u}(0)|\mathbf{R}|T_{1g}) \tag{31}$$

We now use the Wigner–Eckart theorem, and instead of calculating the above matrix element, we evaluate the reduced matrix element

$$(T_{2g}(0)|\mathbf{L}_M|T_{1g}(-M)) \tag{32}$$

The functions $T_{2g}(0)$ and $T_{1g}(M)$ are taken as the appropriate spherical harmonics quantized along the trigonal axis (7). They are

$$T_{2g}(0) = \frac{1}{\sqrt{2}}(Y_3{}^3 + Y_3{}^{-3}) \tag{33}$$

$$T_{1g}(-1) = \sqrt{\tfrac{5}{6}}Y_3{}^2 + \sqrt{\tfrac{1}{6}}Y_3{}^{-1} \tag{34}$$

$$T_{1g}(0) = \tfrac{2}{3}Y_3{}^0 + \tfrac{1}{3}\sqrt{\tfrac{5}{2}}(Y_3{}^3 - Y_3{}^{-3}) \tag{35}$$

$$T_{1g}(1) = \sqrt{\tfrac{5}{6}}Y_3{}^{-2} - \sqrt{\tfrac{1}{6}}Y_3{}^1 \tag{36}$$

For $\mathbf{L}_M$ we take

$$L_1 = \frac{1}{\sqrt{2}}(L_x + iL_y) = \frac{1}{\sqrt{2}}L_+ \tag{37}$$

$$L_0 = L_z \tag{38}$$

$$L_{-1} = \frac{1}{\sqrt{2}}(L_x - iL_y) = \frac{1}{\sqrt{2}}L_- \tag{39}$$

Immediately we get

$$(T_{2g}(0)|L_z|T_{1g}(0)) = \sqrt{5} \tag{40}$$

$$(T_{2g}(0)\left|\frac{1}{\sqrt{2}}L_+\right|T_{1g}(-1)) = (T_{2g}(0)\left|\frac{1}{\sqrt{2}}L_-\right|T_{1g}(1)) = \frac{\sqrt{5}}{2} \tag{41}$$

The intensity ratio of I_π to I_σ is then given as

$$\frac{I_\pi}{I_\sigma} = \frac{4}{1} \tag{42}$$

The $^4A_2 \to {}^4T_1$ band is evidently going to show considerable dichroism; experimentally the intensity ratio is found (57) to be about two. It is clear that we cannot expect that any measured intensity ratio should be exactly one to four. First of all, the expansion of V_{hem} is truncated. Secondly, we have only considered one excited state of T_{2u} symmetry to provide intensity. And thirdly, vibronic contributions to the total band intensity may obscure the picture. The first two error sources are of course not present in the former treatment.

IX. The Zeeman Effect

Measurements of the Zeeman effect offers unique opportunities to study whether the ideas one has about the nature of the excited states are in accord with reality. In the study of the spectra of the rare earths it has indeed been indispensable, since without Zeeman studies very few assignments could have been made at all definitely. Also, the new technique of Zeeman–Raman experiments (*34*) looks very promising here.

In order to observe the Zeeman splittings the lines of the system must be very sharp, preferably with a half-width of not more than a couple of wave numbers. This limits the number of *nd*-complexes that can be studied to those that have bands assigned to transitions between states having the same electronic configurations. All such bands are spin forbidden (*1*). The classic example is again the ruby lines $^4A_{2g} \to {}^2E_g$ found in the d^3 systems. However, the d^2 system in V^{3+} complexes has also received some attention (*40, 45*) and d^8 systems with a very high value of Dq should be a possibility.

The pioneer work in this field is that of Spedding and Nutting (*51*), who studied the Zeeman effect on the absorption lines of $KCr(SO_4)_2 \cdot 12H_2O$. The interpretation of these results was given by Van Vleck (*61*), who proved that the lines that Spedding and Nutting had investigated were due to an intersystem combination $^4A_2 \to {}^2X$. In a following paper Van Vleck (*26*) made the identification $^2X = {}^2E$.

Consider again the case of Cr^{3+} in a trigonal crystal field (*57, 59, 60, 64*). As before, we take the triply degenerate representations T_1 and T_2 to behave like pseudo P states and associate a pseudo crystal quantum number $M_L = 1, 0, -1$ to the three components. In the same sense, one can associate the quantum numbers $M_L = \pm 1$ with the two components of a doubly degenerate E state. Because of the lower field, ^2E is split into two Kramers doublets, which we characterize in the notation (M_L, M_s)

$$(1, \tfrac{1}{2}) \quad (-1, -\tfrac{1}{2}) \qquad \text{and} \qquad (1, -\tfrac{1}{2}) \quad (-1, \tfrac{1}{2})$$

The first set of states have $M_J = M_L + M_S$ equal to $\pm \tfrac{3}{2}$, the second to $\pm \tfrac{1}{2}$. In $Cr(en)_3^{3+}$ complexes they are separated by about 18 cm^{-1}.

We now couple the ^2E states to the nearby-lying $^4T_2$ state via spin–orbit coupling. Remembering that M_J remains a good quantum number in this type of coupling, and that because of the quantization around the threefold axis, $M_L = \pm 2$ equals $M_L = \mp 1$, we get, using first-order perturbation theory

$$^2E(1, \tfrac{1}{2}) + \lambda\,{}^4T_2[(1, \tfrac{1}{2}), (0, \tfrac{3}{2}), (-1, \tfrac{1}{2})]$$
$$^2E(-1, -\tfrac{1}{2}) + \lambda\,{}^4T_2[(-1, -\tfrac{1}{2}), (0, -\tfrac{3}{2}), (1, \tfrac{1}{2})]$$
$$^2E(1, -\tfrac{1}{2}) + \lambda\,{}^4T_2[(1, -\tfrac{1}{2}), (0, \tfrac{1}{2}), (-1, -\tfrac{3}{2})]$$
$$^2E(-1, \tfrac{1}{2}) + \lambda\,{}^4T_2[(-1, \tfrac{1}{2}), (0, -\tfrac{1}{2}), (1, \tfrac{3}{2})]$$

So far we have not used the actual C_{3v} symmetry of the complex. Going from O_h to C_{3v} we have

$$O_h(E, \pm 1) \text{ irreducible representation } E \text{ of } C_{3v}$$
$$(T_2, \pm 1) \qquad\qquad\qquad E$$
$$(T_2, 0) \qquad\qquad\qquad A_1$$
$$(A_2, 0) \qquad\qquad\qquad A_2$$

The electric-dipole operator transforms in C_{3v} as $A_1(Z)$ and $E(X, Y)$.
    Rewriting the perturbed ^2E wave functions as

$$(E_1, \tfrac{1}{2}) + \lambda[(E_1, \tfrac{1}{2}), (A_1, \tfrac{3}{2}), (E_{-1}, -\tfrac{1}{2})]$$
$$(E_{-1}, -\tfrac{1}{2}) + \lambda[(E_{-1}, -\tfrac{1}{2}), (A_1, -\tfrac{3}{2}), (E_1, \tfrac{1}{2})]$$
$$(E_1, -\tfrac{1}{2}) + \lambda[(E_1, -\tfrac{1}{2}), (A_1, \tfrac{1}{2}), (E_{-1}, -\tfrac{3}{2})]$$
$$(E_{-1}, \tfrac{1}{2}) + \lambda[(E_{-1}, \tfrac{1}{2}), (A_1, -\tfrac{1}{2}), (E_1, \tfrac{3}{2})]$$

and taking for the ground state the set

$$(A_2, \tfrac{3}{2}), (A_2, \tfrac{1}{2}), (A_2, -\tfrac{1}{2}), (A_2, -\tfrac{3}{2})$$

we can see the selection rules allow only the transitions $(A_2, M_s'') \to (E, M_s')$ with $M_s'' = M_s'$. The transitions are all $\sigma(X, Y)$ polarized. Had we, on the other hand, also mixed ^2E with the not too distant $^4T_1$ state, Z polarizations would also have been orbitally allowed, since $^4T_1$ in C_{3v} symmetry transforms as $^4E + ^4A_2$.
    Applying a magnetic field parallel to the C_3 axis of the compound will add the term

$$H^{(1)} = \beta H(L_z + 2S_z)$$

to the Hamiltonian. The orbital momentum is quenched in an A or E state, and we may take $H^{(1)} = 2\beta H S_z$. $g_\parallel$ in both the ^2E and the $^4A_2$ state are therefore expected to be close to two. The selection rules and the absorption pattern is pictured in Fig. 2.
    If we want to calculate the relative intensities as given in the drawing of the Zeeman spectrum (Fig. 2) we can proceed as follows. To get the coupling coefficients between ^2E and $^4T_2$, we can write the proportionality equation:

$$(^2E M, M_S |\mathbf{L} \cdot \mathbf{S}| ^4T_2 M', M_S')$$
$$= C \cdot (^2E M, M_S |\mathbf{L}| ^2T_2 M', M_S)(^2T_2 M', M_S |\mathbf{S}| ^4T_2 M'M_S') \quad (43)$$

We now calculate the matrix elements in Eq. (43) using the operators L_z, $(1/\sqrt{2})L_+$, $(1/\sqrt{2})L_-$, S_z, $(1/\sqrt{2})S_-$, and $(1/\sqrt{2})S_+$. Using the Wigner–Eckart theorem we take as basis set for E and T_2 the trigonally quantized set of d-orbitals. To couple a spin doublet with a spin quartet we write down a set of spin wavefunctions made up of two particles. These functions are constructed so as to span $S = \tfrac{3}{2}$ and $S = \tfrac{1}{2}$. Operating on these wavefunctions with *either* $S(1)$ *or* $S(2)$, where (1) and (2) refer to the spin coordinates of particle one and two, gives the coupling coefficients $(^2T_2 |S| ^4T_2)$. Multiplying the coupling coefficients for $\mathbf{L}$ and $\mathbf{S}$ together and squaring the result gives the required *relative* intensities.
    In the case where the splitting of the ground state is large compared with kT, we must calculate the Boltzmann distribution on the Zeeman levels and

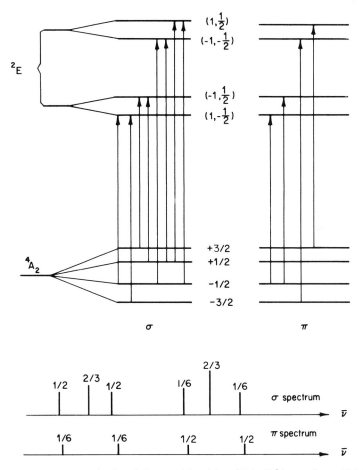

FIG. 2. Zeeman pattern in C_{3v} of the transition $^4A_2 \rightarrow\, ^2E$ in Cr^{3+} complexes $H_0 \parallel C_3$.

weight the relative intensities with these factors if we want to compare experimental and calculated intensities. Also, if the investigated crystal is not diluted, couplings may occur between the chromophores that can completely alter the calculated intensities and mock the selection rules.

Pure compounds for which the Zeeman effect has been studied include (20) $V(urea)_6(ClO_4)_3$ and (4) $2Cr\ en_3Cl_3 \cdot KCl \cdot 6H_2O$. The optical Zeeman splittings of the $Re^{4+}(5d)^3$ ion in single crystals of K_2PtCl_6 and Cs_2ZrCl_6 at 4°K has been studied by Dorain and Wheeler (21). Because of the strong spin–orbit coupling present in Re^{4+} complexes all states are best classified using the double group quantum numbers (1, 28) Γ_6, Γ_7, and Γ_8. Many extremely narrow lines are seen between 8500 and 36,000 cm^{-1}. All the

$(0, 0, \ldots, 0)\ {}^4A_{2g}(\Gamma_8) \to (0, 0, 0, \ldots, 0)\ {}^2T_{2g}(\Gamma_7),\ {}^2T_{2g}(\Gamma_8)$, and ${}^2E_g(\Gamma_8)$ states are identified and their Zeeman patterns elucidated. All of the narrow lines appear to be coupled with odd vibrations; evidently we observe the very rare case of a true vibronic coupling system (21), where the various "odd" contributions to the intensity may be sorted out.

Some very nice work dealing with the fluorescence of the d^3 systems V^{2+} and Mn^{4+} in Al_2O_3 have been carried out by Sturge (53, 54). The emission of the ruby lines were investigated at low temperatures while applying both a magnetic and an electric field. The observation of the "pseudo" Stark splitting of the ${}^2E$ states is of particular interest (54) as one of the very few instances where this technique has been applied to inorganic complexes.

X. The Jahn–Teller Effect

In the case of electronic degeneracy the motions of the nuclei couple with the low-frequency electronic motions via the term (4) in the Hamiltonian. These vibronic couplings may completely destroy the expected vibrational pattern associated with a given potential surface. Consequently the band contour of a transition terminating on a Jahn–Teller distorted state may be highly irregular, and we talk about the state experiencing a dynamic Jahn–Teller effect. As an example, various expected vibrational patterns for an $A \to E$ transition in an octahedral molecule can be seen in a paper by Longuet–Higgins et al. (37).

More papers have been written dealing with this rather specialized topic than one should expect possible. For an excellent summary of the theoretical and experimental features we recommend the review paper by Sturge (56). Here we shall limit ourselves to a few remarks dealing with the experimental aspects of the dynamic Jahn–Teller effect.

Evidence for the operation of a dynamic Jahn–Teller effect in the ${}^2T_{2g}$ excited state of V^{2+} in MgO has been given by Sturge (55). The V^{2+} ion is at a perfect O_h site in MgO. The lowest (0–0) line in the transition ${}^4A_{2g} \to {}^4T_{2g}$ can therefore be a magnetic-dipole transition, an expectation that experiments confirm. However, the (0–0) line is observed as a doublet, split some 40 cm^{-1}, which can be shifted and polarized by applying a uniaxial stress to the crystal. Such a behavior is inconsistent with a cubic symmetry of ${}^4T_{2g}$. Assuming a dynamic Jahn–Teller effect in which the ϵ_g and τ_{2g} vibrations are supposed to be active in the coupling of the electronic components of ${}^4T_{2g}$, the line pattern may, however, be explained.

Even nicer experimental work is furnished by Scott and Sturge (48) in their high-resolution spectral measurements of the electronic origin of the $^3T_1 \to {}^3T_2$ of V^{3+} in Al_2O_3. As we observed when we treated the properties of the spin-forbidden bands in V^{3+} dissolved in Al_2O_3, the octahedral ground state $^3T_{1g}$ is split into $^3A_2 + {}^3E$ in the trigonal field of aluminum oxide. The experimental splitting is about 1000 cm^{-1}. Using also the other observed excited states, the crystal field parameters may be evaluated, and the calculations predict that the octahedral $^3T_{2g}$ excited state should exhibit a combined spin–orbit and trigonal splitting of some 400 cm^{-1}. However, experimentally the expected 400 cm^{-1} splitting is seen to be only 9 cm^{-1}.

The explanation of this effect was given by Ham (29), who pointed out that " off-diagonal" operators such as orbital momentum, spin–orbit coupling, or trigonal fields can be partially or totally quenched according to the strength of the Jahn–Teller coupling. What has happened in the $^3T_2$ state (48) is therefore presumably that the Jahn–Teller coupling has completely quenched the spin–orbit coupling and greatly reduced the trigonal field.

The work of Nelson et al. (44, 65) on the d^1 systems Ti^{3+} and V^{4+} in Al_2O_3 is interesting in that it shows how electronic transitions placed in the far infrared can be used to demonstrate the operations of the dynamic Jahn–Teller effect. Their observations showed that the trigonal splitting of the octahedral $^2T_{2g}(d^1)$ ground state was much smaller than would be expected (41). For Ti^{3+} in a trigonal field, polarized infrared absorption spectroscopy revealed the ground state to be $E_{3/2}(^2E)$ with $E_{1/2}(^2E)$ at 28 cm^{-1}, followed by $E_{1/2}(^2A_1)$ at 53 cm^{-1}. The total splitting of $^2T_{2g}$ is, in other words, only 53 cm^{-1}. To explain this, Macfarlane et al. (38) had again to invoke the operation of a dynamic Jahn–Teller effect together with the " normal" first- and second-order spin–orbit coupling and trigonal field. Including the Zeeman interactions, they could likewise explain the g factors. The measured splittings could only be accounted for by taking a Jahn–Teller energy of about 300 cm^{-1} using an active ϵ_g vibration of 200 cm^{-1}. Both Ti^{3+} and V^{4+} in Al_2O_3 exhibit, therefore, a very pronounced Jahn–Teller effect.

The dynamic Jahn–Teller effect in the excited ^2E_g state of d^1 systems have been considered by McClure (41) and Dingle (16). The last author looked at the strongly polarized, double-peaked band seen at about 17,000 cm^{-1} in the 20°K crystal spectrum of $Ti(urea)_6I_3$. The Ti^{3+} ion is at a D_3 site. Consequently we do not expect the excited ^2E_g state to show other splittings than those due to spin–orbit coupling. This can at most amount to a few hundred wave numbers. Yet the observed splitting is some 2500 cm^{-1}. It seems reasonable in order to explain the polarizations to assume that a strong dynamic Jahn–Teller coupling is operating inside the ^2E_g manifold, and it is tentatively proposed that the Jahn–Teller active vibration is of such a nature as to destroy the quantitization of the ^2E_g state around the C_3 axis.

XI. General Conclusions

As demonstrated, the phenomena one can study making use of crystal spectra are highly varied. Ranging from simple evaluations of crystal field parameters to penetrating studies of Zeeman and Jahn–Teller effects, "inorganic" spectroscopy is certainly not lacking in excitement and variance. It is unfortunately true that relatively few molecules show pronounced band structures. However, the deep insight one can gain into the electronic structures of the excited states makes it really worthwhile to search for those molecules that do show band structures.

One such class of molecules are the tetrahedral d^0 systems. In the last few years a combination of experiments (*22, 24, 31, 64*) and theory (*5, 6, 10–12*) has provided an understanding of the electronic structures of such molecules. The work is not finished; many outstanding questions still remain to be answered. However, it is quite certain that the way forward is to design and execute better and better experiments. Without such, we would have nothing to guide us.

Acknowledgment

I am indebted to Professor S. P. McGlynn for having read the chapter in a preliminary form and for his suggestions for improvements.

References

1. Ballhausen, C. J., "Introduction to Ligand Field Theory." McGraw-Hill, New York, 1962.
2. Ballhausen, C. J., *Theoret. Chim. Acta* **1**, 285 (1963).
3. Ballhausen, C. J., *Mol. Phys.* **6**, 461 (1963).
4. Ballhausen, C. J., and Trabjerg, I., (To be published).
5. Becker, C. A. L., Ballhausen, C. J., and Trabjerg, I., *Theoret. Chim. Acta* **13**, 355 (1969).
6. Becker, C. A. L., and Dahl, J. P., *Theoret. Chim. Acta* **14**, 26 (1969).
7. Bleaney, B., and Stevens, K. W. H., *Rept. Progr. Phys.* **16**, 108 (1953).
8. Born, M., and Huang, K., "Dynamical Theory of Crystal Lattices," Appendix 8. Oxford Univ. Press, London and New York, 1954.
9. Condon, E. U., and Shortley, G. H., "The Theory of Atomic Spectra." Cambridge Univ. Press, London and New York, 1935.
10. Dahl, J. P., and Ballhausen, C. J., *Advan. Quantum Chem.* **4**, 170 (1968).

11. Dahl, J. P., and Johansen, H., *Theoret. Chim. Acta* **11**, 8 (1968).
12. Dahl, J. P., and Johansen, H., *Theoret. Chim. Acta* **11**, 26 (1968).
13. Day, P., and Jørgensen, C. K., *Chem. Phys. Letters* **1**, 507 (1968).
14. Denning, R. G., *Chem. Commun.* 120 (1967).
15. Dingle, R., *Acta Chem. Scand.* **22**, 2219 (1968).
16. Dingle, R., *J. Chem. Phys.* **50**, 545 (1969).
17. Dingle, R., *J. Chem. Phys.* **50**, 1952 (1969).
18. Dingle, R., and Palmer, R. A., *Theoret. Chim. Acta* **6**, 249 (1966).
19. Dingle, R., and Ballhausen, C. J., *Mat. Fys. Medd. Dan. Vid. Selsk.* **35**, no. 12 (1967).
20. Dingle, R., McCarthy, P. J., and Ballhausen, C. J., *J. Chem. Phys.* **50**, 1957 (1969).
21. Dorain, P. B., and Wheeler, R. G., *J. Chem. Phys.* **45**, 1172 (1966).
22. Duincker, J., and Ballhausen, C. J., *Theoret. Chim. Acta* **12**, 325 (1968).
23. Dunn, T. M., *Trans. Faraday Soc.* **57**, 1441 (1961).
24. Dunn, T. M., and Francis, A. H., *J. Mol. Spectry.* **25**, 86 (1968).
25. Ferguson, J., Guggenheim, H. J., Kamimura, H., and Tanabe, Y., *J. Chem. Phys.* **42**, 775 (1965).
26. Finkelstein, R., and Van Vleck, J. H., *J. Chem. Phys.* **8**, 790 (1940).
27. Flint, C. D., *J. Chem. Phys.* **52**, 168 (1970),
28. Griffith, J. S., " The Theory of Transition Metal Ions." Cambridge Univ. Press, London and New York, 1961.
29. Ham, F. S., *Phys. Rev.* **138**, A1727 (1965).
30. Herzberg, G., "Molecular Spectra and Molecular Structure III." Van Nostrand, Princeton, New Jersey, 1966.
31. Holt, S. L., and Ballhausen, C. J., *Theoret. Chim. Acta.* **7**, 313 (1967).
32. Hush, N. S., and Hobbs, R. J. M., *Progr. Inorg. Chem.* **10**, 259 (1968).
33. Kingsley, J. D., Prener, J. S., and Segall, B., *Phys. Rev.* **137**, A189 (1965).
34. Koningstein, J. A., and Mace, G., *Chem. Phys. Letters* **3**, 443 (1969).
35. Kraus, D. L., and Nutting, G. C., *J. Chem. Phys.* **9**, 133 (1941).
36. Lax, M., *J. Chem. Phys.* **20**, 1752 (1952).
37. Longuet-Higgins, H. C., Öpik, U., Pryce, M. H. L., and Sack, R. A., *Proc. Roy. Soc. (London)* **A244**, 1 (1958).
38. Macfarlane, R. M., Wong, J. Y., and Sturge, M. D., *Phys. Rev.* **166**, 250 (1968).
39. McCaffery, A. J., Schatz, P. N., and Lester, T. E., *J. Chem. Phys.* **50**, 379 (1969).
40. McClure, D. S., *Solid State Phys.* **9**, 399 (1959).
41. McClure, D. S., *J. Chem. Phys.* **36**, 2757 (1961).
42. McGlynn, S. P., Azumi, T., and Kinoshita, M., "Molecular Spectroscopy of the Triplet State." Prentice Hall, Englewood Cliffs, New Jersey, 1969.
43. Müller-Goldegg, A., and Voitländer, J., *Z. Naturforsch.* **23A**, 1236 (1968).
44. Nelson, E. D., Wong, J. Y., and Schawlow, A. L., *Phys. Rev.* **156**, 298 (1967).
45. Pryce, M. H. L., and Runciman, W. A., *Discussions Faraday Soc.* **26**, 34 (1958).
46. Orgel, L. E., *Mol. Phys.* **7**, 397 (1964).
47. Sayre, E. V., Sancier, K., and Freed, S., *J. Chem. Phys.* **23**, 2060 (1955).
48. Scott, W. C., and Sturge, M. D., *Phys. Rev.* **146**, 262 (1966).
49. Sleight, T. P., and Hare, C. R., *J. Phys. Chem.* **72**, 2207 (1968).
50. Spedding, F. H., and Nutting, G. C., *J. Chem. Phys.* **2**, 421 (1934).
51. Spedding, F. H., and Nutting, G. C., *J. Chem. Phys.* **3**, 369 (1935).
52. Sponer, H., and Teller, E., *Rev. Mod. Phys.* **13**, 75 (1941).
53. Sturge, M. D., *Phys. Rev.* **130**, 639 (1963).
54. Sturge, M. D., *Phys. Rev.* **133**, A795 (1964).
55. Sturge, M. D., *Phys. Rev.* **140**, A880 (1965).

56. Sturge, M. D., *Solid State Phys.* **20**, 91 (1967).
57. Sugano, S., and Tanabe, Y., *J. Phys. Soc. Japan* **13**, 880 (1958).
58. Sugano, S., and Tanabe, Y., *Discussions Faraday Soc.* **26**, 43 (1958).
59. Sugano, S., and Tsujikawa, I., *J. Phys. Soc. Japan* **13**, 899 (1958).
60. Sugano, S., Schawlow, A. L., and Varsanyi, F., *Phys. Rev.* **120**, 2045 (1960).
61. Van Vleck, J. H., *J. Chem. Phys.* **8**, 787 (1940).
62. Wells, E. J., Jordan, A. D., Alderdice, D. S., and Ross, I. G., *Australian J. Chem.* **20**, 2315 (1967).
63. Wilson, E. B., Decius, J. C., and Cross, P. C., "Molecular Vibrations." McGraw-Hill, New York, 1955.
64. Wood, D. L., *J. Chem. Phys.* **42**, 3404 (1965); **44**, 2221 (1966).
65. Wong, J. Y., Berggren, M. J., and Schawlow, A. L., *J. Chem. Phys.* **49**, 835 (1968).

INORGANIC PHOTOCHEMISTRY

D. R. Eaton

DEPARTMENT OF CHEMISTRY
MCMASTER UNIVERSITY
HAMILTON, ONTARIO, CANADA

I. Introduction

The presence of a chapter on photochemistry in a volume devoted to spectroscopy would seem at first sight to be something of an anomaly. The literature on photochemistry has in fact tended to be somewhat divorced from that on spectroscopy. The increasing interest in the physical and mechanistic aspects of photochemistry is, however, drawing the subjects inevitably closer together, and an increasing number of chemists are finding their research interest in the intervening area. This trend has perhaps progressed somewhat further in organic photochemistry than in inorganic photochemistry. The very rapid development of the chemistry of the triplet states of π-electron molecules has been accompanied and illuminated by a wealth of sophisticated spectroscopic and electron-spin-resonance experiments involving these molecules (*68*).

Inorganic photochemistry has remained somewhat more fragmented, although its close relationship to spectroscopy has always been implicit. Thus the flash photolysis of relatively simple inorganic molecules such as ammonia (*56*) to give short-lived fragments such as NH_2 has long been a fruitful field for molecular spectroscopists. On the other hand the photochemical interests of transition-metal chemists are on the whole of more recent origin, since rational development of the field was not possible until ligand field theory and its application to spectroscopy had reached a fairly high degree of sophistication. However, perhaps the most notable development in the very recent past has been the extensive use of electron-spin resonance to detect and identify transient radicals produced during the photolysis of inorganic compounds. The present chapter is not intended to be anything approaching a comprehensive review of inorganic photochemistry. The aim is rather to illustrate those areas of the subject likely to be of most interest to the spectroscopist by reviewing a selected number of recent papers. Even then the choice of topics and papers has been somewhat arbitrary. Thus the whole area of photosensitization by mercury atoms (*27*) has been omitted on the admittedly weak grounds that the subject is of primary interest to the organic chemist. Similarly not more than fleeting reference will be made to the large volume of very excellent work in synthetic inorganic photochemistry. On the other hand there is rather heavy emphasis on transition-metal photochemistry and on applications of electron-spin resonance, since it is in these areas that contact with spectroscopy is most pronounced.

For a general account of the theoretical background to photochemistry and also for an account of the experimental techniques the reader is referred to the excellent textbook by Calvert and Pitts (*17*). It will suffice for the present purposes to draw attention to one or two points of particular interest to inorganic chemists.

In most general terms photochemistry is the study of the chemistry of excited electronic states of molecules. A prime necessity is therefore to formulate a description of excited states. For this purpose simple one-electron molecular orbital theory often suffices to give a naïve but nevertheless qualitatively satisfactory picture of the situation. An electronic excitation is simply regarded as a process whereby an electron is removed from an orbital with certain bonding characteristics and reinserted in another orbital with different characteristics. At the simplest level orbitals can be described as as either localized at a certain bond or delocalized over a number of bonds and can be characterized as bonding, nonbonding, or antibonding. Based on these simple premises the chemistry of the excited state can be deduced in terms of the weakening and subsequent breaking of certain bonds due to the presence of electrons in the appropriate antibonding orbitals or in terms of polarization of charge arising from electron transfer from one site to another leading to susceptibility to attack by electrophilic or by nucleophilic reagents. Thus it can be simply argued (10) that the ligand-field or d–d transitions in transition-metal complexes in the molecular orbital picture correspond to electron transfer from an essentially nonbonding d orbital to an orbital that is antibonding between the metal and ligand but still predominantly metal in character. Heterolytic fission of the metal–ligand bond and subsequent replacement of the ligand by an alternate ligand is therefore the simple expectation. Similarly the intense bands observed in the blue or near ultra-violet for many metal complexes are associated with either metal-to-ligand ($\pi^* \leftarrow d$) or ligand-to-metal ($d \leftarrow \pi$) charge-transfer processes. The expectation here is therefore one of homolytic bond cleavage or reactivity probably at a point distant from the metal atom.

It is, however, unrealistic to discuss the reactivity of different electronic states without some consideration of the lifetimes of these states. Thus if the lifetime of an excited state is 10^{-8} sec, it is clear that the number of observable chemical reactions will be very limited. If the excited state is strongly anti-bonding at an appropriate point in the molecule, unimolecular dissociation is obviously feasible as a primary reaction, but more complex reactions are very improbable. However, the dissociation products are often themselves very reactive species, and the simplicity of the primary process can be more than counterbalanced by the complexity of the secondary reactions. The disentanglement of the primary and secondary reactions in fact constitutes one of the main themes of much of the more physical work in inorganic photochemistry.

The radiative lifetime of an electronic state is the first consideration in determining its average lifetime. It is related to the Einstein transition probability for absorption; thus states connected to the ground state by intense transitions (those with high probability) such as charge-transfer bands will have short radiative lifetimes, and those corresponding to forbidden bands will have long radiative lifetimes (34). Typically the range is 10^{-8} sec for the radiative lifetime corresponding to a strongly allowed band to 10 sec for a highly forbidden transition. Conventionally, radiation from the first type is referred to as fluorescence and from the second type as phosphorescence. Long radiative lifetimes may be expected to be rare in molecules containing atoms beyond the first two short periods—i.e., in most inorganic compounds of interest. This arises from the increasing importance of spin–orbit coupling in the heavier elements. Such coupling mixes together states of different spin multiplicities and relaxes the selection rule forbidding transitions between such states. Purely from this consideration alone we cannot expect to find in transition-metal photochemistry the variety of reactions exhibited by the relatively long-lived organic triplet states.

Photoluminescence is, however, not the sole mode of deactivation of excited states. Less well understood but probably equally important are radiationless processes grouped together under the heading of internal conversion (37). In this process energy is lost to the surrounding medium as thermal energy, so that deactivation of the excited state is accomplished without luminescence. Internal conversion between excited states of the same multiplicity is usually very efficient (18). The effect of this is that only the lowest excited state of a given multiplicity is available for photoreactions. With transition metal-complexes in particular, internal conversion to the ground state is often a strong competitor with possible photochemical reactions. As a result quantum yields can be quite low. A process related to internal conversion is that of intersystem crossing. This is also a radiationless process by means of which excitation of an electronic state of one multiplicity can lead to the population of a lower energy level of different multiplicity. Thus energy can be absorbed in a strongly allowed singlet-to-singlet transition, but the photoreactivity may be that of a triplet state. The latter may have a relatively long lifetime and hence the opportunity to participate in a variety of chemical reactions. Finally the list of processes by means of which a molecule can lose energy without undergoing chemical reaction would be incomplete without mentioning energy-transfer processes by means of which electronic energy is passed from one molecule to another. Such processes are of great importance in photosensitization and are also thought to occur in biological systems involved in photosynthesis, but there has been little study made of their importance in the relatively simple inorganic processes that constitute the topic of the present chapter.

Photochemical studies can be divided into several categories, depending on the point of the overall process on which interest is focused. Thus for synthetic purposes the overriding interest is in the final products, and mechanistic considerations are less likely to receive attention. At the other extreme the pure spectroscopist is interested only in the primary process of electronic excitation and in any redistribution of energy into other electronic levels that might follow this primary process. Most of the work that will be discussed in the present chapter strikes a balance between these two extremes. The emphasis will be on the identification of the reactive intermediates that result from the primary photochemical process and, stemming from this, the deduction of plausible reaction mechanisms that serve to rationalize the chemistry of the situation. In this area the techniques, if not the objectives, are largely spectroscopic in origin. Thus although the classical techniques of photochemistry—the identification of reaction products, the measurement of quantum yields, the methods of chemical kinetics, and the use of scavengers to detect the occurrence of free-radical intermediates—play a major part in the determination of photochemical mechanisms, the most definitive information on the earlier stages of the process has come by and large from studies using the spectroscopic techniques of flash photolysis and electron-spin resonance. In the succeeding sections of this chapter, therefore, selected papers will be discussed to illustrate the application of spectroscopic techniques to the photochemistry of small inorganic molecules, non-transition-metal compounds, and transition-metal complexes.

II. Small Molecules

The study of the electron-spin-resonance spectra of reactive free radicals presents a number of problems. In solution, which is usually the medium of photochemical interest, such radicals will be short lived and their equilibrium concentrations may well be sufficiently small that ESR detection presents a problem. With the present level of instrumentation it is a widely used rule of thumb that radicals with millisecond lifetimes can be expected to be detectable by ESR but those with microsecond lifetimes will probably be present in too small a concentration to be observed. However, in favorable cases this generalization may have to be modified in the future, since it has recently been realized that radicals produced by breaking chemical bonds may be initially formed with populations of the different spin levels that are far from those corresponding to thermal equilibrium (8). If this is the case, there may be sufficient enhancement of the ESR signal to make very short-lived radicals observable. In any event the observation of the ESR spectrum in solution where possible is usually the most informative experiment as regards the structure of a radical intermediate. Radicals in the solid state have much longer lifetimes, and in principle the ESR spectrum of a single crystal containing free radicals with uniform orientations is as informative or even more so than the solution spectrum. In practice photochemistry in single crystals is relatively restricted, and one often has to be content with experiments in polycrystalline or glassy solids. Detailed ESR information from such experiments is often restricted, but nevertheless may suffice to identify the radicals. A frequently used modification of this method involved irradiation at low temperature (typically 77°K) to identify primary radical products, followed by warming of the matrix to allow diffusion of the primary radicals and subsequent ESR identification of secondary radical products.

The potentialities of ESR solution studies are well illustrated by the work of Livingston and Zeldes (41, 42) on photochemical reactions of hydrogen peroxide. This latter molecule absorbs in the ultraviolet, and the excited state most plausibly corresponds to the transfer of an electron from an O–O bonding to an O–O antibonding orbital. Dissociation of the molecule follows, and the primary products of the photolysis are hydroxyl radicals—i.e.,

$$H_2O_2 \xrightarrow{h\nu} 2OH\cdot$$

These primary products are, however, too short lived for ESR observation, but their presence may be manifested by reaction with suitable radical scavengers. Thus in a typical experiment a 1% solution of hydrogen peroxide in ethyl alcohol was flowed through the ESR cavity and irradiated while in the cavity with an AH6 high-pressure mercury lamp. The resulting ESR spectrum is shown in Fig. 1. The principal radical is unambiguously identified as $CH_3\dot{C}HOH$ resulting from abstraction of an α hydrogen atom—i.e.,

$$CH_3CH_2OH + O\dot{H} \rightarrow CH_3\dot{C}HOH + H_2O$$

The spectrum comprises a doublet of quartets from the alkyl hydrogen atoms and a small splitting from the hydroxyl proton. Measurement of the hyperfine coupling constants provides information on the electronic structure of the radical, and the variation of the hydroxyl splitting with temperature can be interpreted in terms of proton exchange. Inspection of the spectrum in Fig. 1 reveals the presence of a second weaker spectrum, which is assigned to $\dot{C}H_2CH_2OH$ resulting from β hydrogen abstraction. The intensities of the two spectra give a measure of the relative importance of α and β hydrogen

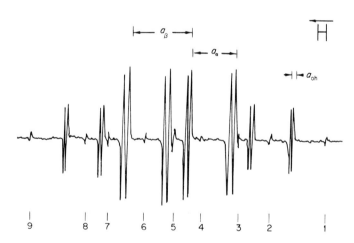

FIG. 1. ESR Spectrum obtained during the photolysis of ethyl alcohol containing hydrogen peroxide at −70°C. The numbered lines arise from $\dot{C}H_2CH_2OH$ and the stronger unnumbered lines from $CH_3\dot{C}HOH$. [*J. Chem. Phys.* **44**, 1245 (1966).]

abstraction. Using this ESR technique the photochemistry of hydrogen peroxide has been extensively explored. At high concentrations of peroxide a
further ESR signal is observed, which is assigned to the HO_2 radical formed
by the reaction

$$\dot{O}H + H_2O_2 \rightarrow H\dot{O}_2 + H_2O$$

It should be noted that the primary reactive species may be an excited electronic
state rather than a short-lived radical. This is the case in the photolysis of
ketones, also studied by Livingston and Zeldes (43), where hydrogen abstraction reactions are attributed to the triplet state of the ketone. A similar photochemical technique using organic peroxides has been extensively developed
by Kochi and Krusic (39).

The second spectroscopic technique of very broad applicability to photochemistry of this type is that of flash photolysis. The use of this technique is
well illustrated by the work of Dogliotti and Hayon (20) on the photolysis
of persulphates. The primary process is again the breaking of the relatively
weak peroxide bond to give in this case $SO_4^{\div}$ radicals—i.e.,

$$[O_3S\!-\!O\!-\!O\!-\!SO_3]^{2-} \xrightarrow{h\nu} 2SO_4^{\div}$$

Radicals are formed by irradiation with a flash of 2537 Å ultaviolet irradiation
from a mercury lamp and detected by measurement of their visible absorption
spectrum observed by means of a second flash after a short interval of time.
A band with a maximum at 455 mμ is observed, and this is assigned to the
sulphate radical. A weakness of this experimental method is that the assignments of the spectra are rarely unambiguous; but on favorable occasions
circumstantial evidence can be fashioned into a very strong case. The lifetime
of the radicals can be measured by varying the duration of the delay between
the exciting flash and the observing flash. Thus the sulphate radical was found
to decay by second-order kinetics and in a typical experiment to have a lifetime of 300 μsec. The sulphate radical is the only transient species observed
at low pH's, but at high pH's (above 8.5) it is rapidly decomposed by the
reaction

$$SO_4^{\div} + H_2O \rightarrow \dot{O}H + SO_4^{2-} + H^+$$

Under these conditions of high pH the resulting hydroxyl radicals readily
react with oxygen to give ozonide ions—i.e.,

$$\dot{O}H + OH^- \rightarrow O^{\div} + H_2O$$
$$O^{\div} + O_2 \rightarrow O_3^{\div}$$

This radical is again identified by its spectrum. Typical reactions of the sulphate radical studied by flash photolysis include hydrogen abstractions from
alcohols and from the bicarbonate ion. Thus in the case of the latter ion the

455-mμ band was rapidly replaced by one at 600 mμ assigned to the carbonate radical—i.e.,

$$SO_4^{\bar{.}} + HCO_3^- \rightarrow HSO_4^- + CO_3^{\bar{.}}$$

In the case of the photolysis of persulphate ions there is ESR evidence to support the contention that sulphate radicals are the primary products. Thus Barnes and Symons (9) have exposed single crystals of potassium persulphate to ultraviolet irradiation and have analyzed the ESR spectrum of the resulting trapped radicals. Unfortunately ^{16}O has no magnetic moment to cause hyperfine splitting, but consideration of the anisotropy of the g value and of the possibility of ^{33}S hyperfine splitting leads to a convincing assignment. One of the more interesting features of this analysis is that the radicals are apparently trapped in the lattice in pairs (as is very reasonable in view of their dissociative method of formation) and remain sufficiently close together to have a significant dipolar interaction. An average separation between members of the radical pair of 15.8 Å was deduced.

As indicated previously the majority of ESR work of interest to photochemists has been carried out in polycrystalline or glassy matrices rather than in solution or single-crystal environments. Although the information obtainable from the former type of experiment is less detailed than that obtained from the latter, nevertheless a great deal of useful work has been carried out in this area, as is indicated by the following example. Stiles et al. (62) have used ESR to examine the photolysis of H_2S, D_2S, and H_2S_2 both in the solid state and in inert-gas matrices. These materials are all decomposed by short-wavelength (λ 2200–2800 Å) ultraviolet light. An example of the type of ESR spectrum obtained is shown in Fig. 2. Comparison with Fig. 1 illustrates the typical difference between liquid-phase and solid-phase spectra. The most intense features of Fig. 2 are attributed to the HS˙ radical. Their absorption frequencies correspond to the three principal g values of the anisotropic radical. Such g values are quite characteristic for a given radical. In addition hyperfine splitting from a single proton can be discerned, and the value of $\sim$7 G is in reasonable agreement with the value of 5.4 G reported for HS˙ in the gas phase. This splitting disappears on deuteration. The primary photolytic process is thus thought to be

$$H_2S \xrightarrow{h\nu} H˙ + HS˙$$

Hydrogen atoms are not observed, presumably because of their shorter lifetime. On warming the matrices a second spectrum appears, which is assigned to polymeric sulfur radicals (S_n)˙, which are secondary reaction products. This spectrum can be observed weakly in Fig. 2. However, changes in the ESR spectrum on warming are not necessarily due to secondary-radical reactions. Thus in this case the spectrum associated with HS˙ changes on warming, but

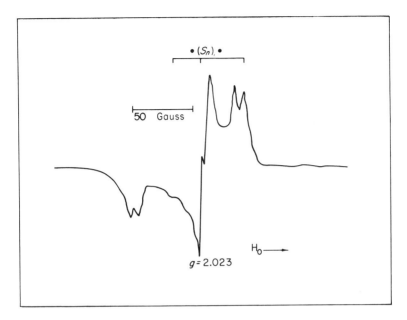

FIG. 2. ESR Spectrum obtained by 24-hr irradiation of hydrogen sulphide. [Reproduced by permission of the National Research Council of Canada from *Can. J. Chem.* **44**, 2149–2155 (1966).]

this is ascribed to physical effects connected with the onset of rotation of the radicals in the lattice. Some caution must obviously be exercised in the interpretation of temperature effects.

A second example of this type of study is provided by the work of Hayon and Saito (*31*) on the photochemistry of nitric acid. Ultraviolet irradiation (2537 Å) at 77°K yields ESR spectra arising from three distinct radicals. The first of these, a triplet with an isotropic ^{14}N splitting of 57.2 G and a smaller anisotropic contribution, has been identified as the NO_2 radical formed by the primary process

$$HNO_3 \xrightarrow{h\nu} NO_2 + OH^{\cdot}$$

The second product of this reaction is not observed, since it is removed by the fast secondary reaction

$$OH^{\cdot} + HNO_3 \rightarrow H_2O + NO_3^{\cdot}$$

The nitrate radical arising from this reaction gives rise to the second ESR signal, which is a rather broad singlet. Experiments with nitric acid isotopically enriched in ^{15}N lead to the replacement of the single line by a partly resolved doublet, confirming the presence of nitrogen in the radical. The absorption spectrum associated with the radical also agrees with that assigned to NO_3 in γ-irradiation studies. It was also shown that substances such as bromide ion, which are known to be hydroxyl radical scavengers, suppress the signal assigned to $NO_3^{\cdot}$, in agreement with the above scheme. This signal can be photobleached by light with $\lambda < 6240$ Å, and on this basis the further photo-reaction:

$$NO_3^{\cdot} \xrightarrow{hv} NO + O_2$$

is postulated. The third ESR signal increases in intensity on photobleaching and is probably associated with the NO molecule.

In the photoreactions considered thus far the initial reaction has been the breaking of a chemical bond. There is a second rather large class of photo-chemical reactions of small inorganic molecules that involve photoionization as an initial step. Early work in this area is discussed in a review by Uri (66). Hayon and McGarvey (32) have used the flash-photolysis technique to study electron-transfer reactions of sulphate, carbonate, and hydroxyl ions. All of these ions absorb only in the vacuum ultraviolet (below 2000 Å), and some special experimental techniques are necessary. In each case the initial reaction is one of ionization—e.g.,

$$SO_4^{=}(aq) \xrightarrow{hv} SO_4^{-}(aq) + e(aq)$$

In the case of the sulphate radical two transient species are observed. The first, absorbing at 455 mμ, is identified as the sulphate radical. The band position is the same as that observed in the photolysis of persulphates and discussed previously. The second band at 240 mμ is assigned to the O_2^{-} radical produced by the reaction

$$O_2 + e(aq) \rightarrow O_2^{-}$$

Both of these radicals decay with second-order kinetics. At high pH the OH$^{\cdot}$ radical arising from the reaction

$$SO_4^{\pm} + H_2O \rightarrow SO_4^{2-} + OH^{\cdot} + H^{+}$$

was observed. Similar experiments with the carbonate ion gave two transient bands at 600 mμ ($CO_3^{\pm}$) and 240 mμ ($O_2^{\pm}$). In these particular experiments the lifetimes of the solvated electron were too short for detection, but in ice matrices it can be observed both optically or by ESR. Halmann and Platzner (29) have studied the photochemistry of phosphate solutions irradiated at 1849 Å. The primary process is again one of electron transfer—i.e.,

$$HPO_4^{2-} \xrightarrow{hv} HPO_4^{\pm} + e(aq)$$

The presence of solvated electrons was demonstrated by their reactions with scavengers such as nitrous oxide and methanol.

The intermediates in all of the reactions discussed above have been free radicals. Such species are a natural consequence of the bond-breaking or single electron-transfer processes that occur on the photolysis of most simple inorganic molecules. It should be pointed out though that not all photolytic reactions involve free radicals. Thus Burak and Treinin (16) have investigated the photochemistry of aqueous solutions of azide ions irradiated at 254 mμ. The overall reaction observed at low azide concentrations can be expressed as

$$N_3^- + 2H_2O \xrightarrow{h\nu} N_2 + NH_2OH + OH^-$$

The excitation of the azide ion is to a short-lived $^1A''$ state, which is thought to dissociate by reaction with an adjacent solvent molecule—i.e.,

$$N_3^{-*} + H_2O \rightarrow NH + N_2 + OH^-$$

The NH formed is in a $^1\Delta$ state, which is relatively long lived, since fluorescence to the $^3\Sigma$ ground state is forbidden by both spin and orbital selection rules. The final products are obtained by reactions of the $^1\Delta$ NH, which is not a free radical. Thus the principal reaction is

$$NH + H_2O \rightarrow NH_2OH$$

giving hydroxylamine, but side reactions with excess azide ion lead to the formation of nitrogen, hydrazine, hydrogen, and ammonia. Addition of ammonia boosts the yield of hydrazine by the reaction

$$NH + NH_3 \rightarrow N_2H_4$$

III. Non-Transition-Metal Compounds

Photochemical research has naturally not been limited to the simple inorganic molecules considered in the previous section. However, the majority of the work on more complex inorganic molecules, with the exception of transition-metal complexes, which will be considered separately, has been oriented more toward studies of products than of mechanisms. What mechanistic work has been carried out has yielded results not dissimilar to those obtained with simpler molecules. These generalities can be illustrated by reference to a few recent papers.

Langmuir and Hayon (40) have reported flash-photolysis studies of mercuric halides and their complexes, undertaken with a view to establishing the primary photolytic reaction. Photolysis of $HgCl_2$, $HgCl_4^{2-}$, $HgBr_2$, $HgBr_4^{2-}$, and HgI_4^{2-} yielded transient species identified as $Cl_2^{\cdot-}$, $Br_2^{\cdot-}$, or $I_2^{\cdot-}$, depending on the halide present in the starting material. In these experiments short-wavelength ultraviolet light was removed by filters to eliminate the photolysis of uncomplexed halide ions. Several possible primary processes were considered. The participation of hydroxyl radicals formed by charge transfer from a solvent water molecule to a mercury ion and the subsequent reaction of the hydroxyl radical with halide ion was disproved by studies of the pH dependence of the radical yields and by the lack of effect of OH-radical scavengers. Two alternative processes are considered; namely,

$$HgX_2 \rightarrow HgX + X \cdot \quad \text{followed by} \quad X \cdot + X^- \rightarrow X_2{}^{\pm}$$

or

$$HgX_2 \rightarrow [Hg^+X_2{}^-] \rightarrow X_2{}^{\pm}$$

It was not found possible to distinguish between these two possibilities. In the photolysis of HgI_2 a transient species absorbing at 330 mμ that is not $I_2{}^{\pm}$ was observed. This could be either $I \cdot$ or $HgI \cdot$. The latter is perhaps favored by the observation that mercurous iodide is produced in this reaction. Flash-photolytic experiments of this kind provide a very convenient method of producing free radicals such as $X_2{}^{\pm}$ and of measuring their rates of reaction with various substrates. Some comparative data obtained in this way for various inorganic radicals and collected together by Langmuir and Hayon are shown in Table I. The general order of reactivity is $OH \cdot > SO_4{}^{\pm} > Cl_2{}^{\pm} > NO_3 \cdot > Br_2{}^{\pm} \gg I_2{}^{\pm}$. It should be noted that $I_2{}^{\pm}$ reacts very slowly—e.g., its rate of reaction with alcohols is $< 10^2 M^{-1} sec^{-1}$.

The photochemistry of organometallic compounds of mercury has been extensively reviewed by Bass (12). In general there seems to be little doubt that the primary process is the breaking of a mercury–carbon bond and that

TABLE I

COMPARISON OF THE RATE CONSTANTS OF REACTION OF INORGANIC OXIDIZING RADICALS WITH SOME SIMPLE ORGANIC AND INORGANIC COMPOUNDS IN NEUTRAL AQUEOUS SOLUTIONS [a]

Compound	OH	$SO_4{}^-$	$Cl_2{}^-$	NO_3	$Br_2{}^-$
CH_3OH	4.8×10^8	$2.5 \pm 0.4 \times 10^7$	$3.0 \pm 1.0 \times 10^7$	$1.0 \pm 0.1 \times 10^6$	$1.5 \pm 0.2 \times 10^4$
CH_3CH_2OH	1.1×10^9	$7.7 \pm 2.2 \times 10^7$	$2.9 \pm 0.2 \times 10^7$	$3.9 \pm 0.3 \times 10^6$	$1.4 \pm 0.2 \times 10^5$
i-PrOH	3.9×10^9	$8.3 \pm 3.0 \times 10^7$	$6.1 \pm 0.7 \times 10^7$	$3.6 \pm 0.2 \times 10^6$	$3.0 \pm 0.5 \times 10^5$
HCOOH	1.0×10^9	$1.4 \pm 0.2 \times 10^6$	—	$2.1 \pm 0.1 \times 10^5$	—
CH_3COOH	1.4×10^5	$8.8 \pm 0.2 \times 10^4$	—	$4.6 \pm 0.4 \times 10^4$	—
H_2O_2	4.5×10^7	—	$7.4 \pm 0.3 \times 10^7$	—	$3.9 \pm 0.2 \times 10^7$
Ce^{3+}	2.2×10^8	$1.4 \pm 0.3 \times 10^8$	—	$3.7 \pm 0.1 \times 10^5$	—
Tl^+	9.0×10^9	$1.7 \pm 0.2 \times 10^9$	—	$3.5 \pm 0.1 \times 10^7$	—
$HCO_3{}^-$	2.8×10^3	$9.1 \pm 0.4 \times 10^6$	—	—	—

[a] $k(R_{ox} + S)$, $M^{-1} sec^{-1}$.

subsequent events depends entirely on the reactivity of the resulting organic radicals. Thus, for example, the phenyl radicals obtained from mercury diphenyl readily react with isopropyl alcohol:

$$C_6H_5{\cdot} + (CH_3)_2CHOH \rightarrow (CH_3)_2C{\cdot}OH + C_6H_6$$

whereas the resonance-stabilized benzyl radicals from mercury dibenzyl undergo few reactions other than dimerization. Of greater spectroscopic interest is the report by Winkler *et al.* (*72*) of the generation of free-radical anions by the photolysis of aromatic hydrocarbons in the presence of phenyl lithium. Thus irradiation of anthracene and phenyl lithium in diethyl ether leads to a solution showing a well-resolved ESR spectrum of the lithium salt of the anthracene anion. The large hyperfine coupling with ^7Li clearly shows that the lithium is associated with the paramagnetic species. The mechanism of this photolysis is uncertain. The authors suggest that it proceeds by electron transfer from the " anion of the organolithium reagent " to the hydrocarbon. With anthracene there is a slow thermal reaction, but the photoreaction is faster by a factor of 10^5. Photochemical electron-transfer reactions are of course of great biological significance in photosynthesis, and this has provided the impetus for a number of studies involving inorganic analogues. Thus Quinlan (*55*), for example, has observed photochemical electron transfer from zinc tetraphenyl porphine to *p*-benzoquine, using ESR to detect the resulting semiquinone.

Finally we note two examples in which some deductions regarding mechanism have been made from product analysis. These are typical of a great amount of photochemical work and illustrate some of the problems that await more detailed mechanistic studies. Kochi and Bethea (*38*) have studied the photolysis of Tl(III) carboxylates. They attribute the high yields of dimers from primary carboxylic acids to high local concentrations of radicals due to the instability of the Tl(II) species generated in an excited state. Flash-photolytic or ESR evidence for Tl(II) compounds has not yet been brought forward. In a similar vein Williams *et al.* (*71*) have investigated the photolysis of tetrarylborates. Biphenyls are the principal products, and it was shown that their formation proceeds by an intramolecular reaction around the boron atom and that the coupling between the phenyl rings occurs at the carbons originally bonded to boron. Details of the mechanism would obviously be of considerable chemical interest.

IV. Transition-Metal Complexes

There has recently been a renewed interest in the photochemistry of transition-metal complexes. The subject has been reviewed by Wehry (*69*). It is usual to distinguish three different classes of photochemical reactions;

namely, ligand-substitution processes, isomerizations, and oxidation–reduction reactions. The simple expectation would be that ligand substitution or rearrangement would result from excitation of electronic energy levels corresponding to d–d transitions, and oxidation or reduction reactions from irradiation of charge-transfer bands. Experimental results show that this is a rather drastic oversimplification of the situation (*11*). This is perhaps not surprising, in view of the various processes alluded to in the introduction that can lead to the rapid transfer of energy from one excited state to another, and of the fact that only in purely ionic compounds can the absorption bands be classified as purely d–d or purely charge-transfer transitions. Since most of the chemically more interesting transition-metal complexes have at least some covalent character in their bonding, complexity in their photochemistry can be anticipated on these grounds alone. There are also practical limitations on the complexes for which photochemical ligand-exchange reactions can be studied: if the thermal-exchange reaction is fast, there are obvious difficulties in studying the photochemical reaction. For this reason many of the studies of first-row transition complexes have been concerned with Cr(III) and Co(III) compounds. These two series of complexes differ in that photoreduction occurs with Co(III) but not with Cr(III). This difference no doubt mostly reflects the much reduced stability of Cr(II) compared to Co(II). Initially we will therefore consider some photochemical studies of complexes of these metals.

A. Cr(III) Complexes

The ground state of octahedral Cr(III) complexes has $^4A_{2g}$ symmetry and arises from the $(t_{2g})^3$ configuration. Commonly three fairly strong absorptions are observed in the visible and near-ultraviolet spectrum of chromium complexes and are assigned to transitions to the theoretically expected $^4T_{2g}$, $^4T_{1g}(^4F)$, and $^4T_{1g}(^4P)$ excited states. All these excited states are associated with the $(t_{2g})^2(e_g)$ configuration in the strong-field limit. The lowest excited state is ^2E_g (from the $(t_{2g})^3$ configuration), which is metastable. The main point at issue in studies of the photosubstitution of chromium complexes is the assignment of the reactive intermediate. Arguments have been advanced in favor of the quartet excited states, the doublet state, and even vibrationally excited ground-state levels.

Perhaps the simplest substitution reaction is the photochemical exchange of water with $Cr(H_2O)_6^{3+}$. This has been studied by Plane and Hunt (*53*), using water isotopically enriched in ^{18}O. Their principal observations were firstly that the quantum yield for the reaction was low (~ 0.01) and decreased with decreasing temperature, secondly that the yield was the same for all three allowed absorption bands, and thirdly that no fluorescence could be observed. This last observation eliminates the possibility that the low quantum yields are due to competition between the photochemical reaction and fluorescence. The observation that the quantum yield was independent of wavelength is not unexpected, since rapid internal conversion (lifetime $\sim 10^{-11}$ sec) of the second and third excited quartets to the lowest excited quartet is anticipated. A possible explanation of the low quantum yield is that internal conversion to the vibrationally excited ground state is also efficient. In fact the reactive species may be vibrationally excited ground-state molecules. Alternatively there could be interstate crossing to the ^2E_g, with this state being responsible for the photoreactivity. On this model the low probability of interstate crossing accounts for the low quantum yields.

Quantum yields for other substitution reactions are significantly higher— e.g., ~ 0.3 for the photoaquation of $Cr(NH_3)_6^{3+}$. Edelson and Plane (23) have studied this reaction and concluded that the ^2E_g state is the reactive intermediate on the basis of finding a quantum yield of unity on irradiating at a wavelength of 650 mμ, which corresponds to the very weak quartet-to-doublet transition. However, there are experimental difficulties in carrying out actinometry at these long wavelengths, and a more recent study by Wegner and Adamson (67) gave quantum yields that were identical within experimental error for quartet and for doublet absorption. From a study of the photochemical behavior of a number of Cr(III)-complex ions Schläfer (57) concluded that the ^2E_g was the photochemically important level. He argues that the lifetimes of the quartet states ($1/\tau$ estimated $> 10^7$ sec^{-1}) are too short for chemical reactivity and that population of the doublet state is demonstrated by the observation of phosphorescence from a variety of chromium complexes. Further he points out that although the ^2E_g state arises from the $(t_{2g})^3$ configuration, the additional interelectron repulsion energy compared with the $^4A_{2g}$ ground state will reduce the ligand-field activation energy required to reach a pentagonal bipyramid (for a bimolecular reaction) or a square pyramid (for a unimolecular reaction) intermediate.

In a later paper Adamson (2) has reviewed the photochemistry of Cr(III) complexes and systematically examined the photoreactions and quantum yields of a series of mixed aquo-ammine compounds. One important conclusion from this and similar studies is that the products of thermal and photochemical reactions can be different. This observation argues strongly against the suggestion that reaction occurs from vibrationally excited ground-state levels. On the basis of the mixed aquo-ammine studies Adamson has formulated the following empirical rules for chromium photochemistry:

(1) Each ligand can be assigned a ligand field strength according to its position in the spectrochemical series. The six ligands are considered to be arranged in three pairs corresponding to the three axes of the octahedron.

The axis with the weakest average ligand field strength is labilized. Furthermore the total quantum yield is about the same as would be observed for a symmetrical (O_h) complex with the same average field.

(2) If the labilized axis contains two different ligands, the one with the greater ligand field strength is displaced.

(3) The above preferences are more pronounced for irradiation of the longer-wavelength quartet bands than for irradiation of the shorter-wavelength bands.

Thus, for example, in $Cr(amine)_5H_2O^{3+}$ the amine molecule trans to the water is displaced, in $trans$-$Cr(amine)_4(H_2O)_2^{3+}$ one of the waters is displaced, and in cis-$Cr(amine)_4(H_2O)_2^{3+}$ an amine trans to a water is displaced. At present the interpretation of these rules is not clear, but a detailed description of the photochemical process will obviously need to account for these observations.

B. Co(III) Complexes

The photochemistry of cobalt is somewhat more complicated than that of Cr(III), since both oxidation–reduction and ligand-exchange reactions can occur. It would appear that the two processes are not independent but are quite closely related. Thus Adamson (*1*), for example, has studied the photolysis of a series of Co(III) acido pentammines of the type $Co(NH_3)_5X^{2+}$. He found that their behavior varied from 100% aquation reaction with $X = Cl^-$ to 100% redox reaction with $X = I^-$. The compounds with $X = Br^-$, NO_2^-, SCN^-, and N_3^- gave mixtures of aquation and oxidation products. The obvious correlation is with the ease of oxidation of X^-. The quantum yields are highest for those compounds that photoreduce rather than photoaquate. However, the quantum yields fall off with increasing wavelength, the decrease being greater for photoreduction than for photoaquation. It was suggested that the primary process in both the photoreduction and the photoaquation reactions is the homolytic cleavage of the Co—X bond to give a Co(II) compound and X· atoms or radicals. When initially produced these radicals will be surrounded by a cage of solvent molecules and if they are unable to escape will most probably recombine to give the initial starting materials. It is supposed that the probability of escape from the cage will depend on the energy provided in excess of that needed to effect the charge-transfer process. Thus on this picture absorption of a quantum of high-energy radiation provides a sufficient excess of kinetic energy to separate the charge-transfer products and give a high probability of further reaction. In this manner the wavelength dependence of the quantum yield is rationalized. It is obvious how this primary process can lead to the products of photoreduction of the cobalt complex but a little less obvious how the photoaquation products arise. The suggestion made is that failure of the primary products to escape from the solvent cage, rather than resulting in simple recombination, may be followed by a thermal electron-transfer reaction leading to a five-coordinate Co(III) pentammine molecule. This will then complete its coordination shell by acquiring a ligand from the solvent shell. Since the actual aquation process is thermal rather than photochemical, this mechanism implies that aquation will occur only if there is a significant tendency to undergo aquation as a dark process. In general this prediction appears to be upheld.

More recent work by Penkett and Adamson (51) has provided confirmation of the charge-transfer nature of the primary process. Thus flash photolysis of solutions containing $[Co(NH_3)_5I]^{2+}$ shows the spectra of transient iodine atoms. Bromine atoms formed from the analogous bromide tend to react with the ammonia. However, this simple charge-transfer mechanism cannot be generalized to all transition-metal halides, even when reduction of the metal ion is a plausible process. Thus flash photolysis of $PtBr_6^{2-}$ does not lead to bromine atoms but appears to be best represented by the two-electron reduction:

$$PtBr_6^{2-} \xrightarrow{h\nu} [PtBr_4^{2-}] + Br_2$$

Formally at least, this is a "reductive elimination" reaction that mirrors the oxidative addition reactions common in these Group-VIII complexes, and there may well be mechanistic similarities.

Endicott and Hoffmann(24) have carried out extensive studies of the photolysis of Co(III) complexes with ultraviolet (2537 Å) light with the aim of investigating the primary process more thoroughly. At this wavelength photoreduction is the dominant process. Specifically they studied the pH dependence of the quantum yields and argued from this that protonation of the excited state is an important step in the mechanism. They deduced from this that the reactive state must be relatively long lived and hence of different multiplicity from the ground state. As with chromium complexes the rate of intersystem crossing is therefore a factor of importance. However, Wehry (69) has pointed out that excited-state proton-transfer reactions can be very fast and has questioned the necessity of postulating intersystem crossing to a metastable state. It is apparent that the primary photochemical processes in these Co(III) compounds are still far from being unequivocally established.

There has been some application of ESR to the study of cobalt photochemistry. Thus Eaton and Suart (22) have investigated the photolytic production of hydrogen atoms and free radicals from frozen aqueous solutions of cobalt compounds. With simple Co(II) salts the primary process appears to be photooxidation of the cobalt by a solvent molecule, giving hydrogen atoms which are easily identified by ESR—i.e.,

$$Co^{2+} + H_2O \xrightarrow{h\nu} Co^{3+} + OH^- + H^{\cdot}$$

Similar photooxidations of Fe^{2+} and I^- have been reported by Moorthy and Weiss (48). It was observed that in the presence of oxalate anions the production of hydrogen atoms is about 600 lines more efficient than with a variety of other anions. It was suggested that in this case the electron transfer can proceed through the conjugated π system of the oxalate ion—i.e.,

This kind of electron-transfer mechanism is familiar in thermal reactions from the work of Taube (64). Rather more surprisingly the principal product from the photolysis of trisoxalatocobaltate in ice at liquid-nitrogen temperature is also hydrogen atoms. Since this must be a process involving the photoreduction of Co(III) to Co(II) with the concurrent oxidation of the ligand, hydroxyl radicals rather than hydrogen atoms might be anticipated. It was shown, though, that hydrogen atoms are not a true primary product of this reaction. Thus for short irradiation times (less than a minute) the most intense ESR signal is a line close to $g = 2$ assigned to the oxalate radical. With longer times of irradiation the hydrogen-atom signal becomes relatively more intense. A plot of the relative intensities of the hydrogen and oxalate radicals is shown in Fig. 3. The suggested mechanism involves formation of an oxalate radical followed by its thermal decomposition—i.e.,

The oxalate radicals after a short time of irradiation reach a steady-state concentration, at which point the rate of production by photolysis is balanced by the rate of loss due to decomposition and possibly recombination. The hydrogen atoms, on the other hand, are stable in ice at liquid-nitrogen or lower temperatures and accumulate. In both the Co(II) and Co(III) systems a variety of secondary radicals are observed if the ice is allowed to warm. The products obtained in the presence of carboxylic acids can be rationalized by postulating addition of the hydrogen atom to the carbonyl carbon of the

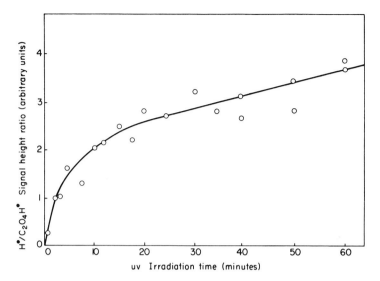

FIG. 3. Ratio of hydrogen-atom signal strength to oxalate-radical signal strength as a function of ultraviolet irradiation time. [Reprinted from *J. Phys. Chem.* **72**, 400 (1968). Copyright by the American Chemical Society. Reprinted by permission of the copyright owner.]

carboxylic acid followed by either homolytic or heterolytic fission of the carbon–carbon bond—i.e.,

$$
H^\bullet + \underset{\underset{R}{|}}{\overset{\overset{O}{\parallel}}{C}}-OH \longrightarrow \left[H-\underset{\underset{R}{|}}{\overset{\overset{O^\bullet}{|}}{C}}-OH \right]
\begin{array}{l}
\xrightarrow{\text{heterolytic}} HCO^\bullet + \left. \begin{array}{l} OH^- \\ R+ \end{array} \right\} ROH \\[2ex]
\xrightarrow{\text{homolytic}} HCOOH + R^\bullet
\end{array}
$$

It is possible that addition to the carbon rather than addition to the oxygen occurs because the oxygens are involved in complexing with the metal ions.

Finally we might take note of a somewhat analogous electron-transfer reaction observed by Feitelson and Shaklay (*25*). They report that the fluorescence of hydroquinone is strongly quenched by $Co(CN)_6^{3-}$. The cobalticyanide does not react with hydroquinone in its ground state, but apparently there is a rapid electron transfer with the singlet excited state. Flash-photolysis experiments showed an increase in the yield of semiquinone, as predicted by this interpretation, on addition of cobalticyanide to the hydroquinone. Thus in this case we have electron transfer from an excited state of a potential ligand to the metal rather than from an excited state of the metal ion to a ligand.

C. Complexes of Other Transition Metals

Photochemical reactions of a large number of complexes of transition metals other than chromium and cobalt have been reported, but detailed mechanistic studies have been attempted in relatively few cases. For many complexes the thermal ligand-exchange reactions are fast, and this places a substantial restriction on the study of photochemical ligand exchange. This is a somewhat less restrictive factor for oxidation–reduction reactions.

An obviously desirable objective would be to compare the photochemistry of a series of complexes containing the same ligand but different transition-metal ions. Some progress in this direction has been made with oxalate complexes. Thus Porter *et al.* (*54*) have measured the quantum yields for the photolysis of Fe(III), Mn(III), Co(III), and Cr(III) oxalates as a function of irradiating wavelength. Their results are shown diagrammatically in Fig. 4. The Cr(III) complex cannot be photolytically reduced, but photoracemization is possible (see below). For the other complexes the quantum yields diminish as the nature of the absorption process changes from being predominantly charge transfer to predominantly *d–d* transitions. The order of decreasing

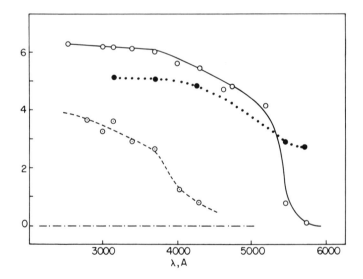

Fig. 4. Quantum yields of transition-metal oxalates as a function of wavelength. Primary: solid line, $Fe(C_2O_4)_3^{3-}$; broken line, $CO(C_2O_4)_3^{3-}$. Quantum-yield spectra: dotted line, $Mn(C_2O_4)_3^{3-}$; dot-dash line, $Cr(C_2O_4)_3^{3-}$. [Reprinted from *J. Am. Chem. Soc.* **84**, 4027 (1962). Copyright by the American Chemical Society. Reprinted by permission of the copyright owner.]

quantum yields is Fe > Mn > Co > Cr, which is the order of increasing energy of the charge-transfer excited state. The quantum yields are always less than unity, and it is pointed out that this is consistent with a mechanism involving intersystem crossing to a metastable excited state. It is considered that the nature of the wavelength dependence precludes the low quantum yields being explained as a cage recombination effect. At long wavelengths the quantum yields are higher than would be expected from absorption in the tail of the charge-transfer band, and rapid internal conversion between ligand-field states and charge-transfer states is suggested. It is interesting that the order of photochemical reactivity Fe > Mn > Co > Cr in these oxalate complexes is confirmed by the ESR studies of Shagisultanova et al. (59). They observed the formation of hydrogen atoms and free radicals during the ultraviolet photolysis of these oxalate complexes (as has been described in more detail for the cobaltioxalate above) and noted the above sequence of radical concentrations for the different metal ions.

The stability of the d^6 complexes of Pt(IV), Ir(III), and Rh(III) makes them attractive compounds for photochemical work, and a number of studies in this area have been reported. In contrast to Co(III) in the first transition series, ligand exchange is the exclusive reaction and redox reactions do not occur. Thus Moggi (46), for example, finds that irradiation of $[Rh(NH_3)_5Cl]^{2+}$ at 254 or 365 mμ gives only photoaquation. Similarly Balzani et al. (5) have irradiated both $PtBr_6^{2-}$ and PtI_6^{2-} in both ligand-field and charge-transfer bands and find exclusive photoaquation in all cases. They deduce that rapid internal conversion leads to reactions involving the same excited state regardless of the frequency of irradiation and propose that the primary reaction involves heterolytic fission of the metal–ligand bond. Bauer and Basolo (13) have utilized this selectivity in photoreactivity to synthesize some otherwise rather inaccessible complexes. Thus the compounds trans-$[M(en)_2X_2]^+$, where en = ethylene diamine, M = Rh or Ir, and X = Cl, Br, or I, can be photoaquated quantitatively. In the presence of excess halide ion, halide exchange rather than aquation occurs, and this reaction was used to synthesize trans-chlorobromo ethylenediamine and trans-chloroiodo ethylenediamine complexes. These clean substitution reactions may be contrasted with the report of Shagisultanova and Il'yukevich (58) that the ultraviolet photolysis of aqueous solutions of Cu(II) ethylenediamine complexes leads to oxidation of the ligand, with carbon dioxide being the principal product. Indeed Cu(II) ions sensitize the photodecomposition of ethylenediamine to carbon dioxide and ammonia, with the relative amounts of these two products depending on the concentration of copper ions.

Finally in this section we might note the interesting photoreaction shown in the diagram, which has been reported by Hata et al. (30).

The presence of the metal hydride was demonstrated by NMR, and the reaction is reversible under ethylene pressure. Thermal reactions involving the migration of the ortho hydrogen of an aryl phosphine ligand to the metal atom are now well recognized (14), but the occurrence of such a photolytic reaction has certainly interesting implications regarding the electronic structure of the excited state and suggests the possibility of rather sophisticated photochemical syntheses with complexes of this type.

V. Cyanide and Carbonyl Complexes

The bonding in most of the transition-metal complexes discussed above involves predominantly electron donation from the ligand to the metal. There is, however, an important class of compounds for which the electron accepting properties of the ligands are of considerable importance. Cyanides and particularly carbonyl compounds are typical of this class of substance. Cyanides always correspond to the strong-field case in the ligand-field formulation, and carbonyls are best regarded as covalent compounds. In these circumstances distinctions between d–d and charge-transfer transitions begin to lose their validity, and it is of interest to inquire what effect this will have on the photochemistry.

Moggi, et al. (47) have investigated the photochemistry of a number of complex cyanides and have also summarized the results of earlier workers in this area. The overall picture is not too much different from that obtained with non-π-bonding ligands. In aqueous solution both photoaquation and redox reactions are possible. In most cases extensive mechanistic studies have not been carried out, but the mode of reaction has been deduced from the final products. Briefly summarized, the conclusions are that the d^1 complexes $Mo(CN)_8^{3-}$ and $W(CN)_8^{3-}$ undergo a redox reaction; d^2, d^3, d^4, d^5, and d^6 complexes exemplified by $Mo(CN)_8^{4-}$, $W(CN)_8^{4-}$, $Cr(CN)_6^{3-}$, $Mn(CN)_6^{3-}$, $Fe(CN)_6^{3-}$, $Fe(CN)_6^{4-}$, and $Co(CN)_6^{3-}$ undergo a primary photoaquation reaction; and the d^8 complexes $Ni(CN)_4^{2-}$, $Pd(CN)_4^{2-}$, and $Pt(CN)_4^{2-}$ are photolytically inactive. Reduced reactivity in the cyanide complexes is not surprising in view of the high metal–ligand bond strengths. It should be noted that in complexes such as $Fe(CN)_6^{3-}$ products resulting from the reduction of the iron are also observed. Moggi et al. consider that their results are most consistent with the primary reaction being one of photoaquation, but are unable to reach a definitive conclusion on this point. Similar remarks apply to $Fe(CN)_6^{4-}$. Thus Balzani et al. (3, 4) have studied the photochemical substitution of bipyridyl and phenanthroline ligands for cyanide and observed reactions involving only substitution. These studies involved 365-mμ irradiation. On the other hand Ohno (50), using 254-mμ irradiation of $Fe(CN)_6^{4-}$, observed redox reactions. The hydrated electron is considered to be the primary product, and secondary reactions with a variety of electron scavengers were studied. Typical of these reactions are

$$e^- + N_2O \rightarrow N_2 + OH^. + OH^-$$
$$e^- + H_2PO_4^- \rightarrow H^. + HPO_4^{2-}$$

The first reaction is suggested as a clean way of producing hydroxyl radicals free from the $HO_2^.$ that is always present in hydrogen peroxide photolyses. Dainton and Airey (19) have considered the primary process in the photo-oxidation of ferrocyanide in rather more detail. They conclude that in neutral or alkaline solution hydrated electrons are the primary product, and that these react competitively with the product $Fe(CN)_6^{3-}$ or with added electron scavengers. In acid solutions either electrons or hydrogen atoms may result from the reactions

$$H\,Fe(CN)_6^{3-} \xrightarrow{hv} H^+ + Fe(CN)_6^{3-} + e$$

or

$$H\,Fe(CN)_6^{3-} \xrightarrow{hv} H^. + Fe(CN)_6^{3-}$$

The most usual photoreaction of carbonyl compounds is substitution. Such photochemical reactions are of synthetic importance, and this area has been reviewed by Strohmeier (63). Thus typically irradiation of chromium hexacarbonyl in the presence of donor ligands such as aniline, pyridine, acetonitrile, benzonitrile, or dimethylsulfoxide leads to the synthesis of compounds of the type $Cr(CO)_5L$. It is notable that ^{14}CO will exchange photochemically with metal carbonyls (such as $Cr(CO)_6$), which are inert to thermal exchange. It is assumed that the substitution takes place by an SN1 mechanism with a five-coordinated intermediate. Herberhold (33) has synthesized $Cr(CO)_5\,TCNE$ (TCNE = tetracyanoethylene) by a similar reaction, and Fields et al. (26) have prepared fluoroolefin complexes of iron by ultra-violet-induced reactions with iron pentacarbonyl.

The presence of radical intermediates during the irradiation of bimo-lecular carbonyl complexes has been demonstrated by the experiments of

Bamford *et al.* (*7*). Bamford and his collaborators had previously demonstrated that polymerization of methylmethacrylate can be initiated by trichloromethyl radicals. In the experiments referred to they irradiated solutions of $Mn_2(CO)_{10}$ and $Re_2(CO)_{10}$ in methylmethacrylate and then added carbon tetrachloride to the mixtures. They interpreted the observed formation of polymer as arising from $CCl_3{\cdot}$ radicals resulting from reaction of a product of the metal carbonyl photolysis with carbon tetrachloride. Their results were consistent with the presence of two types of metal containing intermediates, one of which in the case of the rhenium carbonyl was relatively long lived. It is tempting to suppose that the metal–metal bond has been broken during the photolysis. The observation of Beveridge and Clark (*15*) that tetrafluoroethylene reacts photolytically with $Co_2(CO)_8$ to give an insertion product formulated as $(OC)_4CoCF_2Co(CO)_4$ suggests a similar mechanism.

VI. Photoisomerization of Transition-Metal Complexes and Selection Rules

Organic photochemistry has received considerable stimulation of late from the development by Hoffmann and Woodward (*35*) of the concept of selection rules for chemical reactions. The crux of their method involves the correlation of the electronic configurations of the reactants and of the products together with the plausible supposition that if the ground state of the reactants correlates with an excited state of the products the thermal reaction will require a large energy of activation. However, for a photochemical reaction the correlation involves an excited state of the reactants and the selection rules may be quite different. There are indeed a substantial number of examples known in organic chemistry where the thermal reactivity and the photochemical reactivity are qualitatively different, and it is the consistency of these examples with the rules deduced by Woodward and Hoffmann that has been largely responsible for the enthusiastic acceptance of this approach to chemical reactivity. In transition-metal photochemistry it is much more difficult to find clear-cut examples, but it seems probable that the selection-rule concept will have some utility. Selection rules are most easily applied to reactions involving " concerted " mechanisms, and for this reason the most likely area of application is that of intramolecular rearrangements or isomerizations. A number of photolytic isomerizations and racemizations of transition-metal complexes have been reported, but in many cases it is not clear whether the reaction proceeds by an intermolecular or by an intramolecular mechanism. Thus Noack (*49*), for example, in a study of the infrared spectrum of *cis*-$Fe(CO)_4I_2$, found three carbonyl stretching bands assignable to the cis isomer and a fourth band that only appeared on exposure to light and that he concluded was associated with the trans isomer. Johnson *et al.* (*36*) confirm this observation of photolytic cis → trans isomerization. They also have added the further observation that if the solutions remain in the dark, the band due to the trans isomer disappears but there is not a corresponding enhancement of the cis bands. It would appear, therefore, that this is not a simple case of a reversible cis ⇌ trans isomerization but rather that the trans isomer is an intermediate in the photodecomposition of the cis isomer. This would seem to argue against a simple dissociative mechanism for cis–trans isomerization and in favor of an intramolecular rearrangement, but the argument is by no means unambiguous. In the same paper Lewis *et al.* report studies of the exchange reaction of *cis*-$Fe(CO)_4I_2$ with $C^{18}O$, which presumably occurs by a dissociative mechanism, but do not mention whether trans isomer is produced if the exchange is carried out in the dark. Such an experiment would determine whether the ligand-exchange and isomerization reactions have a common intermediate.

Cis–trans isomerization of square planar complexes has been more extensively studied. Thus Haake and Hylton (*28*) have reported the photolytic isomerization of *cis*- and *trans*-bis(triethylphosphine)-dichloroplatinum(II). Absorption in the *d–d* bands is responsible for the isomerization. They suggest that a possible mechanism would involve a triplet-state intermediate, since the lowest triplet state of d^8 complexes has tetrahedral geometry and a tetrahedral intermediate is required for an intramolecular isomerization process. Further examples of photolytic cis–trans isomerization have been reported by Perumareddi and Adamson (*52*) and by Balzani and Carassiti (*6*). The latter authors have studied the photoisomerization of $Pt(gly)_2$ (gly = glycine) and demonstrated by using labeled glycine that no ligand exchange occurs—i.e., that the isomerization is intramolecular. In this case there is no competing thermal reaction, and quantum yield of 0.13 was found for 313-mμ irradiation, this quantum yield being independent of the presence of excess ligand. Their favored mechanism is again one involving a tetrahedral intermediate.

Eaton (21) has considered the application of symmetry-based selection rules to both isomerization and substitution reactions of transition-metal complexes. The approach adopted is based on crystal field theory. The five *d* orbitals of the transition metal comprise a basis set upon which crystal field Hamiltonians H_A and H_B corresponding to the initial and final geometries operate. The resulting orbitals are correlated according to the symmetry common to H_A and H_B. This simple theory leads to the prediction that cis-trans isomerization of d^8 complexes will be thermally forbidden but photochemically allowed, in accord with experiment. Similar considerations show that exchange of axial and apical ligands in a d^8 trigonal bipyramid is thermally allowed but that this process would be forbidden for spin-free d^2 or spin-paired d^3 and d^4 complexes. Whitesides (*70*) has pointed out some of the limitations of this simple crystal-field approach and emphasized the advantages of correlating electronic states rather than orbitals. There is no doubt that the application of symmetry-based selection rules to transition-metal compounds is more difficult than it is to simple organic compounds because of the necessity of taking into account both the covalent bonding forces and the interelectron repulsion, but the prospects for qualitative utility would seem quite good. Mango and Schachtschneider (*44, 45*) have applied the Woodward–Hoffmann rules rather more directly to transition-metal compounds. They consider expanded correlation diagrams in which the metal orbitals are also included and show that this can result in some reactions that were originally forbidden becoming allowed. Thus this approach can be used to rationalize some examples of transition-metal catalysis. Photolytic reactions have not been explicitly considered in this context. An example of such a reaction is perhaps provided by the photochemical dimerization of norbornene in the presence of copper halides reported by Trecker *et al.* (*65*).

$$\text{Cu halide} \atop h\nu$$

Finally we might note the possibility of racemization reactions. Spees and Adamson (*60*) have reported the photoracemization of trioxalatochromate ions and concluded that it proceeded by a substitution mechanism. More intriguing is the recent report of Stevenson and Verdieck (*61*) that irradiation of racemic mixtures of this and related compounds with circularly polarized light leads to optical resolution of the complexes. With continued photolysis an equilibrium optical activity is attained.

VII. Conclusion

The subject of inorganic photochemistry is a rapidly developing field, particularly that portion dealing with transition-metal complexes. Many of the tools used in this field are those developed by the spectroscopist, and as the attention of the photochemist is focused increasingly on the primary photolytic processes the language employed becomes that of spectroscopy. It is hoped that a chapter on this subject in a book devoted to spectroscopy will help to draw attention to these close ties and encourage further interest in this area by spectroscopists.

References

1. Adamson, A. W., *Discussions Faraday Soc.* **29**, 163 (1960).
2. Adamson, A. W., *J. Phys. Chem.* **71**, 798 (1967).
3. Balzani, V., and Carassiti, V., *Ann. Chim.* **54**, 251 (1964).
4. Balzani, V., and Carassiti, V., *Ann. Chim.* **54**, 103 (1964).
5. Balzani, V., Manfrin, F., and Moggi, L., *Inorg. Chem.* **6**, 354 (1967).
6. Balzani, V., and Carassiti, V., *J. Phys. Chem.* **72**, 383 (1968).
7. Bamford, C. H., Hobbs, J., and Wayne, R. P., *Chem. Commun.* 469 (1965).
8. Bargon, J., and Fischer, H., *Z. Naturforsch.* **22a**, 1556 (1967).
9. Barnes, S. B., and Symons, M. C. R., *J. Chem. Soc.* A, 66 (1966).
10. Basalo, F., and Pearson, R. G., "Mechanisms of Inorganic Reaction," 2nd ed., p. 654. Wiley, New York, 1967.
11. Basalo, F., and Pearson, R. G., "Mechanisms of Inorganic Reactions," 2nd ed., p. 658. Wiley, New York, 1967.
12. Bass, K. C., *Organomet. Chem. Rev.* **1**, 391 (1966).
13. Bauer, R. A., and Basolo, F., *J. Am. Chem. Soc.* **90**, 2437 (1968).
14. Bennett, M. A., and Milner, D. L., *Chem. Commun.* 581 (1967). Parshall, G. W., *J. Am. Chem. Soc.* **90**, 1669 (1968).
15. Beveridge, A. D., and Clark, H. C., *J. Organomet. Chem.* **11**, 601 (1968).
16. Burak, I., and Treinin, A., *J. Am. Chem. Soc.* **87**, 4031 (1965).
17. Calvert, J. G., and Pitts, J. N., "Photochemistry." Wiley, New York, 1966.
18. Calvert, J. G., and Pitts, J. N., "Photochemistry," p. 296. Wiley, New York, 1966.
19. Dainton, F. S., and Airey, P. L., *Nature* **207**, 1190 (1965).
20. Dogliotti, L., and Hayon, E., *J. Phys. Chem.* **71**, 2511 (1967).
21. Eaton, D. R., *J. Am. Chem. Soc.* **90**, 4272 (1968).
22. Eaton, D. R., and Suart, S. R., *J. Phys. Chem.* **72**, 400 (1968).
23. Edelson, M. R., and Plane, R. A., *Inorg. Chem.* **3**, 231 (1964).
24. Endicott, J. F., and Hoffmann, M. Z., *J. Am. Chem. Soc.* **87**, 3348 (1965).
25. Feitelson, J., and Shaklay, N., *J. Phys. Chem.* **71**, 2582 (1967).
26. Fields, R., Germain, M. M., Haszeldine, R. N., and Wiggins, P. W., *Chem. Commun.* 243 (1967).
27. Gunning, H. E., and Strausz, O. P., *Advan. Photochem.* **1**, 209 (1963).
28. Haake, P., and Hylton, T. A., *J. Am. Chem. Soc.* **84**, 3774 (1962).
29. Halmann, E., and Platzner, I., *J. Phys. Chem.* **70**, 2281 (1966).
30. Hata, G., Kondo, H., and Miyake, A., *J. Am. Chem. Soc.* **90**, 2278 (1968).
31. Hayon, E., and Saito, E., *J. Chem. Phys.* **43**, 4314 (1965).
32. Hayon, E., and McGarvey, J. J., *J. Phys. Chem.* **71**, 1472 (1967).
33. Herberhold, M., *Angew. Chem. Intern. ed.* **7**, 305 (1968).
34. Herzberg, G., "Spectra of Diatomic Molecules," p. 20. Van Nostrand, Princeton, New Jersey, 1950.
35. Hoffmann, R., and Woodward, R. B., *Acct. Chem. Res.* **1**, 17 (1968).
36. Johnson, B. F. G., Lewis, J., Robinson, P. W., and Miller, J. R., *J. Chem. Soc.* A, 1043, (1968).
37. Kasha, M., *Discussions Faraday Soc.* **9**, 14 (1950).
38. Kochi, J. K., and Bethea, T. W., *J. Org. Chem.* **33**, 75 (1968).
39. Kochi, J. K., and Krusic, P. J., *J. Am. Chem. Soc.* **90**, 7155, 7157 (1968).
40. Langmuir, M. E., and Hayon, E., *J. Phys. Chem.* **71**, 3808 (1967).

41. Livingston, R., and Zeldes, H., *J. Chem. Phys.* **44**, 1245 (1966).
42. Livingston, R., and Zeldes, H., *J. Am. Chem. Soc.* **88**, 4333 (1966).
43. Livingston, R., and Zeldes, H., *J. Chem. Phys.* **45**, 1946 (1966).
44. Mango, F. D., and Schachtschneider, J. H., *J. Am. Chem. Soc.* **89**, 2484 (1967).
45. Mango, F. D., and Schachtschneider, J. H., *J. Am. Chem. Soc.* **91**, 1030 (1969).
46. Moggi, L., *Gazz. Chim. Ital.* **97**, 1089 (1967).
47. Moggi, L., Bolletta, F., Balzani, V., and Scandola, F., *J. Inorg. Nucl. Chem.* **28**, 2589 (1966).
48. Moorthy, P. N., and Weiss, J. J., *J. Chem. Phys.* **42**, 3121 (1965).
49. Noack, K., *Helv. Chim. Acta* **45**, 1847 (1962).
50. Ohno, S., *Bull. Chem. Soc. Japan* **40**, 2035 (1967).
51. Penkett, S. A., and Adamson, A. W., *J. Am. Chem. Soc.* **87**, 2514 (1965).
52. Perumareddi, J. R., and Anderson, A. W., *J. Phys. Chem.* **72**, 414 (1968).
53. Plane, R. A., and Hunt, J. P., *J. Am. Chem. Soc.* **79**, 3343 (1957).
54. Porter, G. B., Doering, J. G. W., and Karanka, S., *J. Am. Chem. Soc.* **84**, 4027 (1962).
55. Quinlan, K. P., *J. Phys. Chem.* **72**, 1797 (1968).
56. Ramsey, D. A., and Dressler, K., *Phil. Trans. Roy. Soc. London* **251A**, 553 (1959).
57. Schläfer, H. L., *J. Phys. Chem.* **69**, 2201 (1965).
58. Shagisultanova, G. A., and Il'yukevich, L. A. *Russ. J. Inorg. Chem. (English Transl.)* **11**, 510 (1966).
59. Shagisultanova, G. A., Neokladnova, L. N., and Poznyak, A. L., *Dokl. Akad. Nauk SSSR* **162**, 1333 (1965).
60. Spees, S. T., and Adamson, A. W., *Inorg. Chem.* **1**, 531 (1962).
61. Stevenson, K. L., and Verdieck, J. F., *J. Am. Chem. Soc.* **90**, 2974 (1968).
62. Stiles, D. A., Tyerman, W. J. R., Strausz, O. P., and Gunning, H. E., *Can. J. Chem.* **44**, 2149 (1966).
63. Strohmeier, W., *Angew. Chem. Intern. ed.* **3**, 730 (1964).
64. Taube, H., *Advan. Inorg. Chem. Radio Chem.* **1**, 1 (1959).
65. Trecker, D. J., Henry, J. P., and McKeon, J. E., *J. Am. Chem. Soc.* **87**, 3261 (1965).
66. Uri, N., *Chem. Rev.* **50**, 375 (1952).
67. Wegner, E. E., and Adamson, A. W., *J. Am. Chem. Soc.* **88**, 394 (1966).
68. Wagner, P. J., and Hammond, G. S., *Advan. Photochem.* **5**, 21 (1967).
69. Wehry, E. L., *Quart. Rev.* **21**, 213 (1967).
70. Whitesides, T. H., *J. Am. Chem. Soc.* **91**, 2395 (1969).
71. Williams, J. L. R., Doty, J. C., Grisdale, P. J., Regan, T. H., Happ, G. P., and Meier, D. P., *J. Am. Chem. Soc.* **90**, 53 (1968).
72. Winkler, H. J. S., Winkler, H., and Bollinger, R., *Chem. Commun.* 70 (1966).

MATRIX ISOLATION SPECTROSCOPY

J. W. Hastie,* R. H. Hauge, and J. L. Margrave

DEPARTMENT OF CHEMISTRY
RICE UNIVERSITY
HOUSTON, TEXAS

* Present address, Inorganic Materials Division, National Bureau of Standards, Washington, D.C.

I. Introduction

Matrix isolation in its most commonly used form is the preparation of a rigid sample of effectively isolated molecules or reactive species in an inert matrix medium. The well-known technique of treating solid samples by diluting with KBr, for example, and forming a compressed disk of finely dispersed solid in a KBr matrix may be crudely considered as a progenitor of the present form of matrix isolation. Usually matrix isolation is achieved by the co-condensation of a gaseous material with a large excess (10^2–10^4 mole ratios) of inert material such as a rare gas or alkali halide. The trapped species do not undergo translational motion—i.e., diffusion—and usually are prevented from rotating, but may vibrate with frequencies that are within a few percent of the gas-phase values. Thus, the spectra of the material in these matrices are frequently much simpler than those obtained for any other state of matter, particularly if low temperatures are used.

The use of matrix isolation for spectroscopic studies of inorganic species at low temperatures was first demonstrated by Whittle *et al.* (*129*) and Becker and Pimentel (*14*). The emphasis of this early work was on the stabilization of free radicals, and many of the current techniques were developed under the NBS free-radical program. A comprehensive summary of this early development is given by Bass and Broida (*13*) and others (*26, 85*). More recent reviews have been given by Milligan and Jacox (*54*) and Mile (*73*) for radicals, Weltner (*122, 127*) for high-temperature inorganic species, and Pimentel (*95*) for applications to chemical physics. Also very recent surveys on ir spectroscopy at subambient temperatures have been carried out by Hermann (*51*) and also by Hallam (*42*).

In one of its most diverse forms matrix isolation of inorganic materials requires the use of high-temperature Knudsen cells for the production of molecular species and cryogenically cooled surfaces for the trapping of these species in an inert-gas solid. Alternatively, pseudo-high-temperature conditions may be created by subjecting material to radiation by high-energy sources such as microwave or electric discharges, uv lamps, or gamma rays, or by electron or ion bombardment.

Matrix-isolation spectroscopy may be applied to the study of inorganic species and their reactions. These species may be transient radicals such as SiF_3 and OH, hydrogen-bonded molecules of low stability such as $(H_2O)_2$ and $(HCl)_2$, or species present in a high-temperature environment such as C_3, TiO_2, SiCN, SiO and LaF_3. Alternatively, new species—e.g., XeF_2, $XeCl_2$, SiN, etc.—may be produced by reaction of the elements, formed either on the matrix surface during condensation or within the matrix by controlled diffusion. More complex molecules such as $(SiO)_n$ where $n = 1$–5, $(LaF_3)_2$, or $(SF_4)_2$ may be formed by diffusion of the monomers in the matrix. Similarly matrix reactions such as

$$O_2 + Li \rightarrow LiO_2$$
$$O_2 + Na \rightarrow O_2^- + Na^+$$
$$CH_3X + Li \rightarrow CH_3 + LiX$$
$$SiF_2 + LiF \rightarrow LiSiF_3$$
$$SiF_2 + Na \rightarrow NaSiF_2$$
$$Na + H_2O \rightarrow Na^+ + e + H_2O$$

may be studied and the products identified spectroscopically.

One also has the ability to stop a matrix reaction at any stage of completion merely by recooling to a low enough temperature to prevent diffusion. Hence reaction mechanisms and reaction intermediates may be spectroscopically observed.

A variety of fundamental molecular properties can be obtained such as vibrational frequencies (ir), geometry (ir), and electronic levels (near-ir, visible, uv, vacuum-uv), but usually no rotational structure.

The main advantages of matrix over gas-phase spectroscopy at present are

(1) The ability to observe at one's leisure normally unstable or reactive species such as radicals, ions, and high-temperature species.

(2) The ability to reduce and eliminate thermal-rotational and -vibrational excitation, particularly for species produced at high temperatures. As a result the vibrational bands are sufficiently sharp to allow accurate frequency measurement and the use of a wide range of isotopes. Also the sensitivity is enhanced.

(3) The ability to build up sufficient concentration to observe species of low abundance or spectral features with low absorption coefficient.

(4) By diffusion experiments, to follow the production of polymers, possibly to the point of crystal formation, and thereby obtain information bridging the behavior of the molecular and solid crystalline states.

II. Matrix-Isolation Techniques

A prime consideration in matrix-isolation spectroscopy is that of sample preparation. The choice of matrix material is somewhat dependent on the nature of the molecule to be isolated and the wavelength of radiation to be transmitted through the matrix. The original work of Becker and Pimentel (14) demonstrated the isolation efficiencies of solids such as N_2, the rare gases, and other effectively inert materials such as CO_2 and CCl_4. From the appearance of ir-absorption bands due to associated species it was determined that for efficient isolation the ratio of matrix to active molecules (M/A) can vary from several hundred to several thousand. For HN_3 with $M/A = 325$–340 the effective isolation improved from N_2 to Ar to Xe at $20°K$. With ESR spectroscopy radical concentrations of $\sim 10^{-8}$ molar can be detected. Raman and infrared spectroscopy can usually detect species of $\sim 10^{-5}$ and $\sim 10^{-7}$ molar concentration, respectively.

Light scattering from the matrix surface is also an important consideration in the choice of matrix material, deposition rate, and trapping temperature, particularly when optimum resolution is required. For visible spectroscopy glass-forming materials are desirable, whereas for infrared and especially the far-infrared regions crystal-forming materials such as the alkali halides and the rare gases are suitable. In the latter case the matrices are considered to be a collection of many microcrystals, each separated by grain boundaries. Even for infrared spectroscopy light loss from rare-gas crystals can be prohibitive, particularly if the matrix is formed rapidly and at too low a temperature. However, a practical upper limit to trapping temperatures is determined by the isolation effectiveness, which falls off with increasing temperature. Thus the choice of trapping temperature is often the result of a compromise between a gain in light level and a loss in the degree of isolation. A sufficient variety of inorganic species has been studied in various matrices that for new studies an empirical choice of matrix conditions can usually be made, and some of the examples discussed in Section IV will demonstrate typical matrix conditions.

The matrix surface dimensions are such as to allow complete passing of the spectrometer light beam; that is, about 1×4 cm. Matrix thickness t may conveniently be determined from counting n interference fringes in the infrared, using the expression

$$t = \frac{n}{2(\Delta \nu)}$$

where $\Delta \nu$ is the frequency interval. Thicknesses of 0.1–1 mm are typical. Thicker matrices can have a tendency to peel off the cryogenic surface. Also light loss by scattering tends to increase with matrix thickness.

The rate of deposition of a matrix can vary widely—e.g., 10^{-3}–10^{-6} mole/hr. For example, in some experiments, such as the isolation of relatively unreactive molecules in a rare-gas solid, a suitable matrix can be obtained after several minutes of trapping, whereas in others, such as those involving surface reactions, several hours to a day may be required. Matrices can also be deposited in a matter of seconds using a controlled pulse technique (105); this pseudo-matrix-isolation method has been demonstrated to have applications to infrared spectroscopic analysis of gases.

There are at present no means for choosing the most suitable isolating medium. However, the ever-present possibility of extraneous matrix effects, to be discussed in the following section, requires one to form matrices with a variety of isolating media and deposition conditions in order to reduce or at least identify these effects.

A. Preparation of a Matrix

The apparatus required for matrix preparation depends largely on the degree of volatility of both the matrix material and the substance to be isolated.

Case 1: Both Matrix and Inorganic Material Volatile. For this case standard gas-handling techniques are employed, and one may either mix the gases in the desired proportion prior to spraying onto the cryogenic surface or cocondense the materials from two separate jets. Many studies of this type are given in the early literature and most of the readily obtainable inorganic gases such as H_2O, NO_2, HCl, BF_3, and SO_2 have been studied spectroscopically in matrices.

Case 2: Matrix Material Volatile and Inorganic Nonvolatile. This combination is potentially the most important, as the majority of known inorganic species are volatile only at elevated temperatures. Thus a mating of cryogenic and high-temperature technology is required. The production of a molecular beam of material with the use of Knudsen cells is well documented (*31*), as is the production of cold trapping surfaces (*13*). The only special requirement is that the cold trapping surface be reasonably well shielded from the radiation of the high-temperature source. A physical separation of approximately 12 cm is usual; a schematic of a typical system is given in Fig. 1. With this precaution it is even possible to isolate C_3 species produced from a source at 2700°K, in solid neon at ∼4°K.

The matrix concentration is readily controlled by the rate of effusion dw/dt from the Knudsen cell as given by the expression (*31*)

$$\frac{dw}{dt} = AP\left(\frac{M}{2\pi RT}\right)^{1/2}$$

and also by the solid angle subtended by the trapping surface and the Knudsen orifice. A is the area of the Knudsen orifice, P the pressure in the Knudsen cell, R the gas constant, M the molecular weight, and T the cell temperature. Typically only a few percent of the actual effusate reach the cold surface. Similarly the matrix gas is not entirely collected by the cold surface, as evidenced for example by the background-pressure rise during trapping—to ∼10^{-4} Torr as compared with a normal pressure of 10^{-6}–10^{-7} Torr. Thus only

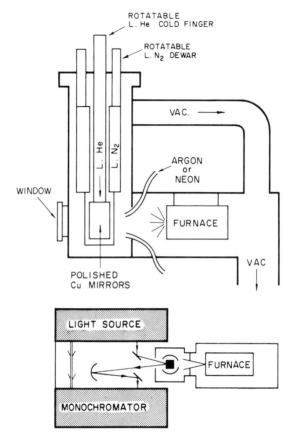

FIG. 1. Apparatus for trapping high-temperature species for matrix spectroscopic studies. [From J. W. Hastie, R. H. Hauge, and J. L. Margrave, *High Temp. Sci.*, **1**, 76 (1969).]

an approximate calculation of the matrix concentration can be made, and one usually tests for the degree of isolation by searching for spectra due to associated species formed on condensation.

A versatile Knudsen effusion assembly, represented by Fig. 2, is capable of operating at 2200°C and is useful for production of lower-valence species of inorganic salts.

For the case where the matrix material is involatile and the inorganic reactant is volatile—e.g., trapping a vapor in an alkali halide matrix—one may use similar procedures to the above case.

Case 3: Both Matrix and Inorganic Material Nonvolatile. Where each material is involatile at room temperature, separate Knudsen sources are required. This would seem to be a useful technique for studies of doped thin

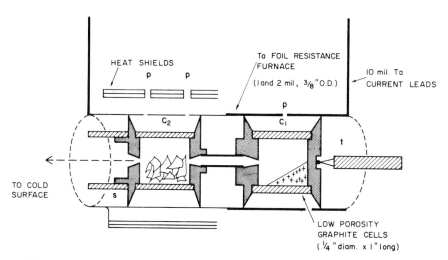

FIG. 2. A Knudsen effusion reactor assembly, used for the equilibrium production of lower-valence species: p, pyrometer sighting holes; t, thermocouple; C_1, cooler cell with salt; C_2, hot cell with metal; S, graphite sheath. [From J. W. Hastie, R. H. Hauge, and and J. L. Margrave, *J. Chem. Phys.* **51**, 2648 (1969), by permission of the American Institute of Physics.]

films, such as rare-earth metals in a CaF_2 matrix. A useful apparatus that may be adapted for trapping from multiple sources is demonstrated by the rotating cryostat used in ESR and ir studies (*19*, *72*). This apparatus has proved useful for the production of negative ions such as O_2^-, CO_2^-, CS_2^-, and H_2S^- at the condensing surface, as described in Section IV,C.

B. Cryogenic Considerations

The major cryogenates in use are liquid He (bp = 4.2°K), liquid hydrogen (bp = 20.4°K), and liquid N_2 (bp = 77.4°K). Pumping on the refrigerant can reduce these temperatures to 1.2°K and 15°K for He and H_2, respectively. Temperatures usually required for matrix formation are around 20 ± 5°K for Ar, Xe, Kr, and N_2, and 5°K for Ne. At these temperatures the solids have vapor pressures of about 10^{-10} Torr. Cooling of the trapping surface is usually achieved by introducing a controlled leak of liquid or gaseous refrigerant to a cold block that is in good thermal contact with the trapping surface. Cryotips that utilize the Joule–Thompson effect in the expansion of He or H_2 gas at high pressure are also commonly in use. Small liquid Dewars with a capacity for several liters of coolant and situated above the cold block are most economical for lengthy experiments that do not require repeated breaking of the vacuum for such purposes as cleaning of the trapping surface or loading a sample in a Knudsen cell. For short-term experiments of a few hours duration it has been found more convenient to pass the liquid coolant from the main reservoir, usually a 25–50-liter Dewar, by means of an efficient transfer line at a low rate sufficient to just maintain the cold-block temperature. This eliminates the inefficiency of cooling down and rewarming an *in situ* Dewar each time it is required to break vacuum.

C. Choice of Trapping Surface and Optical System

The most suitable trapping surfaces for the various spectroscopic regions are the crystalline forms of quartz or Si (far-ir), CsI, KBr, NaCl (ir), quartz, LiF (to 1050 Å), CaF_2 (visible, uv) and sapphire rod or other nonconducting material for ESR spectra. Most studies have utilized these transmission windows, which unfortunately have relatively low thermal conductivity. Metal surfaces cover a wider spectral region as well as providing very good thermal and light-reflection properties. Despite the description of a technique utilizing reflection off a cooled metal surface for the study of thin films as early as 1940 (*37*), only a few workers (*45*, *72*) have since utilized the method for matrix-isolation spectroscopy. The schematic of Fig. 1 demonstrates an optical system used for obtaining spectra from matrices on metal surfaces.

III. Matrix Effects

For the interpretation of spectra of inorganic species in a matrix one has to identify the special perturbation effects caused by the matrix environment (*96*). Such effects, which are usually manifested as small frequency shifts, multiplet structure, and variations in absorption intensities (*118*), are often a complicating factor in the initial spectral interpretation. However, once a better understanding of this perturbation is at hand, additional information may be derived from the matrix-isolated molecule.

A. Matrix Shifts

The most common and readily identified matrix effect is the frequency shift from the gas-phase values. This effect has been observed for rotational, vibrational, and electronic transitions. For the majority of known rotational and vibrational matrix spectra of inorganic species the shift is to lower frequencies and by only a few percent of the gas-phase values.

1. PERTURBATION OF ROTATION IN MATRICES

Of the many inorganic species that have been studied in matrices, only a few small molecules, such as H_2O, HF, HCl, HBr, and their deuterated forms CH_4, NH_3, and CO have shown a tendency to rotate in rare-gas matrices.

Even for these cases the rotation is apparently hindered by the solid environment. Studies in higher-temperature matrices, such as those provided by zeolites, alkali halide crystals, H_2O crystal, etc., could be expected to show a higher prevalence of rotational effects. In fact some studies are benefited when the matrix-isolated species is in a position to rotate freely in a trapping site, but still is not permitted to move translationally in the matrix. For example, better-resolved ESR spectra are possible for a freely rotating radical.

For the diatomic hydrogen halides, both vibration–rotation structure in the normal infrared and pure rotation–rotation transitions in the far infrared have been observed.

Table I shows the effect of various matrices on the $J = 1 \leftarrow J = 0$ transition in HF and DF, where it has been noted (104) that the matrix shift varies approximately linearly with the rare-gas atomic polarizability. These blue shifts are thought to be due to the collapse of the lattice around the small HF molecule, an effect that should increase, as observed, with the larger rare-gas atoms (92).

Direct observations of pure rotational transitions of matrix isolated H_2O have also been made (103) in the far infrared and support the rotational–vibrational interpretation of the normal infrared spectra.

A number of models, based on hindered rotation, planar or octahedral barriers, and rotation–translation coupling, have been proposed to explain the observed rotational structure in various matrices (14, 51). If one traps HCl in an SF_6 martix, no rotation is observed and a large matrix shift occurs, from which it is suggested that the HCl molecule is hydrogen-bonded to the SF_6 matrix (130).

Several interesting examples of ortho–para nuclear-spin conversion, enhanced by the presence of O_2 in the matrix are known (100), e.g., H_2O and CH_4. The H_2O study, which utilized matrices of Ar, Kr, and Xe, concluded

TABLE I

THE $J = 1 \leftarrow J = 0$ ROTATION TRANSITION IN
RARE-GAS MATRICES[a]

	HF	DF
Gas phase	41.9 cm^{-1}	22.0 cm^{-1}
Ne matrix	39.8	22.8
Ar	44.0	27.3
Kr	45.0	30 (± 1)
Xe	50.5	32.8

[a] Data taken from Robinson and Von Holle (104); accuracy ± 0.4 cm^{-1}.

that the H_2O molecules were trapped in single-atom vacancy substitutional sites with very little hindrance to free rotation. This is not unexpected, as the H_2O molecular diameter is certainly less than that of a single rare-gas atom. On the other hand, the hindered rotation of HCl has been attributed to angle-dependent interactions of HCl with the surrounding rare-gas atoms [see Keyser et al. (64) for further discussion]. It is surprising that the OH radical does not appear to rotate in rare-gas matrices.

It would seem feasible to study the rotation of almost any species in a matrix-isolated state, given the correct choice of matrix material and temperature. For example, larger species such as NO_2 and NO_3 have been observed by ESR spectroscopy to rotate freely in a Linde 13-X molecular-sieve matrix at 150°K and 85°K, respectively (131). Similarly HCN^- rotates freely in KCl at 77°K (91).

2. MOLECULAR ORIENTATION IN MATRICES

The observation of free or hindered rotation in matrices also suggests the possibility of preferred orientation for conditions where rotation is physically not possible. A preferred orientation has been observed for several species, such as $Cu(NO_3)_2$ and CuF_2 (60), AlF_3 (112), VO (61), and CO_2^- (16) and it is not likely that this is a rare occurrence. AlF_3 tends to orient with its plane parallel to the matrix trapping surface, while VO locates with its internuclear axis perpendicular to the matrix surface. The question of whether or not one is isolating a molecule in a random orientation is important, as a preferred orientation of species will result in an observed discrimination against certain symmetry transitions. Ideally one should test for the presence of nonrandom orientations in order to interpret matrix spectra correctly.

3. MATRIX SHIFT EFFECT ON VIBRATIONAL SPECTRA

The effect of a matrix environment on a molecular vibration may be viewed initially as a dielectric effect where the red frequency shift is predicted by the Kirkwood–Bauer–Magat relation (14)

$$\frac{v_0 - v}{v_0} = C\left\{\frac{(D-1)}{(2D+2)}\right\}$$

where C is a constant characteristic of the molecule, D is the dielectric constant of the matrix, v_0 is the gas-phase frequency, and v is the expected matrix frequency. This relation fits the observed shifts of LiF, LiBr, and LiCl in argon, krypton, and xenon matrices but notably not in nitrogen (107). More specifically the two main factors leading to a red shift appear, from the work of Linevsky (65), to be the attractive-dipole–induced-dipole and the dispersive effects (induced-dipole–induced-dipole). The implication of a red shift, which is most commonly observed, is that the $v = 1$ state experiences more of an attractive force from the matrix than the $v = 0$ state. This force should vary as $\alpha/r,^6$ where α is the polarizability and r the radius of the matrix atom (53). In fact treating the attractive forces in terms of the semiempirical Lennard–Jones potential leads to a total attractive potential of (34)

$$-\left(\frac{\mu_1^{\,2}\alpha_1}{r_{12}^6}\right) - \left[\frac{3}{2}\left(\frac{\alpha_1\alpha_2}{r_{12}^6}\right)\left(\frac{I_1 I_2}{I_1 + I_2}\right)\right]$$

where μ_1, α_1, and I_1 are the dipole moment, polarizability, and ionization potential, respectively, for the trapped species; α_2 and I_2 are the polarizability and ionization potential, respectively, for the nearest-neighbor matrix molecule; and r_{12} is the distance between the trapped and nearest-neighbor matrix species. The first term of the above expression is the dipole–induced dipole interaction, and the second term the induced-dipole–induced-dipole interaction. Alternatively, the dispersive interactions resulting in a red shift may be given by the expression of Longuet-Higgins and Pople (68)

$$\text{red shift} \simeq \frac{3\alpha_1\alpha_2\,zE}{8r_{12}^6}$$

where z is the number of nearest-neighbor matrix molecules at a mean distance r_{12}, and E is the energy for the transition. At present there is a sufficient lack of data for both the trapped molecule and the matrix environment that very few useful tests can be made of the models proposed for treatment of the matrix-shift effect. However, certain useful empirical predictions may be made on the basis of existing experimental matrix-shift data.

We find that the following relations appear to hold for inorganic species:

(1) The red vibrational shift increases with increasing matrix polarizability —e.g., from Ne, Ar, Kr, to Xe, as shown in Fig. 3. Shifts in N_2 are still greater.

(2) The red vibrational shift increases with the expected charge separation along the vibrating bond, as given crudely by electronegativity differences or, in the case of diatomic species, by dipole moments—e.g., the examples in Fig. 3 have the following relative order of dipole moments: AlF $\ll$ LiF $\simeq$ LiBr $<$ LiCl.

(3) From the data of Table II the gas-phase vibrational frequency v_g is, to a good approximation (± 6 cm^{-1}), related to the frequencies observed in neon and argon matrices by the expression

$$v_g \simeq v_{Ne} + (0.8 \pm 0.4)(v_{Ne} - v_{Ar})$$

The species listed in Table II represent a wide range of bond types and, since most other inorganic species should fall within this range, their approximate

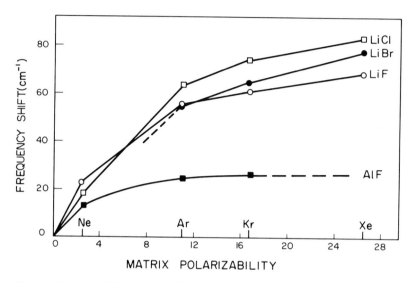

FIG. 3. Relation of frequency shift (gas–matrix) to matrix polarizability (values as given in Table IV). [Frequency data from Hastie *et al.* (*50*), Schlick and Schnepp (*107*), and Table II.]

gas-phase stretch frequencies may be obtained from the above expression. From the few examples quoted it appears that the symmetric stretch and bending frequencies have less of a matrix shift and may even in some instances be shifted slightly to the blue. Clearly, both experimental and theoretical work is needed on the problem of relating matrix vibrations to the gas-phase counterparts.

Figure 4 indicates the known matrix shift that occurs between Ne and Ar matrices for the antisymmetric stretches of various families of inorganic species. The trends are usually in the direction of expected increased ionic character of the vibrating bonds. In fact it is hoped that a clearer understanding of matrix shifts may eventually lead to new knowledge about charge distribution within a molecule. From the examples of Fig. 4 and Table II it appears that one can usually expect matrix vibrational shifts in neon of no more than 4%, and so approximations such as the above expression should lead to predicted gas-phase values that are reliable to within a few percent.

One should note that inorganic (as well as organic) species having a relatively low bond polarity, such as CO_2, NO, N_2O, H_2O, and NH_3, can have both positive and negative vibrational frequency shifts. However, for these cases the shifts are less than 1% of the frequency, as compared to red shifts of up to 4% for more polar bonding species such as those indicated in Fig. 4.

The limited data available for the lowest-frequency modes of polyatomic inorganic species suggest that larger matrix perturbations are likely for these cases. In fact Pimentel and Charles (96) proposed a "cage" model in which low-frequency vibrations feel the effect of a "tight cage" and could tend to show positive (blue) shifts, whereas the higher frequencies should feel a "loose cage" and show negative (red) shifts. The larger amplitudes for lower

TABLE II

A COMPARISON OF MATRIX AND GAS-PHASE VIBRATIONS

| Molecule | Vibration | Frequency (cm^{-1}) | | | $\dfrac{\nu_g - \nu_{Ne}}{\nu_{Ne} - \nu_{Ar}}$ |
		Ar matrix	Ne matrix	Gas	
SO_2 [a]	ν_3	1356.2	1359.3	1360.5	0.4 ± 0.3
SO_2 [a]	ν_1	1153.4	1152.0	1151.4	—
		1148.4	—	—	—
SO_2 [a]	ν_2	520.4	519.6	517.8	—
		518.6	—	—	—
GeF_2 [b]	ν_3	648	655	663	1.0 ± 0.3
GeF_2 [b]	ν_1	676	685	700 ± 5	—
GeF_2 [c]	ν_2	262	—	263 ± 5	—
SiF_2 [d]	ν_3	852.9	864.6	872 ± 3	0.6 ± 0.2
SiF_2 [d]	ν_1	842.8	851.0	855 ± 5	—
TeO [e]	ν	787.2	791.0	796	1.2 ± 0.3
^7LiF [f]	ν	835	868	890	0.7
$^7Li^{35}Cl$ [f]	ν	578	623	641	0.4
AlF [g]	ν	776	785	798	1.4 ± 0.4
AlF [h]	ν	772.3	—	—	$\sim 1.0 \pm 0.3$
^{69}GaF [h]	ν	591.5	609.0	~ 610 [i]	—

[a] Data taken from Hastie *et al.* (45). Spectra for Kr matrices have also been recorded (5) where ν_1 and ν_2 were essentially the same as for argon and ν_3 decreased to $1350.7 cm^{-1}$.

[b] See Hastie *et al.* (43).

[c] Gas-phase value from R. Hauge, V. M. Khanna, and J. L. Margrave, *J. Mol. Spectry*, **27**, 143 (1968); matrix value unpublished data by the present authors.

[d] Gas-phase data from V. M. Khanna, R. Hauge, R. F. Curl, and J. L. Margrave, *J. Chem. Phys.* **47**, 5031 (1967); matrix data from Hastie *et al.* (47).

[e] See Muenow *et al.* (88).

[f] See A. Snelson and K. S. Pitzer, *J. Phys. Chem.* **67**, 982 (1963); G. L. Vidale, *ibid.* **64**, 314 (1960); W. Klemperer, W. G. Norris, A. Buchler, and A. G. Emslie, *J. Chem. Phys.* **33**, 1534 (1960), and unpublished data by the present authors.

[g] Snelson (112).

[h] Hastie *et al.* (50).

[i] Calculated from data given in R. W. B. Pease and A. G. Gaydon, "The Identification of Molecular Spectra," 3rd ed. J. Wiley, New York, 1963; limit of uncertainty not known but probably within $\pm 5 cm^{-1}$.

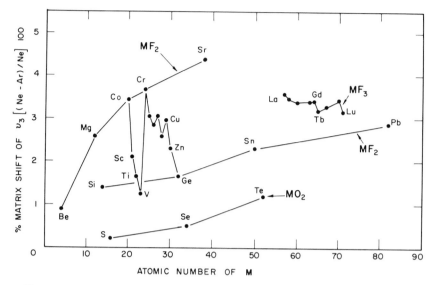

Fɪɢ. 4.   Variation of relative matrix shifts for the antisymmetric stretch frequency in Ne and Ar matrices. [Group IIA data taken from A. Snelson, *J. Phys. Chem.* **72**, 250 (1968). All other data obtained by the present authors (*50*).]

vibrations are not considered to be of major importance in determining matrix shifts. The species SnF_2 shows a higher v_2 for neon (201.5 ± 0.5) than argon (198.0 ± 0.5), and the gas-phase value is known to be approximately 185 ± 15 cm^{-1}. Also KrF_2, which shows (*36*) a red shift for the antisymmetric stretch (588 cm^{-1} gas phase to 580 cm^{-1} in Kr matrix), has a blue shift in the bending frequency (232.6 cm^{-1} gas phase to 236 cm^{-1} in Kr matrix). On the other hand, the $(HCl)_2$, H-bonded species shows (*63*) pronounced red shifts with increasing matrix polarizability for the B_u mode (at approximately 200 cm^{-1}). The large shifts—e.g., 25% from neon to xenon —are consistent with a strong matrix perturbation for a motion associated with a weak force constant (about 0.03 mdyn/Å in this case) but do not lend support to the expected blue shift of a tight-cage effect.

There does not appear to be any evidence of a measurable matrix effect on the vibrational anharmonicities, at least for the lowest vibrational levels. The species LiF, LiCl, SiO, and SO_2 apparently have the same anharmonic constants in rare-gas matrices as for the gas phase, even though the frequency shifts can be appreciable. Also it is predicted (*29*), at least for diatomic species, that the solvent shift is constant for various isotopic species. This means that isotopic frequency ratios observed in matrices may be assumed to represent those in the gas phase, even though the absolute frequencies differ. If this prediction holds for polyatomic species, then it follows that measured matrix isotopic frequency ratios may be used to yield gas-phase structures of molecules, as has been found to be the case. More discussion on this point appears in Section IV.A.

4, MATRIX EFFECTS ON ELECTRONIC STATES

The matrix perturbation effects that one observes on the electronic states of a trapped species are shifts in the energy of the electronic state, changes in the vibrational structure of that state, and changes in the hyperfine structure (electron-spin–nuclear-spin coupling) of ESR spectra.

A discussion of the theoretical aspects of treating matrix shifts for electronic as well as other spectra has been given by Robinson (102) and by McCarty and Robinson (69). The theories essentially treat the matrix as a solvent for the trapped species, and the interaction is described by a Lennard–Jones potential. The theory does not appear to hold for Ne matrices.

Blue shifts appear to be more common for electronic than vibrational spectra. The important question of a possible reversal of electronic ground and low-lying excited states in a matrix also arises. The C_2 spectrum was thought to be an example of this, but recent work has shown that the interpretation was erroneous because of the presence of C_2^-. At the present time there is no evidence for reversal of electronic levels in matrices. Table III indicates typical matrix shifts for electronic transitions. As with vibrational transitions the spectra shift to lower energies with increasing matrix-atom polarizability— i.e., from Ne, Ar, Kr to Xe.

Matrices can also have the useful effect of eliminating perturbations that are present in the gas-phase spectra. For example, the vibrational spacings for the $B\,^2\Pi$ and $B'^2\Delta$ states of NO are irregular in the gas phase because of interaction with Rydberg states; but in Ne, Ar, or Kr matrices, where the Rydberg states are unobservable, the spacings are regular (106). The importance of the different excited NO electronic configurations in relation to matrix effects is indicated by the observation of sharper spectra in Ne than Ar for the $B'^2\Delta$ and $G^2\Sigma^-$ but not the $B^2\Pi$ states.

Blue matrix shifts of considerable magnitude may also be possible for Rydberg series of organic species, since the first member of acetylene, at 8.155 eV in the gas phase, shifts to 8.67 eV and 9.01 eV in Kr and Ar matrices, respectively (98). The limitations of the theories for predicting the matrix shift of Rydberg levels are also discussed by Pysh et al. (98).

An enhancement of phosphorescence in the $a\,^3B_1$–X^1A_1 electronic transition for SO_2 by various matrices and temperatures has been indicated, and an interaction of the SO_2 triplet state with the lattice vibration has been suggested to explain this effect. In this study both blue and red shifts were found for the v_1 and v_2 vibrational modes (94).

TABLE III

COMPARISON OF MATRIX SHIFTS FOR ELECTRONIC TRANSITION ENERGIES (cm^{-1}) [a]

Species	Electronic transition	Ar	Kr	Xe	Gas
Na [b]	$^2P_{3/2,\,1/2} \leftarrow {}^2S_{1/2}$	$+60$ $+1400$	—	—	16965
Hg [b]	$^3P \leftarrow {}^1S_0$	$+1281$	$+796$	$+30$	39412
NH [b]	$A\,^3\Pi \leftarrow X\,^3\Sigma^-$	-192	-264	-370	29773
NH_2 [b]	$^2A_1(080) \leftarrow {}^2B_1$	$+28.7$	-15.8	-40.3	15902.3
In [c]	$5\,^2P_{1/2} \leftarrow 5s5p^2,\,{}^2S_{1/2}$	-100	-1350	-3580	1757Å
SiO [d]	$A\,^1\Pi \leftarrow X\,^1\Sigma^+$				
	$v' = 0$	$+109$	-232	-802	42657
	$v' = 1$	$+65$	-245	-826	43499
	$v' = 2$	$+79$	-225	-850	44328
	$v' = 3$	$+69$	-236	-862	45139
	$v' = 4$	$+40$	-239	-945	45943
	$v' = 5$	$+98$	-251	—	46694
Li_2 [e]	$^1\Pi_u \leftarrow {}^1\Sigma_g^+$	$+220$	-360	-1020	20439
S_2 [f]	$^3\Sigma_u^- \leftarrow {}^3\Sigma_g^-$	-340	-790	-910	31690

[a] A positive matrix shift—i.e., a blue shift—means that the matrix spectra fall at higher energies than those in the gas phase.

[b] Hg, NH, NH_2, and Na data as given by McCarty and Robinson (69). The orbital degeneracy is removed by the matrix effect for the case of Na.

[c] Indium, data of W. O. Duley and W. R. S. Garton, *Proc. Phys. Soc.* **92**, 830 (1967).

[d] SiO, data of J. S. Shirk and A. M. Bass, *J. Chem. Phys.* **49**, 5156 (1968).

[e] Li_2, L. Andrews and G. C. Pimentel, *J. Chem. Phys.* **47**, 2905 (1967).

[f] S_2, L. Brewer, G. D. Brabson and B. Meyer, *J. Chem. Phys.* **42**, 1385 (1965); L. Brewer and G. D. Brabson, *ibid.* **44**, 3274 (1966).

The effect of matrices on the $B\,^3\Sigma_u^-$ state of S_2 is such that the potential energy for excited vibrational levels, at constant internuclear separation, decreases in the order, Ar > gas > Ne > Xe > Kr (93). That is, the potential curves widen in this same order.

For atomic species one usually observes the largest blue shifts for Ar and the smallest for the most polarizable matrix (24). Most electronic spectra are sensitive to matrix temperature, and both blue and red shifts may be observed on warming the matrices.

The theory derived for treating the hyperfine-structure matrix shift in nonpolar matrices has also been tested on the polar H_2O matrix for Cu, Ag, and Au atoms (132); it was concluded that the main matrix effect is not a function of the polarity of the matrix molecules. To explain an apparent breakdown in the theory for the shifts of Cu in solid H_2O it was suggested that chemical interaction occurred and tended to delocalize the unpaired electron of Cu.

B. Nature of Rare-Gas Matrix and Site Effects

In addition to the actual perturbation of rotational, vibrational, or electronic states by the matrix, one also often observes a splitting or multiplet structure in the spectra. This is currently believed to be due to the trapped species being located in several distinct sites, and therefore the phenomenon is stongly dependent on the relative dimensions of the isolated species and the surrounding matrix atoms.

Site splitting is usually a few wave numbers in magnitude and is sometimes difficult to resolve. In many cases the splitting can either be reduced or eliminated by a so-called annealing of the matrix, which is an irreversible process. This is to be contrasted with the reversible temperature-dependent behavior of rotational structure.

Prior to a consideration of the actual nature of these trapping sites it will be useful to review briefly the nature of the pure rare-gas solid and the effect of impurities on the solid structure.

1. NATURE OF RARE GAS SOLIDS

A number of sources provide a detailed account of the nature of rare-gas solids (*20, 38, 97*). We shall mention only the properties that appear to be most relevant to the problems of matrix isolation.

The structure of these pure rare-gas solids is one of cubic close packing with six nearest neighbors per atom. The unit cell is face-centered cubic and the lattice parameter typically increases by several thousandths of an angstrom per degree rise in temperature for the regions used in matrix-isolation experiments. As the treatments of matrix perturbation potentials involve terms of the inverse sixth power of the distance r separating the trapped species and a nearest-neighbor rare-gas atom, then a small change of r, caused by a temperature change, for example, is magnified by at least ten times in terms of the interaction potential. Thus this sensitivity to temperature explains, in part at least, the remarkable effect of small temperature changes on the occupation of matrix sites and on actual matrix shifts.

As previously indicated, the intermolecular potential energy of two rare-gas atoms is given by the Lennard–Jones approximation

$$V(r) = 4E\left[\left(\frac{\sigma}{r}\right)^{12} - \left(\frac{\sigma}{r}\right)^{6}\right]$$

where E may be taken as the depth of the potential well and σ as a measure of the effective molecular diameter of the atom. Values of σ are given in Table IV, together with other rare-gas parameters.

The effect of impurities on the nature of these solids has not received much

TABLE IV

PROPERTIES OF RARE-GAS SOLIDS[e]

Molecule	Atomic size[c] (Å)	σ^a (Å)	α^b	Triple point[a](°K)	Density[c] 0°K (g/cm³)	Lattice parameter[c] 0°K (Å)	Substitutional site[d] (Å)	Interstitial diameter[d] (Å) (octahedral)
Ne	3.147	2.67	2.663	24.66	1.508	4.462	3.16	1.30
Ar	3.866	3.41	11.08	83.81	1.769	5.312	3.76	1.55
Kr	4.056	3.66	16.73	115.77	3.094	5.644	3.99	1.65
Xe	4.450	3.97	27.29	161.36	3.782	6.131	4.34	1.79

[a] Data given by Boato (20).

[b] Calculated polarizability data (40), units of α in 1.482×10^{-25} cm³.

[c] Data listed by Pollack (97).

[d] From Hallam (42); tetrahedral sites are about one-half of the corresponding octahedral values.

[e] Substitutional and interstitial site diameters for N_2 correspond closely to those of Kr.

attention. There is evidence that argon tends to crystallize in a hexagonal lattice when impurities are present (12). Also the presence of a small N_2 impurity in rare-gas matrices is sufficient to quench rotation (21). It is of note that N_2 packs more closely than the rare gases. N_2 matrices form a face-centered cubic lattice structure (below 35°K) with octahedral sites that can accommodate species in the size range 3.2–4.2 Å, hexagonal sites for species in the range 1.3–1.8 Å, and tetrahedral sites that will accept species in the range 0.5–1.0 Å. However, many of the polyatomic metal halides studied are larger than any of the available sites provided by N_2 or the rare gases. Thus matrices of these species are probably very disordered, and the observation of multiple site splitting for these matrices is in accord with this view.

Grain boundaries between microcrystals have been suggested as possible matrix sites that are fairly insensitive to lattice environment, while the main sites are considered to be substitutional (22).

From Table IV it is apparent that only a very-small-diameter trapping site is available in Ne. However, as noted earlier this matrix provides the least apparent matrix perturbation of vibration and rotation. Most likely this useful property of Ne is due to a fortuitous near cancellation of attractive and repulsive forces.

It is believed that for small molecules both substitutional and interstitial sites can be simultaneously occupied, as, for example, in the case of CO in Xe and Kr matrices, where the site splitting is approximately 10 cm⁻¹, less than 0.5% of the total frequency (33). As indicated by Table IV, substitutional sites are more likely for larger species.

Another important aspect of introducing impurities into rare-gas or N_2 solids is the prospect of activating phonon bands of the solids, which are expected in the far infrared. Several strong absorptions have been reported (62) in the region of 33–100 cm^{-1} for rare-gas solids containing 1 mole % of HCl or HBr; these are attributed to either localized modes or phonon bands. Similarly strong absorptions have been observed (50) in nitrogen matrices, irrespective of the impurity introduced but dependent on the matrix temperature at 220 cm^{-1} and 48 cm^{-1}. These effects are at present not well understood and are causing difficulty in obtaining matrix ir spectra in this region of the far infrared.

2. MATRIX-SITE EFFECTS

In choosing a matrix material for isolation of an inorganic species one should ideally take into consideration the molecular size of the trapped species in relation to that of the substitutional sites. Certain combinations of these dimensions may lead to the matrix orientation effects that have so far been observed.

Because the molecular size of most polyatomic inorganic species exceeds the available space of a substitutional or interstitial site, the probability of the trapped species locating in a site where the cage also contains one molecule of the same species is increased. This cage will clearly perturb the trapped species by a different amount than a cage formed entirely by matrix molecules and thus will produce another effect known as an aggregation site effect. This is not to be confused with the formation of dimer species, as actual chemical combination does not necessarily occur.

An example has been observed where new metastable sites are formed when the matrix ScF in Ne is exposed to visible light (70). Another unusual visible-light-sensitive phenomenon has been observed for the colored neon and argon matrices of ScO, YO, and LaO, where the colors are bleached by exposure to light (128). This effect has been interpreted as being due to diffusion and reaction of excited M—O molecules in the matrix, presumably to form $(MO)_n$ polymers, which are not uncommon in matrices. Table V presents a comparison of matrix-splitting and matrix-shift data for the v_3 stretch of the first-row transition-metal difluorides (50). Clearly this behavior is very sensitive to only small changes in molecular size and, as yet, cannot be predicted *a priori*.

The effect of matrix sites on the electronic spectra of atoms has received considerable attention recently. It would appear that some direct structural observations on matrices are needed to fully understand the matrix-site phenomenon.

It is very probable that matrices remove the degeneracy of vibrations to

TABLE V

MATRIX SHIFT AND SPLITTING EFFECTS FOR THE
FIRST-ROW TRANSITION-METAL DI-FLUORIDES
$(cm^{-1})^a$

	$\nu_3(Ne) - \nu_3(Ar)$	Splitting of ν_3 in neon matrix
Ca	20	—
Sc	—	—
Ti	12.6	2.5
V	9.6	—
Cr	25.7	4.0
Mn	22.0	5.8, 11.2
Fe	21.5	8.9
Co	22.7	10.8, 18.9
Ni	20.6	7.2, 12.4
Cu	22.6	14.9, 28.2
Zn	18.0	9.6

[a] Hastie *et al.* (*50*)

provide yet another source of matrix splitting. This appears to be the case for the first-row transition-metal difluorides, where the doubly degenerate ν_2 frequency is split to the extent of several cm^{-1} in neon or argon matrices (*50*).

IV. Applications of Matrix-Isolation Spectroscopy to Inorganic Chemistry

In the following sections, dealing with specific examples, we hope to indicate the ability of matrix-isolation spectroscopy to deal with a wide range of problems in inorganic chemistry. Some of the applications include determinations of geometry and electronic and vibrational structure for radicals, high-temperature species, ions, and polymeric species.

A. Determination of Vibrational Frequencies and Geometrical Structures

Over the past decade, with the use of the Knudsen effusion–mass-spectrometric technique, hundreds of new inorganic species have been identified. From these studies it appears that a high-temperature environment provides a unique opportunity to produce simple inorganic molecules containing bonds between almost any two elements of nature. For example, metal–carbon bonds in species such as SiC, SiCN, CeC_2; metal–nitrogen bonds in species such as Si_2N, LaN, ThN; metal–hydroxyl bonds in species such as $Be(OH)_2$, LiOH, $(LiOH)_2$, OBOH; strong metal–metal bonds from LaAu, CuAu, ZrAu, ZrPt; and the numerous metal halides, metal chalcogenides, and metal oxyhalides, etc., are known. A glance at the periodic classification of the elements and a knowledge of only the species identified thus far suggests an almost endless variety of combinations.

Progress in the structural area for high-temperature species has been hampered by thermal excitation, which is particularly severe for heavy-element molecules and also those with considerable ionic character. For example, most triatomic molecules of this nature will have very low bending frequencies v_2, and, as indicated from the Boltzman population factor, this will result in a high population of excited levels together with a large amplitude for bending. The overall effect is a washing out or extreme broadening of vibrational spectra and a difficulty in observing the atomic coordinates by methods such as electron diffraction. Matrix isolation of the species at low temperatures effectively eliminates this problem but also reduces the number of available spectroscopic tools that may be utilized.

Fortunately the narrow bands characteristic of infrared spectra for matrix-isolated species allow a wider use of isotopic effects than would be possible in the gas phase. Isotopes may be used to verify the species and frequency assignments and for the actual determination of geometry. Selection rules and force-field analyses can also be used to determine geometry, though less definitively.

Table VI indicates some typical results obtained only in recent years. One should note the agreement, within experimental uncertainty, for the bond angles of SO_2, SeO_2, and SiF_2 with the accurately known gas-phase values. To date there is no evidence that geometries of species produced in the gas phase and isolated in matrices differ significantly from those of the gas phase. However, new species produced during the trapping process may have isomeric forms not known in the gas phase, for example, $(NO_2)_2$ (41) and $(AX)_2$.

1. GEOMETRY OF TRIATOMIC SPECIES

The theory relating bond angle with vibrational frequencies and isotopic mass is well documented (52). Essentially the bond angle 2α for symmetrical XY_2 species may be obtained from isotope-shift measurements of the antisymmetric stretch v_3 by the relation

$$\sin \alpha = \frac{M_X M_X^i [M_Y(\omega_3)^2 - M_Y^i(\omega_3^i)^2]}{2M_Y M_Y^i [M_X^i(\omega_3^i)^2 - M_X(\omega_3)^2]}$$

TABLE VI

GEOMETRIES OBTAINED FROM MATRIX IR SPECTRA

Molecule	Structure (angles in degrees)	Comments and references
$(LiF)_2$	Isomers: cyclic V_h, linear $D_{\infty h}$	Li isotopes (1)
$(LiO)_2$	cyclic V_h	Li, O isotopes (10)
LiO_2	116	Li, O isotopes (10)
Li_2O	$D_{\infty h}$	Li, O isotopes (10)
LiO_2	C_{2v}, isosceles triangle $\theta(OLiO) = 90 \pm 6$	Li, O isotopes (10)
LiO_2Li	D_{2h}, rhombus, $\theta(OLiO) = 58$	Li, O isotopes (10)
$(SiO)_2$	cyclic V_h, $\theta(SiOSi) = 93$	Si, O isotopes (6, 48)
$(SiO)_3$	cyclic, D_{3h}	O isotopes (6, 48)
BeF_2	180	absence of ν_1 (30)
MgF_2	158 ± 5	Mg isotopes (30)
CaF_2	140 ± 5	Ca isotopes (30)
SrF_2	108	Sr isotopes, strong ν_1 (30)
BaF_2	(100)	angle estimated, strong ν_1 (30)
SiF_2	100 ± 1	Si isotopes, gas value = 100.9 (47)
GeF_2	94 ± 3	Ge isotopes (43)
TiF_2	130 ± 3	Ti isotopes (49)
SnF_2	92 ± 3	Sn isotopes (50)
SO_2	117 ± 2, 119	S isotopes, gas value = 119 (5, 45)
SeO_2	110 ± 3	Se isotopes, gas value = 113 (45)
TeO_2	110 ± 3	Te isotopes (88)
$SiCl_2$	90–120, most prob. 120	Si, Cl isotopes (83)
$NiCl_2$	180	Ni, Cl isotopes (50)
VCl_2	180	Cl isotopes (50)
$PbCl_2$	94 ± 3	Cl isotopes (50)
CuF_2	$163 < \theta < 180$	Cu isotopes (44)
CrF_2	180	Cr isotopes (50)
CH_3	D_{3h} planar	D isotope analysis (80)
LnF_3 (Ln ≡ rare earths)	C_{3v} nonplanar	Selection rules, observation of ν_1 (50)
$SiCl_3$	C_{3v}, 18 ± 5 deviation from planarity	Si, Cl isotopes (57)
$(MCl_2)_2$ (M = Mn, Fe, Co, Ni, Cu, Zn)	C_{2v}, bridge and terminal M—Cl	Valence force-field analysis of B_{3u} and B_{2u} (116)
$(NiF_2)_2$	C_{2v}, bridge and terminal Ni—F	Valence force-field analysis of B_{3u} and B_{2u} (50)

TABLE VI (Continued)

Molecule	Structure (angles in degrees)	Comments and references
$(MF_2)_2$ $(M = Ge, Sn)$	M—F—M bridges, M—F terminal	Valence force-field analysis of B_{3u} and B_{2u} (50)
$KSnF_3$	C_{3v}, K—F—Sn bridges	Frequency observations
$CsPbCl_3$	C_{3v}, Cs—Cl—Pb bridges	only [Criteria (50)]
MBO_2 $(M = Li, Na, K, Rb, Cs)$	Cs, θ(M—O—B) = 90 (O—B=O) linear	Li, B isotopes, force-constant analysis (110)
$(HCl)_2$	Cyclic C_{2h} (H-bonding)	(63)
CsOH	Linear or near-linear	D isotope analysis (2)
Al_2O	C_{2v}, 145 $\pm$ 10	O isotope analysis (66)
Ga_2S	C_{2v}, 112 $\pm$ 8	S isotope analysis (50)

where $\omega_3 = v_3 - 2x_{33} - \frac{1}{2}x_{13} - \frac{1}{2}x_{23}$ or $\omega_3 \simeq v_3 - 2\beta$, where 2β is taken as the total anharmonic contribution to the fundamental frequency. Also the anharmonic contribution to an isotopic fundamental frequency is given approximately by

$$\frac{\beta^i}{\beta} \approx \frac{(\omega_3{}^i)^2}{(\omega_3)^2} \approx \frac{(v_3{}^i)^2}{(v_3)^2}$$

or alternatively by

$$\frac{\beta^i}{\beta} \approx \frac{(\omega_3{}^i)}{(\omega_3)}$$

These approximations usually result in an uncertainty for α that is less than the experimental error incurred in measuring the isotopic frequency shift $(v_3 - v_3{}^i)$. It can be shown from the original bond-angle expression that neglect or underestimation of the anharmonic contribution results in the calculated bond angle being either a lower limit or an upper limit to the actual value, depending on whether isotopic substitution of X or Y is used. It also follows that the effect of an uncertainty in $(v_3 - v_3{}^i)$ leads to an increasingly large error in α as the angle approaches 180°. In fact with present instruments and techniques a practical upper limit to bond angles that can be determined by this technique is in the region of 160° (49).

In the infrequent event that one can observe all the isotopic features for a species having isotopes of both X and Y, as, for example, in the case of SO_2, it is possible to obtain the anharmonic correction and a precise bond angle (5). For most other cases one must resort to estimation of the anharmonic effect using, for example, the known anharmonicity of the diatomic XY species.

Alternatively, the combined isotopic shifts of v_1 and v_2 may be used to obtain α. For most inorganic species, however, the bending frequency often occurs in the far-infrared with an isotopic shift that is comparable in magnitude with the instrumental resolution and therefore inaccurately determined.

As an example, a spectrum for TeO_2 from which a bond-angle value of $110 \pm 3°$ was obtained is given in Fig. 5. The important effect of annealing out another matrix site, denoted by an asterisk, for yielding the correct

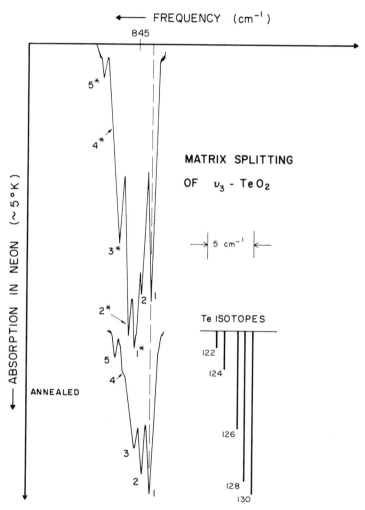

FIG. 5. Infrared spectrum of matrix-isolated TeO_2 showing both matrix splitting and isotopic structure for the v_3 frequency mode. [From Muenow *et al.* (*88*) by permission of the Faraday Society.]

distribution of isotopic intensities is also demonstrated. The absorptions at positions 1–5 in the figure correlate with the naturally occurring Te isotopes of 130, 128, 126, 124, and 122 as shown. Another example, demonstrating the use of both X- and Y-type isotopes, makes use of the v_3 structure of $NiCl_2$ shown in Fig. 6. For this case both the Ni and Cl isotopic features are consistent with a linear geometry and preclude angles of less than 175°. The lower-frequency set of bands, at around 480 cm^{-1}, also leads to the same result and therefore suggests that these absorptions are from the same vibrational species but in another matrix site. It should be noted that the half-bandwidths are as predicted from the instrumental resolution, and also the relative intensities are in accord with those calculated from the natural isotopic abundances.

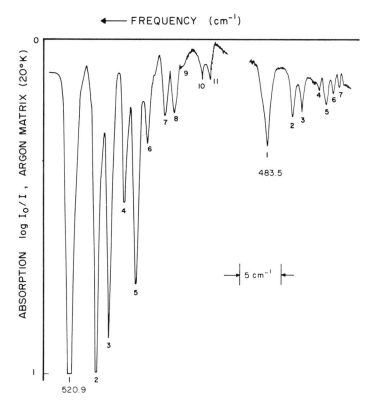

FIG. 6. Infrared spectrum of $NiCl_2$ matrix isolated in solid argon at 20°K, and showing the following isotopic v_3 absorptions: (1) 35-58-35, (2) 37-58-35, (3) 35-60-35, (4) 37-58-37, (5) 37-60-35, (6) 35-62-35, (7) 37-60-37, (8) 37-62-35, (9) 35-64-35, (10) 37-62-37, and (11) 37-64-35, The two distinct sets of data represent different matrix sites (50).

2. Geometry and Vibrations of Polyatomic Species

A frequent occurrence in matrix isolation, which can be considered either desirable or undesirable depending on the aim of the study, is the tendency to form aggregates or polymeric forms of the trapped species. Thus, when it becomes necessary to deposit heavy matrices for the study of weak absorptions, one has to deal with the presence of extra bands due to polymers, which may interfere with the interpretation of the monomer spectra. For most of these polyatomic species it is not possible, at present, to resolve isotopic structure, and therefore one must rely on the use of selection rules and simplified force-field arguments.

To exemplify the problem we consider the species $(MX_2)_2$, which occurs frequently in the study of triatomics. A number of specific examples are cited in Table V. For a C_{2v} structure containing four M—X bridge bonds in a ring and two terminal M—X bonds there are three infrared-active symmetry species, B_{1u}, B_{2u}, and B_{3u}, each giving rise to two vibrations. From the analogy of the alkali halide dimers, $(SiO)_2$, and $(LiO)_2$ the B_{1u} out-of-plane vibrations should fall in the lowest-frequency region of the spectrum. A normal coordinate analysis, neglecting interaction force-constant terms, leads to the following expressions for B_{3u} and B_{2u} species respectively (28):

$$\lambda^2 - (\mu_M + \mu_X)[F_d(1 + \cos \varphi) + F_r]\lambda$$
$$+ F_d F_r\left[(\mu_M + \mu_X)^2(1 + \cos \varphi) - 2\mu_M^2 \cos^2 \frac{\varphi}{2}\right] = 0 \quad (1)$$

or for $\varphi = 90°$

$$\lambda^2 - (\mu_M + \mu_X)(F_d + F_r)\lambda + F_d F_r(\mu_X + 2\mu_M)\mu_X = 0 \quad (2)$$

Similarly, for B_{2u}

$$\lambda^2 - \lambda\left[F_d(\mu_M + \mu_X) + 2\frac{F_\gamma}{r^2}\mu_X\right] + 2\frac{F_\gamma}{r^2}F_d\left[(\mu_M + \mu_X)\mu_X - \frac{\mu_M^2}{4}\right] = 0 \quad (3)$$

The μ's are the reciprocal masses for the atoms; F_r, F_d, and F_γ/r^2 are force-constant terms for outer-bond stretching, bridge-bond stretching, and outer-bond bending respectively; φ is the X–M–X angle in the ring; $\lambda = (2\pi v)^2$; and the solution of Eq. (1) or (2) yields v_{12} and v_{11}, which represent terminal-bond stretch and bridge-bond stretch frequencies, respectively. Solution of Eq. (3) gives v_{10} and v_9, representing terminal-bond bend and bridge-bond stretch frequencies, respectively. The magnitude of v_9 is particularly sensitive to the value of φ. The force constants may be approximated as follows: $F_r \simeq k_1$, the stretch-force constant for the MX_2 monomer; $F_d \simeq F_r/2$ and $F_\gamma/r^2 \simeq F_r/10$. These approximations are not entirely arbitrary, as it is known that the bridge-bond energies are slightly greater than one-half the terminal-bond values. Also for the monomer species the bending-force constants are found to be of the order of and usually less than one-tenth the stretch-force constant values. For the dimer species of this type given in Table V quite good agreement between calculated and observed frequencies has been found, and the above equations certainly serve to distinguish between dimeric and other species in the matrices. Also these equations should be more reliable in predicting isotope shifts where experimental data are available.

From the form of the above expressions it also follows that v_{12} followed by v_9 are the highest frequencies and fall in the region of M–X stretches, while v_{11} and v_{10} fall in the region typical of X–M–X bending frequencies. Also, as the mass of M becomes heavier relative to X, v_{12} has a stronger tendency to decrease relative to v_{11}. An increase in φ also has the effect of increasing v_9.

For $(MX)_2$ species, such as the alkali halide dimers and the Group IVA-oxide dimers, with the neglect of interaction-force constants very good agreement between observed and calculated spectra is obtained. The B_{3u} and B_{2u} modes, assuming a valence force field, have been calculated and are given respectively by

$$\lambda B_{3u} = 2(\mu_M + \mu_X)\left(F_r \sin^2 \theta + \frac{2F_\gamma}{r^2} \cos^2 \theta\right)$$

and

$$\lambda B_{2u} = 2(\mu_M + \mu_X)\left(F_r \cos^2 \theta + \frac{2F_\gamma}{r^2} \sin^2 \theta\right)$$

where 2θ is the X–M–X angle and the two in-plane bending-force constants are assumed to be equal and denoted by F_γ/r^2.

A central force field has also been utilized to describe the potential energies of the alkali halide and LiO dimers. The valence force field would appear to be more generally applicable. The more difficult problem of calculating the remaining ir-active out-of-plane mode B_{1u} has been dealt with using the assumption of free rotation about the M–X bonds and gives a lower limit to the frequency v_6, as follows:

$$v_6 = 0.149\left(\frac{hF_\gamma}{r^4}\right)^{1/3}(\tan^2 \theta + \cot^2 \theta)^{1/3}\left(\frac{\mu_M}{2} + \frac{\mu_X}{2}\right)^{2/3}$$

where h is Planck's constant.

This expression leads to a calculated lower-limit value (6) of 33 cm^{-1} for $(SiO)_2$, compared to an observed value (48) of 79 cm^{-1} in an argon matrix.

These examples of relatively simple polymeric species, where many approximations are used, serve to indicate the difficulty of determining structures of more complex species from matrix ir observations alone and particularly in the absence of observable isotope effects.

3. CORRELATIONS OF VIBRATIONAL PROPERTIES FOR SYMMETRIC MX_2 SPECIES

The ability of the matrix-isolation–ir technique to deal with a wide range of species enhances the possibility of recognizing on a broad basis the factors important to molecular vibrations. One of the questions we wish to answer is: Can we predict the vibrational properties of a whole family of inorganic species from observations on only a few members? Considering symmetric MX_2 as the simplest species to contain both stretching and bending vibrational properties, the question may conveniently be reduced to one of predicting the force constants, k_1 (stretching), k_δ/r^2 (bending), k_{12} (stretch–stretch interaction), and k_{13} (stretch–bend interaction). Usually the last of these interaction constants is not obtainable and is the least significant of the four general valence-field force constants.

For the following discussion we therefore consider the potential energy V of MX_2 to be given by

$$V = \tfrac{1}{2}k_1(\Delta r_1{}^2 + \Delta r_2{}^2) + k_{12}\,\Delta r_1\,\Delta r_2 + \tfrac{1}{2}k_\delta\,\delta^2$$

where Δr terms represent changes in internuclear separations along the M–X bonds and δ is the bond-angle change associated with the bending motion.

Figure 7 demonstrates that k_1 is, to a very good approximation, related directly to the ligand (X) electronegativity (Pauling's values). Thus a knowledge of the antisymmetric frequency (v_3), from which k_1 is obtainable, for a few MX_2 species is sufficient to define the values for X-substitution.

For cases where the central atom M is varied, X being fixed, k_1 also decreases with decreasing electronegativity of M, but not in a linear fashion. The first-row transition-metal difluorides show a reasonably close correlation of $(k_1 - k_{12})$ with the energy required to dissociate the molecule to ions as shown in Fig. 8. Also the general trend of an increasing force constant is in qualitative agreement with the known increase in electronegativity across the row.

A relationship between force constant, internuclear separation r, and dissociation energy D, of the form

$$\frac{k_1 r}{D} = C$$

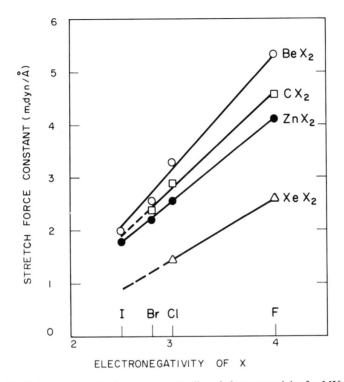

FIG. 7. Relation of stretch force constant to ligand electronegativity for MX_2 species. [BeX_2 data from A. Snelson, *J. Phys. Chem.* **72**, 250 (1968); ZnX_2 data from A. Loewenschuss, A. Ron and O. Schnepp, *J. Chem. Phys.* **49**, 272 (1968); XeX_2 data listed by Nelson and Pimentel (*89*); CX_2 data from references cited in Table VII.]

where C is an empirical constant, has been proposed for diatomic species having similar bond type (*114*). We note that the Group-II and Group-IV difluorides follow the empirical relation

$$\frac{k_1 r}{(D)EN} = C$$

where r is the F–M internuclear separation in Å, D the F–M bond energy of MF_2 in electron volts, k_1 is in mdyn/Å, and EN is electrostatic electronegativity as listed by Rich (*101*). For Group II, $C = 0.8 \pm 0.1$, and for Group IV, $C = 0.73 \pm 0.1$.

Correlations of bending-force constants have historically been difficult, but it is usually thought that k_δ/l^2 (i.e. k_δ/r^2) should decrease with an increasing ionic character in the bonding. It has been observed that for species such as O_3, LiON, OF_2, H_2O, and Al_2O the logarithm of k_δ/l^2 decreases linearly

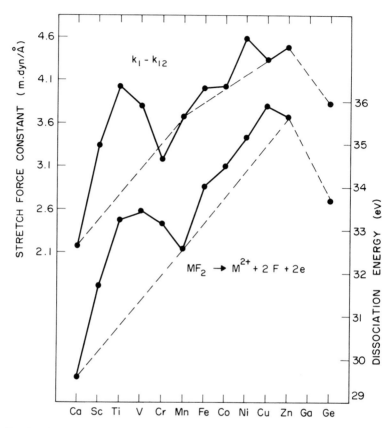

FIG. 8. Correlation of stretch force constants for the first-row transition-metal difluorides with the dissociation energy for the process $MX_2 \rightarrow M^{2+} + 2X + 2e$ (50).

with increasing electronegativity difference (7). However, this correlation does not have general utility, as the species ZnI_2, $ZnBr_2$, $ZnCl_2$, NiF_2, ZnF_2, TiF_2, and CaF_2 follow the reverse trend. There is also no simple relation between $(k_1 - k_{12})$ and k_δ/l^2 for various MX_2 species where X is fixed. However, for the case where M is fixed—for example, the Be halides and Zn halides —the ratio of $k_1 : k_\delta/l^2$ is approximately constant—i.e., 45 ± 3 and 55 ± 5, respectively. Calculations based on an ionic model also yield reasonable values of k_δ/l^2 for these Group-II halides (28).

The quadratic interaction force constant k_{12}, which describes bond–bond interaction in the molecule, is usually much smaller than the stretch-force constant. As pointed out by Linnett and Hoare (67) and more recently by Machida and Overend (71) for species such as NO_2, BO_2, and CF_2, the sign of k_{12} is usually positive if the bonds in MX_2 are longer than those in diatomic MX species. One may logically expect this to include relative dissociation energies; the difference Δr should be positive if the M–X bonds in MX_2 are stronger than that of MX, where Δr is $r_{MX} - r_{MX-X}$.

Linnett and Hoare (67) proposed such a rule based on the observed properties of species such as CO_2, CS_2, SO_2, NO_2, and other covalent molecules. The rationale for this rule would appear to be that bond shortening necessarily implies concomitant strengthening of the bond.

That the magnitude and sign of k_{12} are very sensitive to bond polarity is indicated by the examples of KrF_2 ($k_{12} = -0.2$ mdyn/Å) and XeF_2 ($k_{12} = +0.13$ mdyn/Å). In the former case each F atom has a negative charge of only 0.47, and in the latter, 0.79. The implied extra contribution of a nonbonding resonance structure for KrF_2 was used by Coulson (39) to explain the change of k_{12} from XeF_2 to KrF_2.

4. CORRELATIONS OF VIBRATIONAL PROPERTIES FOR ASYMMETRIC MX_2 SPECIES

From matrix-isolation studies of the type considered in more detail in the following section, force-constant and structural analyses have been carried out by various workers on asymmetric species such as FO_2, ClO_2, HO_2, and LiO_2, as summarized very recently by Andrews (10). The magnitude of the f_{O-O} stretch-force constant has been suggested by Spratley and Pimentel (115) to be inversely related, by reason of the antibonding nature of the next available molecular orbital on O_2,[†] to the amount of negative charge on the O_2 group in MO_2. The extent to which an electron is transferred from M to O_2, or vice versa can be approximately predicted from the electronegativity difference $\Delta EN(M-O)$, using one of the available ΔEN versus ionic character curves.[‡] A very satisfactory linear correlation of f_{O-O} with the estimated charge on O_2 is indicated by Fig. 9. Clearly this correlation should prove useful for predicting the vibrational properties of new MO_2 species.

B. Matrix Reactions and Formation of Reactive Species

Much of the initial interest in matrix isolation was directed to the production and trapping or "stabilization" of free radicals and other reactive species and is well documented. However, this is still one of the most active areas in matrix-isolation spectroscopy, and the following discussion is concerned with the most recent developments in this area.

[†] Good evidence for the antibonding nature of electron addition to O_2 is the fact that the dissociation energy of O_2^- is 1 eV less than that for the neutral O_2 species.

[‡] One such curve that has both an experimental and theoretical basis is given by Hastie and Margrave (46).

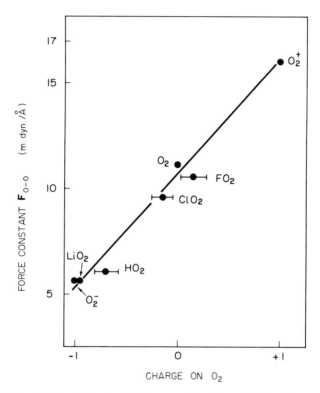

FIG. 9. Correlation of the stretch force constant f_{o-o} with the charge on O_2 in unsymmetrical MO_2 species [f_{o-o} data taken from Andrews (*10*)].

Many of the successful discoveries of new radical species have utilized the principle of a bi- or trimolecular reaction being the most controllable in a matrix trapping environment. A major concern is therefore the method of production and combination of the reactant species. The following techniques listed in Table VII have been used for the production of reactive atoms such as C, F, Cl, Br, H, O, and N. Also listed are examples of species produced with these primary reactants. These studies demonstrate the application of uv and vacuum-uv photolysis, atom abstraction (e.g., with Li), microwave discharge, flash photolysis, gamma radiation, matrix diffusion, and simple chemical combination for the production of new species.

A limiting feature for reactions of this type is the so-called cage effect (discussed by Pimentel in Bass and Broida, *13*, p. 95) which is essentially due to the surrounding matrix atoms hindering the escape of a group split off from a photodissociating species. Thus only the smallest of atoms, such as F or H, can effectively migrate in the matrix at normal trapping temperatures. More efficient production of radicals by photolysis is usually

TABLE VII

MATRIX FORMATION AND ISOLATION OF REACTIVE SPECIES

Product species	Production of reactants	Comments and references
HCO	$HI \xrightarrow{uv} H + I$	CO matrix, 20°K (76)
ClCO	$HCl/CO \xrightarrow{uv} ClCO + HCO$	CO matrix, 14°K (55)
CF$_3$	$Ar/BrCF_3 \xrightarrow{vac\ uv} CF_3 + Br$	0.5% matrix conc. of reactant (79)
CF$_3$	$CO/HCF_3 \xrightarrow{vac\ uv} CF_3 + HCO$	1.2% conc. of reactant
CF$_3$	$Ar/N_3CN \xrightarrow{uv} C$	CF$_3$, C_{3v}
	$t - N_2F_2 \xrightarrow{uv} F$	
	$C + 3F \longrightarrow CF_3$	
CH$_3$	$CH_3X + Li \longrightarrow CH_3 + LiX$	X = halogen, Ar/CH$_3$X = 200–600, Li/CH$_3$X = 1–2; 12 hr trapping
C$_2$H$_6$	$2CH_3 \xrightarrow{\Delta} C_2H_6$	CH$_3$, D_{3h}, Ref. (8)
SiCl$_3$	$HSiCl_3 \xrightarrow{vac\ uv} SiCl_3 + H$	1216 Å uv source, Ar, N$_2$ matrix, 14°K, C_{3v}, 72 + 5° angle (57)
SiCl$_2$	$SiH_2Cl_2 \xrightarrow{vac\ uv} SiCl_2 + 2H$	C_{2v}, 120° angle (83)
CH$_3$	CH$_4$/Ar 14°K, vac uv	D_{3v}, (80)
BrBrCl$_2$	Cl$_2$ + Br$_2$ + inert g., micr. disch.	(90)
CBr$_3$	Li + Ar/CBr$_4$ (200–600)	C_{3v}, (9)
CBr$_2$	Li + Ar/CBr$_4$ (200–600)	C_{2v}, (32)
HNF	F$_2$, uv pyrex filter $\rightarrow$ 2F	Ar/F$_2$, 50–200
HNF$_2$	$F + HN_3$, uv $\longrightarrow HF + N_3$	Ar/HN$_3$, 75–200
	$N_3 \xrightarrow{uv} N_2 + N$	See Jacox and Milligan (56)
	$N + F \longrightarrow NF, NF_2, NF_3$	
	$NH + F \longrightarrow HNF$	
NCN	$Ar/N_3CN \xrightarrow{uv} NCN + N_2$	$\lambda > 2800$ Å source
OH	$Ar/H_2O\ (250) \xrightarrow{vac\ uv} OH + H$	H$_2$ resonance lamp through LiF window (3)
XeCl$_2$	Xe/Cl$_2$ (200–100), micr. disch.	(89)
CaO	$Ca/O_2 \xrightarrow{uv} CaO + O$	(25)
LiON	Li + NO/Ar	Bent, 100 ± 10° (7)
NH$_2$	NH_3 solid $\xrightarrow{\gamma\text{-radiat.}} NH_2$	(117)
CO$_3$	$O_3 \xrightarrow{uv} O + O_2$,	Planar, C_{2v} (86)
	$O + CO_2 \longrightarrow CO_3$	
ClOO	OClO/Ar, uv, 4°K	Isomeric rearrangement (11)
$\begin{bmatrix} HOBr \\ HOCl \end{bmatrix}$	HX − O$_3$/Ar, uv, 4°K	(109)
CF$_2$	Ar/CF$_2$N$_2$ (100), 20°K	13-kJ flash discharge (108)

TABLE VII (Continued)

Product species	Production of reactants	Comments and references
$\begin{bmatrix} GaF \\ GaF_3 \\ Ga_2F_4 \end{bmatrix}$	$Ga + Ar/F_2$ (100), $15°K$	(50)
$NaSiF_2$	$Na + SiF_2 + Ar$, $15°K$	Products decompose on warmup to form Na_2SiF_6
$LiSiF_3$	$LiF + SiF_2 + Ar$, $15°K$	and Li_2SiF_6 at room temperature (50)
$\begin{bmatrix} NH \\ CH_2 \end{bmatrix}$	$CH_3N_3 \xrightarrow{uv} CH_2 + NH + N_2$	(74)
CO_2NH	$NH + CO_2 \longrightarrow CO_2NH$	CO_2 matrix, structure not definitely established (75)

$$O=C{\overset{\displaystyle O}{\underset{\displaystyle\overset{N}{\underset{H}{|}}}{\diagdown\diagup}}}$$

observed during matrix formation rather than within the matrix because of surface reactions occurring in the absence of a cage effect. The use of excess radiation to produce a local temperature increase, and matrices with a relatively high softening ability such as N_2 and Ar are also thought to reduce the cage effect [Minkoff (85), p. 55].

C. Study of Ions by Matrix Isolation

The application of mass spectrometry to inorganic chemistry has led to a knowledge of the identity and stabilities of a large number of positive ions and a far lesser number of negative ions. However, very little spectroscopic or structural information is available for ions, and recent studies indicate matrix-isolation spectroscopy to be a suitable technique for the study of ions.

Ions such as CO_2^- may be produced *in situ* by exposing a matrix containing $NaHCO_3$ isolated in an alkali disk (or even a zeolite matrix in this case) to gamma rays from a cobalt source. Alternatively, cocondensation of CO_2 with atomic Na or K can lead to electron transfer from the alkali metal to produce CO_2^- isolated in its molecular solid (16). CS_2^-, H_2S^-, and O_2^- can be formed by a similar procedure (17–19). The molecular halogen ions Cl_2^- and Br_2^- can be produced in an isolated state by gamma irradiation of H_2SO_4 glasses doped with chloride (27).

One may also produce matrix-trapped electrons (73). For example, Na atoms trapped in ice at 77°K lose their valence electron, which is then stabilized by interaction with six protons from the surrounding H_2O matrix. On the other hand, the delicacy of the stabilization of an electron is indicated by the failure of solid H_2S to act as an electron trap under similar conditions. In this case H_2S^- is formed because of an expected higher electron affinity of H_2S as compared with H_2O. The ammoniated electron is also known.

No direct attempt to matrix-isolate a beam of ions has been reported, but there is very recent evidence that vaporization of graphite produces C_2^- as well as C_2 and other species, which have apparently been isolated in rare-gas matrices and spectra obtained (84).

D. Determination of Thermodynamic Properties

There are several avenues by which matrix-isolation spectroscopy can lead to important thermodynamic data for inorganic species, particularly those produced at high temperatures. It is possible to obtain geometries, vibrational frequencies, and electronic structures (discussed in the following section) of species, and this is sufficient information to allow accurate calculation of heat capacities, entropies, etc. from the well-known statistical relationships. However, this information still does not define completely the thermodynamic nature of a species, and the necessary heats of formation, sublimation, etc. are usually obtained from measurements of the variation of partial pressure with temperature. Currently, the mass-spectrometric–Knudsen-effusion technique is one of the main tools used for obtaining these enthalpy-change quantities. This technique is sometimes limited by an uncertainty in the exact relation of the mass-spectral fragmentation pattern to the neutral molecular precursors, and certainly it does not distinguish between various geometrical isomers.

As shown in Table V, the matrix–ir technique has indicated the existence of isomeric forms of $(LiF)_2$, LiO_2, and possibly $(LiO)_2$, and further studies may indicate that these are not rare cases. Thus one requires for a thermodynamic characterization of such isomers a technique capable of determining the partial pressure versus temperature properties for each isomer. Several recent studies (49, 50, 113) have shown that one can matrix-trap a molecular beam so that the relative concentrations of various species in the matrix are representative of those in the Knudsen effusion cell. In particular, Snelson (113) has demonstrated that the absorption intensity for an infrared band of a matrix-isolated species obeys the Lambert–Beer law. Hence by isolating species in separate matrices at various known Knudsen-cell temperatures and recording the absorption intensities, one can obtain a second-law determination for the heat of sublimation, vaporization, or reaction of a species. In this manner Snelson obtained a value for the heat of sublimation of AlF_3 that was in excellent agreement with reliable values obtained by other methods. Significantly different heats of vaporization for the linear and cyclic forms of $(LiF)_2$ were found—i.e., $\Delta H_v(1200°K) = 61 \pm 3$ kcal/mole (cyclic form) and $\Delta H_v(1200°K) = 76 \pm 4$ kcal/mole (linear form). Also, the energies for the formation of these dimers from the monomers are $\Delta E_0^0 = -56.0$ kcal/mole (cyclic form) and $\Delta E_0^0 = -39.9$ kcal/mole (linear form). This lower stability of the linear form does not preclude its existence in the high-temperature vapor, as the larger entropy of the linear species allows a favorable free energy of formation.

In a similar study (49), the thermodynamic properties of TiF_2 and TiF_3 were evaluated by using matrix spectroscopy to determine the equilibrium concentrations of the gaseous species.

E. Electronic Structure

1. OPTICAL SPECTROSCOPY

Studies of electronic spectra by matrix isolation continue to be of interest especially for species that are unstable at room temperature—for example, free radicals and high-temperature molecules. Previous work has been referenced by Ramsay (99) and Weltner (127). We shall discuss some of the recent studies of electronic spectra that best illustrate the usefulness of the matrix-isolation methods.

The objective in a study of electronic spectra is usually the determination of the nature and relative energy-level spacings of electronic states within a molecule. Properties such as internuclear distances and angular momentum along a molecular symmetry axis of an excited electronic energy state are determined by high-resolution studies of rotational fine structure. Because most molecules do not rotate in matrices, however, one cannot obtain such information, and the emphasis is therefore on the study of electronic and vibrational energy-level spacings.

It is apparent from previous work that a neon matrix usually provides the best gaslike spectra—i.e., the smallest matrix shifts—for most molecules. Matrix shifts for a neon matrix are given in Table VIII for those molecules for which gas-phase data are available. It is seen that most molecular electronic transitions undergo a blue shift of less than 1000 cm^{-1}.

It is interesting to note that a neon matrix affects the A $^2\Pi_{3/2}$ state of Group IIIB oxides differently than the A $^2\Pi_{1/2}$ state, or in other words causes an appreciable change in the spin–orbit coupling constant.

An important advantage in these matrix studies arises from the fact that only the lowest electronic–vibrational energy state is occupied because of the

TABLE VIII

ELECTRONIC TRANSITIONS OF MATRIX-ISOLATED INORGANIC MOLECULES

Molecule	Spectral region (Å)	ν(Ne matrix) $-\nu$(gas) (cm^{-1})	Remarks	References
CH	4304	54	$A\,^2\Delta \leftarrow X\,^2\Pi$	(80)
	3910–3920	−209	$B\,^2\Sigma^- \leftarrow X\,^2\Pi$	
	3149	−45	$C\Sigma^+ \leftarrow X\,^2\Pi$	
CN	3850	—	$B(^2\Sigma^+) \leftarrow X(^2\Sigma^+)$	(78)
O_2	1750–2000	∼50	$B\,^3\Sigma_u^- \leftarrow X\,^3\Sigma_g^-$ Progression in ν'; observed in liquid and solid argon	(13)
	2400–3000	—	Observed only at high oxygen concentrations; possibly $A\,^3\Sigma_u^+ \leftarrow X\,^3\Sigma_g^-$	
C_2	2323	200	$D\,^1\Sigma_u^+ \leftarrow X\,^1\Sigma_g^+$	(81, 119)
Si_2	3461–3974	—	Long progression in ν'; $H\,^3\Sigma_u \leftarrow X\,^3\Sigma_g^-$	(121)
S_2	2500–3000	−10	$^3\Sigma_u^- \leftarrow\,^3\Sigma_g^-$ Long progressions seen in absorption and fluorescence	(23)
ScF	4792	535	$E\,^1\Pi \leftarrow X\,^1\Sigma$	(70)
	6227	−39	$C\,^1\Sigma \leftarrow X\,^1\Sigma$	
	9405	−31	$B\,^1\Pi \leftarrow X\,^1\Sigma$	
ScO	4782	335	$B\,^2\Sigma \leftarrow X\,^2\Sigma$	(128)
	5859	508	$A\,^2\Pi_{3/2} \leftarrow X\,^2\Sigma$	
	5946	372	$A\,^2\Pi_{1/2} \leftarrow X\,^2\Sigma$	
TiO	7104	53	$A\,^3\Phi_2 \leftarrow X\,^3\Delta$	(126)
	6124	133	$^3\Pi_0 \leftarrow X\,^3\Delta_1$	
	5137	120	$C\,^3\Delta_1 \leftarrow X\,^3\Delta_1$	
YO	4646	776	$B\,^2\Sigma \leftarrow X\,^2\Sigma$	(128)
	5697	825	$A\,^2\Pi_{3/2} \leftarrow X\,^2\Sigma$	
	5883	698	$A\,^2\Pi_{1/2} \leftarrow X\,^2\Sigma$	
LaO	5443	529	$B\,^2\Sigma \leftarrow X\,^2\Sigma$	(128)
	7188	410	$A\,^2\Pi_{3/2} \leftarrow X\,^2\Sigma$	
	7706	337	$A\,^2\Pi_{1/2} \leftarrow X\,^2\Sigma$	
	4152	—	$C\,^2\Pi \leftarrow X\,^2\Sigma$	
	4070	—	$C\,^2\Pi \leftarrow X\,^2\Sigma$	
	5592	—	Seen in fluorescence	

TABLE VIII (Continued)

Molecule	Spectral region (Å)	ν(Ne matrix) $-\nu$(gas) (cm^{-1})	Remarks	References
ZrO	3660	163	$^1\Sigma \leftarrow X\,^1\Sigma$	(123)
	5154	—	—	
	5872	—	—	
	6446	—	—	
TaO	8252	55		(35, 125)
	7697	137	$^2\Phi_{5/2} \leftarrow X\,^2\Delta_{3/2}$	
	7297	131	$^2\Delta_{3/2} \leftarrow X\,^2\Delta_{3/2}$	
	6945	25		
	6254	104	$^2\Phi_{5/2} \leftarrow X\,^2\Delta_{3/2}$	
	5980	—	—	
	5552	—	—	
	4804	—	—	
	4787	—	—	
	4518	—	—	
	4476	—	—	
	4273	54	$^2\Pi_{1/2} \leftarrow X\,^2\Delta_{3/2}$	
	4140	87	$^2\Phi_{5/2} \leftarrow X\,^2\Delta_{3/2}$	
	3901	31	$^2\Pi_{3/2} \leftarrow X\,^2\Delta_{3/2}$	
	3803	—	—	
	3737	79	$^2\Delta_{3/2} \leftarrow X\,^2\Delta_{3/2}$	
	3419	—	—	
WO	4968–5835		Extensive perturbations are present in most transitions	(124)
	5262–5784			
	4750–5209			
	4057–4807	−3		
	3706–4648			
	3692–4278			
	4043–4201			
C$_3$	3500–4057	42	$^1\Pi_u \leftarrow X\,^1\Sigma$	(119, 120)
	5856	—	$(^3\Pi_u \rightarrow X\,^1\Sigma)$? Seen in fluorescence; lifetime of excited state $\sim$0.02 sec in neon matrix	
SiC$_2$	4091–4963	58	Complex progression in ν_1', ν_2', and ν_3'	(121)
Si$_3$	4200–4700	—	Identity of absorber not definitely proven; progression in ν_1'; transition not analogous to that of C$_3$ at 4050	(121)

TABLE VIII (Continued)

Molecule	Spectral region (Å)	ν(Ne matrix) $-\nu$(gas) (cm^{-1})	Remarks	References
FCO	2214–3348	—	Extensive progression with 650 cm^{-1} spacing	(77)
CNN	4189 3964			(78)
NCO	4000–4500	< -50	A($^2\Sigma^+$) ← X($^2\Pi$); progression in ν_1, ν_2, and ν_3	(82)
	2400–3200	< -160	B($^2\Pi$) ← X($^2\Pi$); progression in ν_1 and ν_3	
ScF$_2$	2800–3200	—	Progression in ν_1; other bands observed but not assigned	(70)
WO$_2$	7890	—	Long progression in ν_2; short progression in ν_1	(124)
	7806	—	Progression in ν_2	
TaO$_2$	5587–6159	—	Main progression in ν_2; ν_1' also observed	(125)
	7361–8607	—	Progression in ν_2	
CH$_3$	1503	~ 0		(80)
WO$_3$	3253–3453	—	Spectra not definitely assigned to WO$_3$	(124)
Si$_2$C$_3$	5600–6612	—	Spectra complex and assignment uncertain	(121)

low temperatures used. This has proved to be very useful in the study of high-temperature molecules, where it is common to observe gas-state absorption spectra originating from at least two different low-lying electronic states as well as many different vibrational states. Thus a comparison of matrix to gas-phase spectra will establish which of the gas-phase transitions originate from the ground state. The matrix spectra are greatly simplified by the absence of band overlapping arising from absorptions out of excited vibrational states. Table VIII also lists some high-temperature molecules whose ground states have been determined with the matrix method. These results are mainly due to the efforts of W. Weltner, Jr. and D. McLeod, Jr.

The simplicity of the matrix spectra has allowed the determination of many new excited electronic states. In particular, studies of matrix-trapped TaO have demonstrated the existence of sixteen excited electronic states, all observed in absorption from the ground electronic state. A later high-resolution study by Cheetham and Barrow (*35*) of the gas-phase rotational structure for TaO showed good agreement with the neon matrix spectra. In all cases the deviation from gas phase was less than 150 cm^{-1}, as shown by Table VIII. The combined studies have now demonstrated the existence of twenty-seven excited electronic states for TaO.

Judging by the examples of Table VIII, the matrix-isolation method should be very useful in future studies of electronic spectra for high-temperature molecules. This will be particularly true for the transition-metal, lanthanide, and actinide compounds where one expects many excited states with similar properties and small energy separations to exist.

Electronic states of a molecule are sometimes divided into singlet and triplet systems, and it is often the case that the relative energy-level spacings between any two singlet states or any two triplet states are known, but the singlet to triplet separations are unknown. The difficulty in obtaining singlet–triplet energy separations arises from the low transition probabilities for optical transitions between these states. Phosphorescence from an excited triplet state to the ground singlet state has been observed for many organic molecules and a few inorganic molecules. This is usually accomplished by populating a singlet state, which then decays by some nonradiative path to an excited triplet state. Subsequently the triplet state decays to the ground state with light emission. Since the phosphorescence is usually very weak, a high population of the triplet state is required for observation. This is possible by optical means only when the nonradiative decay from the excited singlet to triplet state is very efficient. As a nonradiative transition usually involves a transfer of energy to a third body, one might expect such transitions to occur more readily in a matrix. Phosphorescence from a possible triplet state of C_3 has been reported (*120*) as shown in Table VIII, and for SnS and SnO in various matrices as a $^3\Pi$ to $X^1\Sigma$ transition (*111*). Similar observations on many other molecules would clearly be very helpful in eliminating the persistent problem of relating states of singlet and triplet systems on an energy scale.

2. ESR SPECTROSCOPY

The application of electron-spin-resonance (ESR) spectroscopy to the study of matrix-isolated radicals is expected to grow rapidly in the next few years. There has been considerable study of radicals by other forms of spectroscopy, and there is no apparent reason why these same radicals cannot be studied by ESR spectroscopy.

There are several ways of producing trapped radicals. One often used method is the irradiation of an *in situ* parent molecule by gamma rays (*87*, *87a*) or vacuum ultraviolet radiation (*16*). Some radicals exist at appreciable concentration in equilibrium at high temperatures and can be trapped from a molecular beam (*127*). Another very promising method (*8*, *73*) is the reaction during condensation of two species such as an alkali metal and a halogen-containing molecule to produce a salt and a radical minus the halogen. One can also react an alkali metal with a species of high electron affinity to form an ion pair (*17*).

A recent review by Mile (*73*) summarizes ESR studies of carbon-containing radicals, and a convenient division, with some chemical significance, is given where radicals are classified as either σ or π. In σ-radicals the free-electron orbital is s-p-hybridized, while π-radicals have contributions from only p-orbitals of the zero-valence atoms. Studies of HCO have shown it to be a σ-radical, and since an accurate measure of s–p hybridization is often obtained, it is possible to infer a molecular geometry—125° bond angle in this case—assuming there are no other effects such as d-orbital involvement in the bonding. The trifluoromethyl, CF_3, and other fluorinated methyl radicals form an interesting group, as there appears to be a gradual change from a σ-radical to a π-radical in going from CF_3 to CH_3 with the successive replacement of a fluorine with a hydrogen atom. This is reflected in the change in geometry through the series: CF_3 (nonplanar with a deviation from planarity of 17.8°), CF_2H (12.7°), CFH_2 (near-planar), and CH_3 (planar, as would be expected for a π-radical).

Of the current Dewar designs those used by Kasai *et al.* (*60*) and Bennett and Thomas (*15*) seem to be the most versatile. The rotating cryostat, used by the latter workers, is very useful for the production of free radicals via chemical reactions. It has been used extensively in the studies of alkali-metal–halocarbon (*15*, *73*) reactions and in studies of ion-pair formation with electron-accepting species such as H_2S (*17*), CO_2 (*16*), and CS_2 (*18*). Kasai, Weltner, and Whipple have studied high-temperature radicals such as $Cu(NO_3)_2$ and CuF_2 (*60*), NF_2 (*59*), VO (*61*), ScO, YO, LaO (*128*).

Perhaps the most significant recent study is that of NO_2 (*58*), which demonstrated that molecules could be oriented in relation to the trapping surface. This seems to be frequently true (*128*), as it appears that molecules with a well-defined plane are trapped with the molecular plane parallel to the trapping surface. As a result a matrix formed on a rotatable substrate provides many of the advantages of working with a single crystal. Recent studies of NF_2 (*59*) and VO (*61*) have used this effect to analyze the anisotropic parts of the hyperfine interactions. Molecular orientation is found to occur most often with neon matrices. However, NF_2 was found to orient in argon when it was preheated to 300°C before trapping. It seems that orientation can be expected if local annealing is present during condensation.

Morehouse et al. (87a) have studied the radicals CH_3, SiH_3, GeH_3, and SnH_3 by irradiating the parent MH_4 in situ with gamma rays. Anisotropic g-factors were measured, and for SiH_3 hyperfine interactions of ^{29}Si revealed 22% s-character in the orbital of the unpaired electron. Assuming d-orbitals are unimportant, the radical is predicted to be nonplanar with an H–Si–H angle of 110.6°.

Adrian et al. (4) have recently studied HO_2 produced by the reaction of H atoms, from the in situ photolytic decomposition of HI, with O_2. The g-factor was very anisotropic, and the spectra could be best explained by assuming 71% spin density on the central oxygen and 29% on the end oxygen. The geometry is assumed to be similar to that found in H_2O_2, in agreement with conclusions from the infrared spectra obtained by Milligan and Jacox (cited by Mile, 73).

V. Conclusions and Future Applications

We conclude by pointing out the following variety of inorganic and other chemical problems that have been, or potentially could be, dealt with by matrix-isolation spectroscopy

(1) There exists a broad scope for the identification of species, whether they be atoms, molecules, radicals, ions, or in special cases electrons.

(2) One may obtain molecular and spectroscopic parameters for these species.

(3) One has the ability to artificially introduce perturbations on these species and activate "forbidden" transitions, as shown, for example, by the vibrational transition of S_2 in the infrared.

(4) In principle it is possible to determine reaction energetics, mechanisms, and rates between these species.

(5) The synthesis of new materials can be followed from the atomic to the simple molecular, complex molecular, and eventually solid-state levels.

(6) The ability to orient certain species in matrices should assist in spectroscopic assignments and also possibly in the development of new materials. Subjecting matrices to strong electric or magnetic fields would enhance the general utility of orientation effects.

(7) The ability to study initially high-temperature species in an isolated low-temperature environment gives much additional information on stellar molecules.

Clearly many of these applications will require the interpretation of complex spectra, for which it will be necessary to do the following:

(1) Use high-resolution spectrometers.

(2) Make maximum use of isotopes.

(3) Develop force-field approximations for large molecules, based on the experience gained with smaller species.

(4) Adapt the fullest range of spectroscopic tools—i.e., extend to Raman and Mössbauer spectroscopy, etc.

Now that the techniques for the production of a variety of free radicals have been developed, one can expect future studies to deal more with the reaction chemistry of these radicals with other molecular entities.

As pointed out by Weltner in 1963, very little matrix infrared work had been carried out on high-temperature vapors at that time. More recently, however, such studies have received considerable attention, which should continue in the future. Spectroscopic studies on high-temperature species would seem to be particularly timely in view of the growing knowledge of the stabilities and the extreme lack of accurate entropy or free-energy-function data for these species. The interpretation of spectra obtained from high-temperature species in their thermally excited states can be greatly simplified with supplementary studies in low-temperature matrices where no ambiguity about the molecular ground states can arise. Attempts to determine f-numbers for transitions should be made, particularly for species of stellar importance.

More complex and higher-molecular-weight species will naturally have a high number of fundamental vibrations, many of which can be expected to fall in close proximity to each other. Hence the demonstrated sharpness of matrix spectra appears to be the only means of studying the vibrational properties of such species.

In the area of structural chemistry, reduced entropy effects at cryogenic temperatures should reveal the existence of new, isomeric forms for inorganic species.

It is now apparent that a high degree of symmetry is likely for many inorganic species, thus reducing the proportion of ir-active to total possible vibrations. This, together with the usual absence of rotational structure, suggests that future efforts should also be directed to the use of Raman spectroscopy with matrices.

Before optimum utilization of the matrix technique can be made, much more study of matrix effects and the structure of matrices, perhaps by means of x-ray diffraction, seems necessary. As indicated by Pimentel (*13*) a better structural understanding of the cage effect should lead to the design of experiments in which efficient *in situ* preparation of new species by photolysis can occur. *In situ* photolysis offers more control at every stage of the reaction than the presently more efficient condensation–photolysis approach.

Radiolysis by x-rays or gamma rays is clearly fruitful for the study of ions and even ion–molecule reactions in matrices. The use of short-wavelength radiation for either photolysis or spectroscopy is somewhat limited by light scattering, though the search for a wider range of glassy matrix materials might reduce this difficulty.

The matrix technique is expected to prompt more theoretical work on molecular potential force fields, which should also complement the interpretation of matrix spectra.

Finally, it is interesting to note the prevalence of matrix spectroscopy reports over the years in journals oriented to chemical physics or physical chemistry, but more emphasis on inorganic chemistry is inevitable now that the technique is well developed and its suitability to inorganic studies demonstrated.

Acknowledgments

Research in matrix-isolation spectroscopy is supported at the Rice University by the U.S. Atomic Energy Commission and by the Robert A. Welch Foundation.

References

1. Abramowitz, S., Acquista, N., and Levin, W. I., Infrared matrix spectra of lithium fluoride, *J. Res. Natl. Bur. Stds.* **72A**, 487 (1968).
2. Acquista, N., Abramowitz, S., and Lide, D. R., Structure of the alkali hydroxides II. The ir spectrum of matrix isolated CsOH and CsOD, *J. Chem. Phys.* **49**, 780 (1968).
3. Acquista, N., Schoen, L. J., and Lide, D. R., Infrared spectrum of the matrix-isolated OH radical, *J. Chem. Phys.* **48**, 1534 (1968).
4. Adrian, F. J., Cochran, E. L., and Bowers, V. A., ESR spectrum of HO_2 in argon at $4.2°K$, *J. Chem. Phys.* **47**, 5441 (1967).
5. Allavena, M., Rysnik, R., White, D., Calder, V., and Mann, D. E., Infrared spectra and geometry of SO_2 isotopes in solid krypton matrices, *J. Chem. Phys.* **50**, 3399 (1969).
6. Anderson, J. S., and Ogden, J. S., unpublished data (1969); Anderson, J. S., Ogden, J. S., and Ricks, M. J., Infrared spectra and structures of matrix isolated disilicon, digermanium, and ditin oxides (Si_2O_2, Ge_2O_2, and (Sn_2O_2), *Chem. Commun.*, 1585 (1968).
7. Andrews, W. L. S., and Pimentel, G. C., Infrared spectrum, structure, and bonding of lithium nitroxide, LiON, *J. Chem. Phys.* **44**, 2361 (1966).
8. Andrews, L., and Pimentel, G. C., Infrared detection of methyl radical in solid argon, *J. Chem. Phys.* **44**, 2527 (1966).
9. Andrews, L., and Carver, T. G., Matrix infrared spectrum and bonding in the tribromomethyl radical, *J. Chem. Phys.* **50**, 4223 (1969).

10. Andrews, L., Infrared spectrum, structure, vibrational potential function and bonding in the lithium superoxide molecule LiO_2, *J. Chem. Phys.* **50**, 4288 (1969).
11. Arkell, A., and Schwager, I., Matrix infrared study of the ClOO radical, *J. Am. Chem. Soc.* **89**, 5999 (1967).
12. Barrett, C. S., and Meyer, L., Argon–nitrogen phase diagram, *J. Chem. Phys.* **42**, 107 (1965); Phase diagram of argon–carbon monoxide, *ibid.* **43**, 3502 (1965).
13. Bass, A. M., and Broida, H. P., "Formation and Trapping of Free Radicals." Academic Press, New York, 1960.
14. Becker, E. D., and Pimentel, G. C., Spectroscopic studies of reactive molecules by the matrix isolation method, *J. Chem. Phys.* **25**, 224 (1956).
15. Bennett, J. E., and Thomas, A., The chemical preparation and electron spin resonance spectra of specific trapped hydrocarbon radicals, *Proc. Roy. Soc. Ser. A* **280**, 123 (1964).
16. Bennett, J. E., Mile, B., and Thomas, A., Electron spin resonance spectrum of the CO_2^- radical ion at 77°K, *Trans. Faraday Soc.* **61**, 2357 (1965).
17. Bennett, J. E., Mile, B., and Thomas, A., Preparation and properties of the H_2S negative ion, *Chem. Commun.*, 182 (1966).
18. Bennett, J. E., Mile, B., and Thomas, A., Electron spin resonance spectrum of CS_2^- radical ion at 77°K, *Trans. Faraday Soc.* **63**, 262 (1967).
19. Bennett, J. E., Mile, B., and Thomas, A., Electron spin resonance spectrum of the O_2^- radical ion trapped in non-ionic matrices at 77°K, *Trans. Faraday Soc.* **64**, 3200 (1968); see also earlier papers by this group.
20. Boato, G., The solidified inert gases, *Cryogenics* **4**, 65 (1964).
21. Bowers, M. T., and Flygare, W. H., Vibration–rotation spectra of monomeric HCl, DCl, HBr, DBr and HI in the rare gas lattices and N_2 doping experiments in the rare gas lattices, *J. Chem. Phys.* **44**, 1389 (1966).
22. Bowers, M. T., Concentration-dependent transitions of HCl in rare-gas solids, *J. Chem. Phys.* **47**, 3100 (1967).
23. Brewer, L., and Brabson, G. D., Ultraviolet fluorescent and absorption spectra of S_2 isolated in inert-gas matrices, *J. Chem. Phys.* **44**, 3274 (1966).
24. Brewer, L., King, B. A., Wang, J. L., Meyer, B., and Moore, G. F., Absorption spectrum of silver atoms in solid argon, krypton and xenon, *J. Chem. Phys.* **49**, 5209 (1968).
25. Brewer, L., and Co-workers, unpublished data (1969).
26. Broida, H. P., Stabilization of free radicals at low temperatures, *Ann. N.Y. Acad. Sci.* **63**, 530 (1957).
27. Brown, D. M., and Dainton, F. S., Matrix isolation of unstable halogen radical ions, *Nature* **209**, 195 (1966).
28. Buchler, A., Studies in high temperature physical chemistry, PhD Thesis, Harvard Univ. (1960).
29. Buckingham, A. D., Solvent effects in infrared spectroscopy, *Proc. Roy. Soc.* **A248**, 169 (1958).
30. Calder, V., Mann, D. E., Seshadri, K. S., Allevena, M., and White, D., *J. Chem. Phys.* **51**, 2093 (1969).
31. Carlson, K. D., *in* "The Characterization of High Temperature Vapors" (J. L. Margrave, ed.). Wiley, New York, 1967.
32. Carver, T. G., and Andrews, L., Matrix infrared spectrum and bonding in the dibromomethyl radical, *J. Chem. Phys.* **50**, 4223 (1969).
33. Charles, S. W., and Lee, K. O., Infrared spectra of carbon monoxide in krypton and xenon matrices at 20°K, *Trans. Faraday Soc.* **61**, 614 (1965).

34. Charles, S. W., and Lee, K. O., Interpretation of the matrix-induced shifts of vibration and band frequencies, *Trans. Faraday Soc.* **61**, 2081 (1965).

35. Cheetham, C. J., and Barrow, R. F., Rotational analysis of electronic bands of gaseous TaO, *Trans. Faraday Soc.* **63**, 1835 (1967).

36. Claassen, H. H., Goodman, G. L., Malm, J. C., and Schreiner, F., Infrared and Raman spectra of krypton difluoride, *J. Chem. Phys.* **42**, 1229 (1965).

37. Conn, G. K. T., Lee, E., and Sutherland, G. B. B. M., Investigation of the vibration spectra of certain condensed gases at the temperature of liquid nitrogen, *Proc. Roy. Soc.* **A176**, 484 (1940).

38. Cook, G. A., ed., "Argon, Helium and the Rare Gases." Wiley (Interscience), New York, 1961.

39. Coulson, C. A., Force fields in KrF_2 and XeF_2, *J. Chem. Phys.* **44**, 468 (1966).

40. Dalgarno, A., and Kingston, A. E., Van Der Waals forces for hydrogen and the inert gases, *Proc. Phys. Soc. (London)* **73**, 607 (1961).

41. Fateley, W. G., Bent, H. A., and Crawford, Jr., B., Infrared spectra of the frozen oxides of nitrogen, *J. Chem. Phys.* **31**, 204 (1959).

42. Hallam, H. E., Recent advances in the infrared spectroscopy of matrix isolated species, private communication (1969).

43. Hastie, J. W., Hauge, R. H., and Margrave, J. L., Infrared vibrational properties of GeF_2, *J. Phys. Chem.* **72**, 4492 (1968).

44. Hastie, J. W., Hauge, R. H., and Margrave, J. L., Infrared spectra and geometry of the difluorides of Co, Ni, Cu and Zn isolated in neon and argon matrices, *High Temp. Sci.* **1**, 76 (1969).

45. Hastie, J. W., Hauge, R. H., and Margrave, J. L., Infrared spectra and geometry of SO_2 and SeO_2 in rare gas matrices, *J. Inorg. Nucl. Chem.* **31**, 281 (1969).

46. Hastie, J. W., and Margrave, J. L., *J. Phys. Chem.* **73**, 1105 (1969).

47. Hastie, J. W., Hauge, R. H., and Margrave, J. L., Infrared spectra of silicon difluoride in neon and argon matrices, *J. Am. Chem. Soc.* **91**, 2536 (1969).

48. Hastie, J. W., Hauge, R. H., and Margrave, J. L., Reactions and structures of isolated SiO species at low temperatures from infrared spectra, *Inorg. Chim. Acta* **3**, 601 (1969).

49. For example see Hastie, J. W., Hauge, R. H., and Margrave, J. L., Infrared spectra and geometry of TiF_2 and TiF_3 in rare gas matrices, *J. Chem. Phys.* **51**, 2648 (1969).

50. Hastie, J. W., Hauge, R. H., Ross, M., and Margrave, J. L., unpublished data (1969).

51. Hermann, T. S., Infrared spectroscopy at sub-ambient temperatures, *J. Appl. Spectros.* **23**, 435 (1969).

52. Herzberg, G., "Infrared and Raman Spectra of Polyatomic Molecules." Van Nostrand, Princeton, New Jersey, 1945.

53. Hirschfelder, J. O., Curtiss, C. T., and Bird, R. B., "Molecular Theory of Gases and and Liquids." Wiley, New York, 1954.

54. Jacox, M. E., and Milligan, D. E., Low-temperature infrared studies of the chemistry of free radicals, *Appl. Opt.* **3**, 873 (1964).

55. Jacox, M. E., and Milligan, D. E., Matrix isolation study of the reaction of Cl atoms with CO. The infrared spectrum of the free radical ClCO, *J. Chem. Phys.* **43**, 866 (1965).

56. Jacox, M. E., and Milligan, D. E., Production and reaction of atomic fluorine in solids. Vibrational and electronic spectra of the free radical HNF, *J. Chem. Phys.* **46**, 184 (1967).

57. Jacox, M. E., and Milligan, D. E., Matrix-isolation study of the vacuum-ultraviolet photolysis of trichlorosilane. The infrared spectrum of the free radical $SiCl_3$, *J. Chem. Phys.* **49**, 3130 (1968).

58. Kasai, P. H., Weltner, W. Jr., and Whipple, E. B., Orientation of NO_2 and other molecules in neon matrices at $4°K$, *J. Chem. Phys.* **42**, 1120 (1965).

59. Kasai, P. H., and Whipple, E. B., Electron spin resonance spectrum of the NF_2 radical isolated in a neon matrix at $4°K$, *Mol. Phys.* **9**, 497 (1966).

60. Kasai, P. H., Whipple, E. B., and Weltner, W. Jr., ESR of $Cu(NO_3)_2$ and CuF_2 molecules oriented in neon and argon matrices at $4°K$, *J. Chem. Phys.* **44**, 2581 (1966).

61. Kasai, P. H., ESR of VO in argon matrix at $4°K$: establishment of its electronic ground state, *J. Chem. Phys.* **49**, 4979 (1968).

62. Katz, B., Ron, A., and Schnepp, O., Far-infrared spectra of HCl and HBr in solid solutions, *J. Chem. Phys.* **46**, 1926 (1967).

63. Katz, B., Ron, A., and Schnepp, O., Far-infrared spectrum of hydrogen chloride dimers, *J. Chem. Phys.* **47**, 5303 (1967).

64. For example see Keyser, L. F., and Robinson, G. W., Infrared spectra of HCl and DCl in solid rare gases—I. Monomers, *J. Chem. Phys.* **44**, 3225 (1966).

65. Linevsky, M. J., Infrared spectrum of lithium fluoride monomer by matrix isolation, *J. Chem. Phys.* **34**, 587 (1961).

66. Linevsky, M. J., White, D., and Mann, D. E., Infrared spectrum and structure of gaseous Al_2O, *J. Chem. Phys.* **41**, 542 (1964).

67. Linnett, J. W., and Hoare, M. J., Molecular force fields—IX. A relation between bond interactions during vibrations and on association, *Trans. Faraday Soc.* **45**, 844 (1949).

68. Longuet-Higgens, H. C., and Pople, J. A., Electronic spectral shifts of nonpolar molecules in nonpolar solvents, *J. Chem. Phys.* **27**, 192 (1957).

69. McCarty, M. Jr., and Robinson, G. W., Environmental perturbations on foreign atoms and molecules in solid argon, krypton and xenon, *Mol. Phys.* **2**, 415 (1959).

70. McLeod, D. Jr., and Weltner, W. Jr., Spectroscopy and Franck–Condon factors of scandium fluoride in neon matrices at $4°K$, *J. Phys. Chem.* **70**, 3293 (1966).

71. Machida, K., and Overend, J., Dissociation of polyatomic molecules and the relation of ensuing geometrical change to the force constants, *J. Chem. Phys.* **50**, 4437 (1969).

72. Mamantov, G., Fletcher, W. H., Cristy, S. S., Edwards, C. T., and Morton, R. E., Use of rotary cryostat for infrared studies, *Rev. Sci. Instr.* **37**, 836 (1966).

73. Mile, B., Free radical studies at low temperatures, *Angew. Chem.* **7**, 507 (1968).

74. Milligan, D. E., and Jacox, M. E., Infrared spectroscopic study of the photolysis of methyl azide and methyl-d_3 azide in solid argon and carbon dioxide, *J. Chem. Phys.* **35**, 1491 (1961).

75. Milligan, D. E., Jacox, M. E., Charles, S. W., and Pimentel, G. C., Infrared spectroscopic study of the photolysis of HN_3 in Solid CO_2, *J. Chem. Phys.* **37**, 2302 (1962).

76. Milligan, D. E., and Jacox, M. E., Infrared spectrum of HCO, *J. Chem. Phys.* **41**, 3032 (1964).

77. Milligan, D. E., Jacox, M. E., Bass, A. M., Comeford, J. J., and Mann, D. E., Matrix-isolation study of the reaction of F atoms with CO. Infrared and ultraviolet spectra of the free radical FCO, *J. Chem. Phys.* **42**, 3187 (1965).

78. Milligan, D. E., and Jacox, M. E., Matrix-isolation study of the infrared and ultraviolet spectra of the free radical CNN, *J. Chem. Phys.* **44**, 2850 (1966).

79. Milligan, D. E., and Jacox, M. E., Infrared spectrum of the free radical CF_3 isolated in inert matrices, *J. Chem. Phys.* **44**, 4058 (1966).

80. Milligan, D. E., and Jacox, M. E., Infrared and ultraviolet spectroscopic study of the products of the vacuum-ultraviolet photolysis of methane in Ar and N_2 matrices. The infrared spectrum of the free radical CH_3, *J. Chem. Phys.* **47**, 5146 (1967).

81. Milligan, D. E., Jacox, M. E., and Abouaf-Marguin, L., Vacuum-ultraviolet photolysis of acetylene in inert matrices. Spectroscopic study of the species C_2, *J. Chem. Phys.* **46**, 4562 (1967).

82. Milligan, D. E., and Jacox, M. E., Matrix-isolation study of the infrared and ultraviolet spectra of the free radical NCO, *J. Chem. Phys.* **47**, 5157 (1967).

83. Milligan, D. E., and Jacox, M. E., Matrix-isolation study of the vacuum-ultraviolet photolysis of dichlorosilane. The infrared spectrum of the free radical $SiCl_2$, *J. Chem. Phys.* **49**, 1938 (1968).

84. Milligan, D. E., and Jacox, M. E., Studies of the photoproduction of electrons in inert solid matrices. The electronic spectrum of the species $C_2{}^-$, *J. Chem. Phys.* **51**, 1952 (1969).

85. Minkoff, G. J., "Frozen Free Radicals." Wiley (Interscience), New York, 1960.

86. Moll, N. G., Clutter, D. R., and Thompson, W. E., Carbon trioxide: its production, infrared spectrum, and structure studied in a matrix of solid CO_2, *J. Chem. Phys.* **45**, 4469 (1966).

87. Morehouse, R. L., Christiansen, J. J., and Gordy, W., ESR of free radicals in inert matrices at low temperature P, PH_2 and As, *J. Chem. Phys.* **45**, 1747 (1966).

87a. Morehouse, R. L., Christiansen, J. J., and Gordy, W., ESR of free radicals trapped in inert matrices, *J. Chem. Phys.* **45**, 1751 (1966).

88. Muenow, D., Hastie, J. W., Hauge, R. H., Bautista, R., and Margrave, J. L., Vaporizationthermodynamics and structures of species in the tellurium–oxygen system, *Trans. Faraday Soc.* **65**, 3210 (1969).

89. Nelson, L. Y., and Pimentel, G. C., Infrared detection of xenon dichloride, *Inorg. Chem.* **6**, 1758 (1967).

90. Nelson, L. Y., and Pimentel, G. C., Infrared spectra of chlorine–bromine polyhalogens by matrix isolation, *Inorg. Chem.* **7**, 1695 (1968).

91. Othmer, S., and Silsbee, R. H., Abstract from March 1968 Meeting of Phys. Soc.

92. Pandey, G. K., and Chandra, S., The matrix spectra of the HF molecule, *J. Chem. Phys.* **45**, 4369 (1966).

93. Petropoulos, B., and Herman, L., Effect of rare gas matrices on the spectrum of the S_2 molecule, *Compt. Rend. Acad. Sci. Paris (Ser. B)* **264**, 1196 (1967).

94. Phillips, L. F., Smith, J. J., and Meyer, B., The ultraviolet spectra of matrix isolated disulfur monoxide and sulfur dioxide, *J. Mol. Spectry.* **29**, 230 (1969).

95. Pimentel, G. C., Matrix technique and its application in the field of chemical physics, *Pure Appl. Chem.* **4**, 61 (1962) (IUPAC, 5th Congr. Mol. Spectry).

96. For a review of matrix perturbations on infrared spectra see Pimentel, G. C., and Charles, S. W., Infrared spectral perturbations in matrix experiments, *Pure Appl. Chem.* **7**, 111 (1963).

97. Pollack, G. L., Solid state of rare gases, *Rev. Mod. Phys.* **36**, 748 (1964).

98. Pysh, E. S., Rice, S. A., and Jortner, J., Molecular rydberg transitions in rare gas matrices. Evidence for interaction between impurity states and crystal states, *J. Chem. Phys.* **43**, 2997 (1965).

99. Ramsay, D. A., "Formation and Trapping of Free Radicals" (A. M. Bass and H. P. Broida, eds.). Academic Press, New York (1960).

100. Redington, R. L., and Milligan, D. E., Molecular rotation and ortho–para nuclear spin conversion of water suspended in solid Ar, Kr and Xe, *J. Chem. Phys.* **39**, 1276 (1962).

101. Rich, R., "Periodic Correlations." Benjamin, New York, 1965.

102. Robinson, G. W., Electronic spectra, *in* "Methods of Experimental Physics" (D. Williams, ed.), Vol. 3, p. 155. Academic Press, New York, 1962.

103. Robinson, G. W., Spectra of matrix isolated water in the "pure rotation" region, *J. Chem. Phys.* **39**, 3430 (1963).

104. Robinson, G. W., and Von Holle, W. G., Far infrared spectra of matrix-isolated HF, *J. Chem. Phys.* **44**, 410 (1966).

105. Rochkind, M. M., Infrared pseudo matrix isolation spectroscopy: analysis of gas mixtures, *Science* **160**, 196 (1968).
106. Roncin, J., Damany, N., and Romand, J., Far ultraviolet absorption spectra of atoms and molecules trapped in rare gas matrices at low temperature, *J. Mol. Spectry.* **22**, 154 (1967).
107. Schlick, S., and Schnepp, O., Infrared spectra of the lithium halide monomers and dimers in inert matrices at low temperature, *J. Chem. Phys.* **41**, 463 (1964).
108. Schoen, L. J., and Mann, D. E., Flash-photolysis spectroscopy for matrices, *J. Chem. Phys.* **41**, 1514 (1964).
109. Schwager, I., and Arkell, A., Matrix infrared spectra of HOBr and HOCl, *J. Am. Chem. Soc.* **89**, 6006 (1967).
110. Seshadri, K. S., Nimon, L. A., and White, D., Infrared spectra of matrix isolated alkali metal metaborates, *J. Mol. Spectry.* **30**, 128 (1969).
111. Smith, J. J., and Meyer, B., The absorption and fluorescence spectrum of SnS and SnO: Matrix-induced intersystem crossing, *J. Mol. Spectry.* **27**, 304 (1968).
112. Snelson, A., Infrared spectrum of AlF_3, Al_2F_6 and AlF by matrix isolation, *J. Phys. Chem.* **71**, 3202 (1967).
113. Snelson, A., Heats of vaporization of the lithium fluoride vapor species by the matrix isolation technique, *J. Phys. Chem.* **73**, 1919 (1969).
114. Somayajulu, G. R., Dissociation energies of diatomic molecules, *J. Chem. Phys.* **33**, 1541 (1960).
115. Spratley, R. D., and Pimentel, G. C., The $[p—\pi^*]\sigma$ and $[s—\pi^*]\sigma$ bonds. FNO and O_2F_2, *J. Am. Chem. Soc.* **88**, 2394 (1966).
116. Thompson, K. R., and Carlson, K. D., Bending frequencies and new dimer modes in the far-infrared spectra of transition metal dihalides, *J. Chem. Phys.* **49**, 4379 (1968).
117. Tupikov, V. I., Pshezhetskii, V. S., and Pshezhetskii, S. Ya., Study of transformations and reactions of radicals with molecules in some solid molecular matrices at low temperatures by EPR method, *Elementarnye Protsessy Khim. Vysokikh, Energ Akad. Nauk. SSR, Inst. Khim. Fiz., Tr. Simpoziuma* 215 (1963).
118. Verstegen, J. M. P. J., Goldring, H., Kimel, S., and Katz, B., Infrared spectra of HCl in pure and impure noble-gas matrices, *J. Chem. Phys.* **44**, 3216 (1966).
119. Weltner, W. Jr., Walsh, P. N., and Angell, C. L., Spectroscopy of carbon vapor condensed in rare gas matrices at 4° and 20°K. I, *J. Chem. Phys.* **40**, 1299 (1964).
120. Weltner, W. Jr., and McLeod, D. Jr., Spectroscopy of carbon vapor condensed in rare gas matrices at 4° and 20° K. II, *J. Chem. Phys.* **40**, 1305 (1964).
121. Weltner, W. Jr., and McLeod, D. Jr., Spectroscopy of silicon carbide and silicon vapors trapped in neon and argon matrices at 4° and 20°K, *J. Chem. Phys.* **41**, 235 (1964).
122. Weltner, W. Jr., The matrix isolation of vaporizing molecules, "Condensation and Evaporation of Solids" (E. Rutner, P. Goldfinger, and J. P. Hirth, eds.), p. 243. Gordon and Breach, New York, (1964).
123. Weltner, W. Jr., and McLeod, D. Jr., Ground state of zirconium monoxide from neon matrix investigations at 4°K, *Nature* **206**, 87 (1965).
124. Weltner, W. Jr., and McLeod, D. Jr., Spectroscopy of tungsten oxide molecules in neon and argon matrices at 4° and 20°K, *J. Mol. Spectry.* **17**, 276 (1965).
125. Weltner, W. Jr., and McLeod, D. Jr., Spectroscopy of TaO and TaO_2 in neon and argon matrices at 4° and 20°K, *J. Chem. Phys.* **42**, 882 (1965).
126. Weltner, W. Jr., and McLeod, D. Jr., Spectroscopy of titanium, zirconium, and hafnium oxides in neon and argon matrices at 4° and 20°K, *J. Phys. Chem.* **69**, 3488 (1965).
127. Weltner, W. Jr., Stellar and other high-temperature molecules, *Science* **155**, 155 (1967).

128. Weltner, W. Jr., McLeod, D. Jr., and Kasai, P. H., ESR and optical spectroscopy of ScO, YO, and LaO in neon and argon matrices; establishment of their ground electronic states, *J. Chem. Phys.* **46**, 3172 (1967).
129. Whittle, E., Dows, D. A., and Pimentel, G. C., Matrix isolation method for the experimental study of unstable species, *J. Chem. Phys.* **22**, 1943 (1954).
130. Whyte, T. E. Jr., Infrared spectra of matrix isolated hydrogen chloride in sulfur hexafluoride and hydrogen iodide, PhD Thesis, Howard Univ. (1965).
131. Wood, D. E., and Pietrzak, T. M., unpublished data (1969).
132. Zhitnikov, R. A., and Kolesnikov, N. V., Theoretical consideration of matrix shift of hyperfine structure splitting for Cu, Ag and Au atoms stabilized in polar matrix (H_2O), *Soviet Phys.—Solid State* **9**, 121 (1967).

SPECTROSCOPY OF DONOR–ACCEPTOR SYSTEMS

C. N. R. Rao and A. S. N. Murthy

DEPARTMENT OF CHEMISTRY
INDIAN INSTITUTE OF TECHNOLOGY
KANPUR, INDIA

I. Introduction

The formation of molecular complexes between electron donors and acceptors has been recognized as an important phenomenon for quite some time. Many of the molecular complexes are colored and give rise to new absorption bands in the electronic absorption spectra. These molecular complexes have been investigated by employing a variety of physical methods, such as dielectric measurements, optical spectroscopy, magnetic resonance spectroscopy, and x-ray diffraction. Studies of molecular complexes formed between electron donors and acceptors are of great value in understanding various types of reaction mechanisms, including molecular phenomena in biological systems. Some of the important varieties of molecular complexes involve inorganic compounds, often very simple ones such as iodine. Extensive studies have been carried out on these molecular complexes employing optical spectroscopy. In this chapter we shall discuss the spectroscopic studies of the interaction of electron donors and acceptors in the light of recent theoretical developments, particularly due to Mulliken, restricting our discussion to systems involving inorganic molecules. We shall also briefly deal with some aspects of the studies of radical ions produced through donor–acceptor complexes. Since the spectra of the radical ions are markedly affected by the nature of ionic species in solution, we shall examine some aspects of ion-pair equilibria as well.

Hydrogen bonding is an important aspect in the study of the interaction between electron donors and acceptors, and it is needless to emphasize the importance of this phenomenon in chemistry and biology. The literature abounds in publications on this subject, particularly with respect to spectroscopic studies. We shall presently discuss some of the highlights in the spectroscopic studies of hydrogen bonds in inorganic systems.

Recently several workers have proposed empirical scales for the acidity and basicity of organic and inorganic molecules. Acids and bases have been classified as soft or hard on the basis of equilibrium and kinetic data. We shall comment on these classifications and indicate how they are related to some of the results obtained from spectroscopic measurements on donor–acceptor complexes.

II. Charge-Transfer Complexes

Although a new absorption band in the ultraviolet spectra of solutions of iodine and benzene was recognized quite early as characteristic of the molecular complex between benzene and iodine (*16*), it was only after Mulliken

(*171–179*) propounded the charge-transfer theory that various features of the spectra and other properties of these molecular complexes could be explained. The essential feature in the theory is that a molecular complex is formed between an electron donor and an acceptor; the ground- (ψ_N) and excited-state (ψ_E) configurations of the complex can be written in terms of the no bond (ψ_0) and dative (ψ_1) contributions. The difference between the energies of the ground and excited states

$$\psi_N = a\psi_0(D, A) + b\psi_1(D^+, A^-) \tag{1}$$

and

$$\psi_E = a^*\psi_1(D^+, A^-) - b^*\psi_0(D, A) \tag{2}$$

is equal to the energy of quantum for the characteristic charge-transfer band.

Before discussing the theory of Mulliken, we shall briefly summarize some of the main features of these molecular complexes.

(1) There is a characteristic intense absorption band that is not found in the spectrum of either the donor or the acceptor.

(2) Equilibrium constants and enthalpies of formation of these complexes show a wide range of values depending on the magnitude of interaction between the donor and the acceptor.

(3) Absorption bands due to the acceptor unit are also markedly affected on complex formation. Thus, the iodine band in the visible region shows marked blue shifts in complexes. In some instances, the donor absorption bands are also shifted in complexes.

(4) Depending on the magnitude of donor–acceptor interaction, the dielectric properties change appreciably, the molecular complex generally having a larger dipole moment than either of the components.

(5) When the interaction between the donor and the acceptor is large, almost complete electron transfer may take place in the ground state, giving rise to ion pairs or radical ions. Such radical ions show paramagnetic behavior and large electrical conductivity.

(6) The vibrational spectra of the donors as well as acceptors are affected in the complex, and some new vibrational frequencies may be found in some instances.

In the last few years there have been a large number of papers on the spectroscopy of donor–acceptor systems, and the subject has been reviewed by various authors (*5, 40, 138, 211*) including Mulliken and Person (*179, 179a*). Some of the more recent developments in this area are the classification of donors and acceptors by Mulliken (*180*); determination of vibrational spectra, particularly in the far-infrared region (*5, 37*); estimation of the percent charge-transfer by dipole-moment measurements (*25, 40, 137*); study of radical ions produced by charge-transfer intermediates (*138, 212*); and thermal as well as photochemical reactions involving charge-transfer processes (*138*). The literature on this subject continues to increase in terms of newer and novel systems; for example, tropylium ion (*26*) and iodine atom (*239*) have been recently employed as acceptors. The methods of evaluation of equilibrium constants employing spectroscopic methods are being continually improved.

A. Mulliken's Charge-Transfer Theory

The schematic potential-energy diagram for a charge-transfer complex is shown in Fig. 1. It can be seen from the figure that the charge-transfer transition energy is given by

$$hv_{CT} = I_D - (E_A + E_C + W_0) + R_E - R_N \tag{3}$$

$$\approx I_D - E_A - \Delta \tag{4}$$

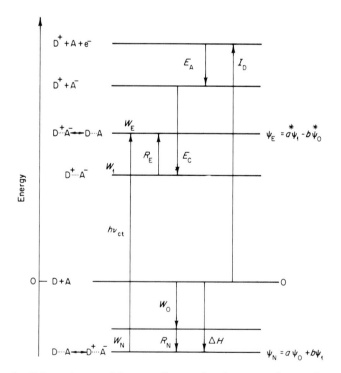

FIG. 1.   Schematic potential-energy diagram for charge-transfer complexes.

and $W_N = \Delta H = W_0 + R_N$, where R_N is the resonance energy. For weak complexes the resonance-energy contribution to $h\nu_{CT}$ can be computed by perturbation theory:

$$R_E - R_N \approx \frac{\beta_0{}^2 + \beta_1{}^2}{W_1 - W_0} \tag{5}$$

where β_0 and β_1 are the matrix elements and W_0 and W_1 are the energies of the pure no-bond and dative structures respectively.

A quantitative relationship based on the frequency of charge-transfer absorption maximum ($h\nu_{CT}$) and the donor ionization potentials (I_D) has been proposed by McConnell et al. (167). The energy associated with the charge-transfer transition is related to the I_D of the donors (for a particular acceptor) by the expression

$$h\nu_{CT} = I_D - E_A - \Delta + \frac{2\beta^2}{(I_D - E_A - \Delta)} \tag{6}$$

The last two terms account for the various interactions that alter the energies of the ground and excited states and hence the absorption frequency. However, when β is small, the approximate expression [Eq.(4)] is made use of and found to hold good in many cases. In the case of hydrocarbon-iodine systems, it has been found by Briegleb (40) that

$$h\nu_{CT} = I_D - C_1 + \frac{C_2}{(I_D - C_1)} \tag{7}$$

where C_1 and C_2 are terms consisting of $(E_A + E_C + W_0)$ and $(\beta_0{}^2 + \beta_1{}^2)$ respectively. W_0 is the sum of several terms, including electrostatic energy (dipole–dipole interaction, etc.). When $h\nu_{CT}$ is plotted against I_D, Biiegleb finds that the data can be fitted by curves of the form given by Eq. (7). Although the relation expressed by Eq. (7) is nonlinear between $h\nu_{CT}$ and I_D, the plots are only slightly curved over the observed range of I_D (Fig. 2). Thus most of the data (on the weak I_2 complexes) can be fitted by a linear relation of the form

$$h\nu_{CT} = 0.87 I_D - 0.36 \tag{8}$$

within the limits of experimental error. Such linear relationships have slopes some what less than unity. The values of the constants in the linear relation have no direct theoretical significance.

Amines, however, do not fall in line with the $h\nu_{CT} - I_D$ relation [Eq. (7)] for weak I_2 complexes (Fig. 2). An equation has been derived for such systems using the variation method (180):

$$(h\nu_{CT})^2 = \left[\frac{W_1 - W_0}{1 - S_{01}^2}\right]^2 \left[1 + \frac{4\beta_0 \beta_1}{(W_1 - W_0)^2}\right] \tag{9}$$

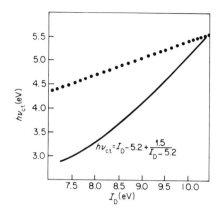

FIG. 2.   Plot of $h\nu_{CT}$ versus the ionization potential of the donor I_D (eV) for complexes of I_2. The dotted line is for amine complexes; the solid line is for other donor complexes [From Mulliken and Person (*179*)].

In Eq. (9) $W_1 - W_0 = I_D - C_1$, and S_{01} is the overlap integral $\int \psi_0 \psi_1 \, d\tau$. It is possible to fit the data for the amines by Eq. (9) with the following parameters: $C_1 = E_A + E_C + W_0 \approx 6.9$ eV, $S_{01} = 0.3$, and $\beta_0 = -2.5$ eV. The values for the weak I_2 complexes are $C_1 = 5.2$ eV, $S_{01} = 0.1$, and $\beta_0 = -0.6$ eV. The difference in these parameters is probably due to a greater overlap of the donor and acceptor molecules in *n*-donor-σ-acceptor complexes than in weak complexes. The increased overlap causes E_C, W_0, S_{01}, and β_0 to become relatively large.

It is evident that Briegleb's relation [Eq.(7)] is not obeyed for donors that are sufficiently different from each other. Usually these differences would result in slight shifts of the curve of Eq. (7), but if the changes in parameters are such that W_1 is nearly equal to W_0, the slope of the curve can be expected to decrease as the resonance interaction becomes more important with respect to the other terms determining $h\nu_{CT}$. In that case Eq. (9) must be used.

A plot of the data using Eqs. (7–9) has shown that Eq. (8) gives a good empirical representation of the other two equations over the practical range for I_D (Fig. 3). The deviation from the unit slope is a result of the nonzero resonance interaction.

The band around 520 mμ of the iodine molecule in the visible region is due to a transition from a σ_g bonding molecular orbital to the antibonding σ_u molecular orbital in the excited state. This band undergoes a shift to lower wavelengths (blue shift) on complexation with a donor. Mulliken (*178*) has explained the blue shift as being mainly due to exchange repulsion between the iodine molecule (in the excited state) and the donor molecules. The exchange repulsion increases the closer the contact between the donor and iodine molecules.

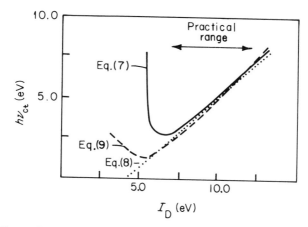

FIG. 3. Comparison of the three equations relating $h\nu_{CT}$ with the ionization potential of the donor $I_D(eV)$. Solid line, Eq. (7) with $C_1 = 5.2$, $C_2 = 1.5$; dashed line, Eq. (9) with $W_1 - W_0 = I_D - 5.2$ and $\beta_0 - \beta_1 = S_{01}(W_1 - W_0)$, where $\beta_0 = -0.6$ and $S_{01} = 0.1$; dotted line, the empirical linear relation of Eq. (8).

Charge-transfer energies in aromatic systems have been interpreted in terms of molecular orbital theory by Dewar and co-workers (70, 71, 146). These workers find that a plot of Hückel energies of the highest occupied π orbitals of the donor against the charge-transfer energy of complexes is fairly linear. This indicates that charge transfer involves the promotion of an electron from the highest occupied orbital of a donor to the lowest unoccupied acceptor orbital. Thus, in a series of π-acceptors, such as *sym*-trinitrobenzene, tetracyanoethylene, and trinitrofluorenone, the charge-transfer energy increases in the same direction as the energy of the lowest unoccupied orbital, *viz.*, trinitrobenzene > trinitrofluorenone > tetracyanoethylene.

B. Classification of Donors and Acceptors

Mulliken (179a, 180, 211) has classified various types of donors and acceptors that give rise to charge-transfer interactions of varying magnitudes (Table I). Donor–acceptor systems vary widely in their energies of formation in the ground state. At one extreme we have the strong Lewis acid–Lewis base addition compounds, and at the other, the weak "contact pairs." The various types of commonly encountered donor–acceptor interactions are shown in Table II along with the approximate energy ranges. The π-donor–σ-acceptor complexes and the n-donor–σ-acceptor complexes have been found to exist as $\infty : \infty$ as well as $1 : 1$ complexes; the $\infty : \infty$ complexes are generally found in solid state and the $1 : 1$ complexes in solution and vapor phases.

The transitory "contact charge-transfer" complexes arise out of the colli-

TABLE I

CLASSIFICATION OF DONORS AND ACCEPTORS

Number of electrons	Functional type	Donor types			Acceptor types	
		Structure	Examples		Structure	Example
Odd	Free radical	R	NO, C_2H_5, H		Q	X, H, OH
Even	(a) Increvalent	n	R_3N, R_2S, R_2O, R_2CO		v	$Br_3, AlR_3, SnCl_4$
	(b) Sacrificial	σ	Aliphatic hydrocarbons		σ	X_2, CCl_4 (X = halogen)
		π	Aromatic and unsaturated hydrocarbons; intramolecular π donor island groups		π	Aromatic and unsaturated hydrocarbons fortified by electronwithdrawing groups; intramolecular π acceptor island groups

TABLE II

DONOR–ACCEPTOR SYSTEMS AND THEIR ENERGY RANGES

Type	Examples	Energy range (kcal/mole)
Contact pairs	Cyclohexane $+ I_2$, benzene $+ CCl_4$	<2
π–σ	Benzene $+ I_2$, phenanthrene $+ I_2$	1–4
π–π	Naphthalene $+ sym$-trinitrobenzene	1–4
n–σ	Carbonyl compounds $+ I_2$, amines $+ I_2$	3–15
n–v	BF_3 + ether	Very strong addition compounds

sions between donor and acceptor species, provided optical absorption takes place during the contact interval. In such contact pairs $a \approx 1$ and $b^* \approx 1$. The new absorption band observed in the case of hydrocarbons in the presence of oxygen was ascribed to contact pairs (252); the enthalpies of formation of these oxygen complexes are very small. The intense absorption bands of contact pairs have been explained by Mulliken (178) on the basis of the relatively greater range of charge-transfer forces compared to the corresponding range of van der Waals forces. The "volume" corresponding to the overlapping of ϕ_D with ϕ_A, known as the electron-acceptation "volume" S_{DA}, can sometimes be much greater than the van der Waals "volume" S_{vdw}, which corresponds to the overlap of the outermost molecular orbitals of the donor and acceptor species in their ground states and in contact. Here S_{DA}^2 will determine the attractive resonance, and S_{vdw}^2 the repulsive exchange forces of the electronic closed shells of D and A. When the donor and acceptor are in contact, or even a little farther apart than corresponding to the van der Waals contact, ϕ_D and ϕ_A can overlap, giving rise to a nonzero S_{DA} and hence a net charge-transfer force of attraction between the components. Murrell (182) has pointed out that the intensity of contact-pair transitions may partly be borrowed from intradonor transitions that can occur because of the partial mixing of states.

When the donor and acceptor species are ionic, then the ground state of the complex, which depends largely on the Coulombic interaction between the components, is influenced markedly by the dielectric constant of the medium. In some n-donor–σ-acceptor complexes, the new bonds that are formed are loosened by environmental assistance and they transform to the "dissociative" inner complex, because of the rupture of the bond. In the presence of nonionizing solvents, the two ions formed by the rupture of the σ bond will be together, while in an ionizing solvent they are separated out. Mulliken (174) has compared the "dissociative" and "associative" donor–acceptor reaction in detail. The "associative" mode helps to form the "outer complex" while the "dissociative" mode helps to form the "inner complex." In the presence of strong environmental conditions the activation energy for the transformation of the outer to inner complex decreases. If the environmental influence is sufficiently strong, the inner complex may be the stable form.

The term charge transfer is often loosely employed in the literature. Charge transfer in most weak complexes takes place only in the excited state. It may, therefore be proper to talk of charge-transfer spectra or transitions and not of charge-transfer complexes, unless the magnitude of charge transfer is appreciable in the ground state; it may be preferable to refer to the molecular complexes as donor–acceptor complexes or preferably name them in terms of the classification discussed earlier.

Hydrogen-bonded complexes may also be considered as specific examples of charge-transfer systems (*180*). In a hydrogen bond, X—H $\cdots$ Y, the electron donor is the Y atom and X—H is the electron acceptor; the hydrogen atom has a formal positive charge. Here again, the donor can be π type or n type. The X—H stretching frequency is considerably decreased on hydrogen bonding, just as is the frequency associated with an acceptor such as I_2 or ICl; the stretching-band intensities are enhanced in both cases. Charge-transfer theory of the hydrogen bond (with regard to the X—H stretching vibration) has been discussed by Puranik and Kumar (*209*) and Friedrich and Person (*96*). The energies of formation of the charge-transfer and hydrogen-bonded complexes are also similar. While hydrogen-bonded complexes with n donors are mostly stabilized by electrostatic forces, charge-transfer forces contribute at least partly to stabilization (*180*). Although there may be some doubt regarding the appropriateness of calling hydrogen-bonded complexes charge-transfer complexes, we can certainly use the general term donor–acceptor complexes for these systems also. A major difference in the properties of the hydrogen-bonded complexes and the halogen complexes is that the latter are characterized by new electronic transitions. New electronic transitions have not been observed in most hydrogen-bonded complexes. Nagakura (*185*) has, however, provided evidence for the charge-transfer transition in the hydrogen bond of the maleate ion. The importance of charge-transfer forces in the hydrogen bond has been nicely pointed out in some recent reviews (*182, 182a, 183*).

Some complexes can be best described as intermediate between n–v and σ–v in type. Thus, the 1 : 1 interaction between MeX and $GaCl_3$ can be represented by the following wave function:

$$\psi = a\,\psi_0(\text{MeX, GaCl}_3) + b\,\psi_1(\text{MeX}^+ - \text{Ga}^-\text{Cl}_3)$$
$$+ c\psi_2(\text{Me}^+\text{X} - {}^-\text{GaCl}_3) + \cdots \qquad (10)$$

Whereas in ψ_1 MeX acts as an n donor, in ψ_2 it acts as a σ donor. In some n–σ complexes we also have symmetrized complexes such as I_3^- and ICl_2^-. In some ways these ions are similar to the bifluoride ion.

In addition to the stable even–even compounds, there are some odd–even reaction intermediates and ion pairs that can be classified as charge-transfer or donor–acceptor systems. Thus many ion pairs (for some of which charge-transfer spectra are known) may be regarded as contact pairs. Even ion pairs, such as $\text{Na}^+\,(\text{TCNE})^-$ having strong interaction, may in some ways be considered as complexes; this is particularly true if such ion pairs could be produced by charge-transfer intermediates as exemplified by the interaction of tetramethylphenylenediamine (TMPD) with tetracyanoethylene:

$$\text{TMPD} + \text{TCNE} \xrightleftharpoons{K_1} \text{TMPD} \cdot \text{TCNE}$$
$$\text{(weak complex)}$$

$$\xrightleftharpoons{K_2} \text{TMPD}^+ \text{TCNE}^- \xrightleftharpoons{K_3} \text{TMPD}^+ + \text{TCNE}^-$$
$$\text{(ion pair)} \qquad\qquad \text{(free ions)} \qquad (11)$$

The second and third equilibria above would depend strongly on the polarity of the solvent.

Charge-transfer transitions are found in many metal complexes, frequently at higher energies than the d–d transitions of the corresponding aquated metal ions. Charge-transfer transitions in inorganic complexes can be of metal $\rightarrow$ ligand type or *vice versa*. A detailed comparison of the inorganic and organic charge-transfer spectra in solution and their analytical applications have been discussed recently by Schenk (*222*).

C. Study of Charge-Transfer Complexes by Electronic Spectroscopy

Electronic spectroscopy is probably the most valuable tool employed in the study of charge-transfer complexes. The important information obtained from electronic spectroscopy is summarized as follows:

(a) *Charge-transfer* (CT) *band*. λ_{CT} (or $h\nu_{CT}$); ϵ_{max}; f_{CT} (oscillator strength); D (transition moment); $\Delta\nu_{1/2}$ (half-bandwidth of the band). A proper characterization of the CT band can be made only when all these features are evaluated. In some instances the CT bands are hidden under donor bands and one has to resort to difference spectra to obtain the position and intensity of the CT band. The strength of interaction between the donor and acceptor can only be assessed if all these data on the CT band are available. In Fig. 4 we have given typical absorption curves obtained in the study of di-t-butylthiourea + iodine complex.

(b) *Acceptor band*. The effect of complexation on the acceptor band provides information on the magnitude of interaction (Fig. 5). Thus, the blue shift, f, ϵ_{max}, and $\Delta\nu_{1/2}$ of the iodine band are very useful in examining charge-transfer complexes with iodine.

(c) CT *equilibria*. By varying the donor or acceptor concentrations, one can evaluate equilibrium constant (K) of interaction. The enthalpy ($\Delta H°$) estimated by determining K at various temperatures provides the strength of interaction. Several methods of obtaining K have been reported in the literature. Special mention must be made of the general equation of Rose and Drago (*219*), which does not make any assumptions regarding the donor and acceptor concentrations. Person (*198*) has recently pointed out the importance of using appropriate donor concentrations while evaluating equilibrium constants by the Benesi–Hildebrand procedure or some other procedure.

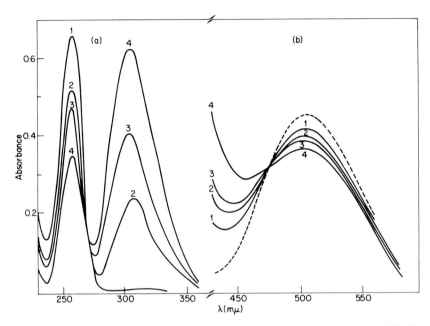

FIG. 4. Absorption spectra of N,N′-di-*t*–butylthiourea-I_2 complex in CH_2Cl_2. (a) Curves 2–4, effect of varying iodine concentration (1 to $2.5 \times 10^{-5}M$) keeping donor concentration constant ($4.9 \times 10^{-5}M$); curve 1, absorption spectrum of pure donor ($4.9 \times 10^{-5}M$). (b) Effect of varying donor concentration (1.2×10^{-5} to $1.5 \times 10^{-4}M$), keeping the iodine concentration constant ($5.1 \times 10^{-4}M$); spectrum of the pure acceptor is shown by the broken curve.

1. CONTACT PAIRS

Halogens such as iodine interact with compounds such as cyclohexane or *n*-butylbromide to give rise to the so-called contact pairs with well-defined absorption maxima (*27*). The λ_{max} in I_2 + cyclohexane or I_2 + methyl cyclohexane or I_2 + *n*-heptyl bromide is around 245 *mμ*. It was earlier believed that the interactions are weak compared to ordinary CT complexes, but stronger than collision pairs. Recent studies by Bhat *et al.* (*27*) have indicated that the energies in such interactions are of the order 1.4–3.0 kcal/mole, a value not too different from the values found for the interaction of iodine with benzene and other π donors (*40*). On the basis of these results, these workers do not find any basis to distinguish between weak CT complexes and contact pairs. However, Julien and Person (*132*) prefer to attribute the additional absorption near 200–250 mμ in the interaction of iodine with *n*-heptane, cyclohexane, and isooctane to contact charge-transfer.

Molecular oxygen has been found to give rise to extra bands on interaction with many donors (*252*); the equilibrium constants and enthalpies are

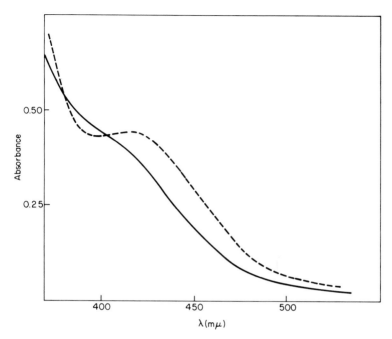

FIG. 5. Effect of high concentration ($\sim 8 \times 10^{-4}M$) on the visible absorption band of iodine (concentration $\sim 1.0 \times 10^{-4}M$): CCl₄ solvent (broken line); CHCl₃ solvent (solid line).

negligible and the interaction appears to be very weak. The oxidation of many aromatic hydrocarbons is expected to involve the initial formation of these contact pairs (20).

The interaction of halide ions with trinitrobenzene, dinitrobenzene, and nitrobenzene (Table III) give rise to characteristic CT bands in the 325–475 mμ region, but the ΔH° is nearly zero in all these systems (41, 79). It appears that these systems may be appropriately classified as contact pairs. Interaction of halide ions with tetracyanoethylene has also been found to give rise to contact pairs with near-zero enthalpy of formation (41). Even though the interaction of halide ions with tetracyanoethylene, trinitrobenzene, or chloranil is very weak, radical anions of these acceptors are readily formed, particularly on irradiation (41, 79).

2. HALOGEN COMPLEXES WITH π AND n DONORS

Molecular complexes of various halogens, particularly of iodine with π and n donors, have been studied exhaustively in the literature (5, 40, 138, 179, 179a, 180, 211). Data on the interaction of a few halogen acceptors with benzene and hexamethylbenzene have been compared with those of a few

TABLE III

HALIDE IONS AS DONORS

Donor	$K^{25°}$ (liters/mole)		
	TNB[a]	DNB[b]	NB[b]
$(C_7H_{15})_4N^+I^-$	11.5	2.2	0.9
$(C_7H_{15})_4N^+Br^-$	13.0	0.5	—
$(C_7H_{15})_4N^+Cl^-$	18.0	—	—

[a] Solvent, CCl_4. [b] Solvent, $CHCl_3$.

TABLE IV

SPECTROSCOPIC AND THERMODYNAMIC DATA ON THE CHARGE-TRANSFER COMPLEXES OF BENZENE AND HEXAMETHYLBENZENE WITH SEVERAL ACCEPTORS

Acceptor	Benzene						Hexamethylbenzene					
	Solvent	$h\nu_{CT}$ (kcal)	K (liters/mole)	Temp. (°C)	$-\Delta H°$ (kcal/mole)	ϵ_{CT}	Solvent	$h\nu_{CT}$ (kcal)	K (liters/mole)	Temp. (°C)	$-\Delta H°$ (kcal/mole)	ϵ_{CT}
Iodine	CCl_4	98.0	0.15	25	1.32	16,400	CCl_4	77.0	1.35	25	3.73	8200
ICl	CCl_4	101.3	0.54	25	—	8,130	CCl_4	85.6	13.2	25	5.32	4000
Br_2	CCl_4	97.8	1.04	25	—	13,400	—	—	—	—	—	—
Cl_2	CCl_4	103.0	0.33	25	—	9,090	—	—	—	—	—	—
TNB	CCl_4	101.0	0.23	20	1.71	0.277[a]	CCl_4	72.0	5.7	20	4.71	0.080[a]
TCNE	CH_2Cl_2	74.6	2.0	22	3.35	3,570	CCl_4	53.7	263	22	7.75	4390
Chloranil	CCl_4	85.0	0.33	20	1.65	—	CCl_4	55.0	9.25	20	5.35	0.093[a]

[a] Oscillator strength.

other acceptors in Table IV. For the interaction of alkylbenzenes with halo-gens, K varies in the order ICl > IBr > I_2 > Br_2 > Cl_2; none of these halogens compares well with a π acceptor such as TCNE, but compares favorably with trinitrobenzene. We have compared in Table V the values of C_1 and C_2 in the Briegleb relation [Eq. (7)] for the interaction of aromatic donors with iodine and other acceptors.

The data on the interaction of iodine with several n donors are given in Table VI. It can be seen that the equilibrium constants and enthalpies in these systems are much larger than with π donors. The blue shifts of the visible band of iodine with n donors also vary with the strength of the interaction. The blue shifts as well as the intensity of the iodine band are larger when the strength of interaction is large, as can be seen from Table VII.

In Table VI we have compared the interaction of n donors with iodine and phenol. It can be seen that amines that are strong donors toward iodine also behave similarly with respect to phenol; aromatics act as weak donors to-ward both iodine and phenol. Such comparisons for the interaction of iodine and phenol with n donors have been made by Singh *et al.* (229).

The pyridine–iodine system has generally been considered as an example of a stable n–σ complex with the lone pair of nitrogen acting as the donor site (21, 30, 58). It was, however, recognized that pyridine could in principle act as a π donor as well. Recently. Mulliken (181) has discussed the possibility of charge transfer from the π molecular orbitals of pyridine to the σ orbitals of iodine in the pyridine–I_2 complex. The slight asymmetry of the charge-trans-fer band with a shoulder on the low-frequency side has been interpreted as involving this π–σ charge transfer.

Charge-transfer interaction between aromatic donors and an iodine atoms produced by flash photolysis of I_2 solutions has been investigated by several workers (104, 238, 239). It has been found that the visible band of I_2 shows a decrease in intensity on flash photolysis, while other regions of the visible spectrum show increased absorption. The equilibrium constants and enthalpy of o-xylene + iodine atom have been found to be higher than those with molecular iodine (239). Formation constants deduced from gas-phase atom-recombination studies are at least an order of magnitude greater than those

TABLE V

Acceptor	C_1 (eV)	C_2 (eV)
Iodine	5.2	1.5
Tetracyanoethylene	6.1	0.54
Chloranil	5.7	0.44
Trinitrobenzene	5.0	0.70

TABLE VI

SPECTROSCOPIC AND THERMODYNAMIC DATA FOR A FEW CHARGE-TRANSFER AND HYDROGEN-BONDED SYSTEMS

Donor	Charge transfer with I_2						Hydrogen bonding with phenol				
	Solvent	K (liters/ mole)	Temp. (°C)	$-\Delta H°$ (kcal/ mole)	$h\nu_{CT}$ (kcal)	f_{CT}	Solvent	K (liters/ mole)	Temp. (°C)	$-\Delta H°$ (kcal/ mole)	$\Delta\nu_{O-H}$ (cm^{-1})
Benzene	CCl$_4$	0.15	25	1.32	98	—	CCl$_4$	0.28	25	1.56	47
Diethyl ether	CCl$_4$	0.87	25	4.2	115	0.17	CCl$_4$	8.83	25	5.41	277
Dimethyl sulfide	Heptane	200	25	8.9	94	0.60	CCl$_4$	14.65[a]	25[a]	4.26[a]	254[a]
Acetone	Cyclohexane	0.80	25	2.5	—	—	CCl$_4$	10.06	30	4.7	230
	Freon	1.70	20	5.8	118	—					
Cyclohexanone	Freon	2.40	20	6.1	113	—	CCl$_4$	16.83	20	4.97	238
Thiocamphor	Cyclohexane	95.0	25	11.0	—	—	CCl$_4$	6.2	25	5.0	—
Pyridine	Heptane	290	25	8.0	122	1.12	CCl$_4$	35.2	30	6.5	465
Triethylamine	Heptane	6000	24.5	12.2	103	0.85	CCl$_4$	58.0	25	9.1	450
Triphenylamine	Cyclohexane	12.7	26	9.5	102;133	—	CCl$_4$	1.54	25	—	460
Triphenylarsine	CHCl$_3$	8330	26	1.5	89	—	CCl$_4$	1.52	25	—	460
Triphenylstibine	CHCl$_3$	1700	26	3.0	99	—		—	—	—	—
1,4-Diselenane	CCl$_4$	7250	25	7.6	—	—	CCl$_4$	11.8[b]	25[b]	3.72[b]	240[b]

[a] The data are with n-butyl sulfide as donor.
[b] The data are with n-butyl selenide as donor.

TABLE VII[a]

BLUE-SHIFTED BANDS OF IODINE

	λ_{max} (mμ)	ϵ_{max}	f
Iodine	520	900	0.012
Iodine + Et_2O	462	920	0.017
Iodine + pyridine	422	1320	0.024
Iodine + Et_3N	414	2030	0.050
Iodine + Et_2S	435	1960	0.035

[a] Solvent, heptane.

for comparable I_2–aromatic complexes in the solution phase (104). Similar observations have not been reported with n donors. Since the electron affinity of the iodine atom is reliably known, some attention has been paid to the relation between ionization potential of donors and $h\nu_{CT}$ in iodine atom complexes. The slope of such a plot was found to be similar to amine–I_2 complexes (104), indicating that the resonance energy is quite large in iodine-atom complexes. An attempt has been made to evaluate the electron affinity of other acceptors by comparing the frequency of the CT band between hexamethylbenzene and iodine atom with the $h\nu_{CT}$ values of hexamethylbenzene complexes with other acceptors (131).

A few workers have examined the interaction of molecular iodine with various π and n donors in vapor phase (141, 213a, 244). The charge-transfer band for the aromatic–I_2 systems are generally found at lower wavelengths in vapor phase than in nonpolar solvents. The $h\nu_{CT}$–ionization-potential relation is valid in the vapor-phase data also, but the slope is different from the solution-phase data; the intensity of CT bands also differ in the two phases. The blue-shift of the iodine bands and $h\nu_{CT}$–I_D relations in vapor phase amine-I_2 complexes have been examined recently (213a). The equilibrium constants and enthalpies of formation are different in the vapor phase from those in the solution phase. There appears to be no doubt that CT forces are operative in vapor phase as well (213a).

D. Complexes of Other Inorganic Molecules

Complexes of alkenes with SO_2, Ag^+, and other metal ions have been investigated extensively in the literature (5). The silver ion in these complexes may serve not only as an electron acceptor but also as an electron source. Apparently, there is π bonding between components, whereby the filled d orbitals of the metal ion overlap with the π^* orbitals of the alkenes. The formation constants of the complexes of alkenes with Ag^+ vary markedly with the structure of the olefin, and the values are generally larger than those for the interaction with iodine. Thus, K for the cyclohexene–Ag^+ complex is 3.6 liters/mole at 40°C, while it is much less than 0.3 liters/mole for the cyclohexene–I_2 complex; the K for cyclohexene–SO_2 complex is still less (<0.05 liters/mole). Aromatic hydrocarbon adducts of Ag^+ are less stable than the olefin complexes. Aromatic-hydrocarbon–SO_2 complexes exhibit K values of the same magnitude as the olefin–SO_2 complexes. While the K values for iodine complexes of benzene and polymethyl benzenes differ by a factor of about 10, the corresponding values for the HF adducts differ by a factor of 10^6; the HF complexes are probably best described as σ complexes in which one of the ring carbons becomes tetrahedral in character. The applications of spectroscopy to the study of Ag^+ or SO_2 adducts with unsaturated compounds are mainly confined to infrared and Raman investigations. Thus, the C=C stretching vibration frequency is considerably lowered on complexation with Ag^+. The interaction of Friedel–Crafts catalysts (Lewis acids) with ketones and other carbonyl compounds produces stable adducts that give rise to large variations in C=O frequency (5, 149). In these complexes there is electron transfer from the donor oxygen to the Lewis acid. Results from infrared studies of these complexes will be discussed at a later stage.

1. *n–v* Systems

It was mentioned earlier that *n–v* systems provide examples of very strong interactions between donors and acceptors. Although a study of the vibrational spectra of such systems has shown marked frequency shifts indicating strong interaction, electronic spectra have not been of much use in studying such interactions. Most of the thermodynamic data in these systems reported in the literature are therefore based on other physical measurements. It would suffice for our discussion to point out the large enthalpies of formation for these adducts (8, 72, 237) (Table VIII). Mössbauer spectroscopy has been employed to examine some of *n–v* addition compounds (125a).

E. Charge Transfer to Solvent (CTTS) Transitions

It has been established for some time that the characteristic absorption spectra of halide ions in various solvents are due to charge-transfer transitions from the halide ions to the solvent cage. Symons and co-workers (106, 231, 232) have proposed that the spectra of halide ions can be interpreted in terms of a simple square-well model for the excited state, where the energy of the transition E_{max} varies inversely with the radius of the solvent cavity r_0, since $E_{max} = I.P. + (h^2/8mr_0^2)$. The excited electron is confined to the first layer of the solvent. In order to establish that the absorption bands of the solvated iodide

TABLE VIII

THERMODYNAMIC DATA FOR A FEW n–v SYSTEMS

Donor	$-\Delta H°$ (kcal/mole)				
	$AlCl_3$	BF_3	BH_3	$B(CH_3)_3$	$Ga(CH_3)_3$
Acetophenone	33.0	—	—	—	—
Benzophenone	29.0	—	—	—	—
Diethyl ether	—	10.9	—	—	—
Diphenyl ether	18.0	—	—	—	—
Tetrahydrofuran	—	13.4	—	—	—
$(CH_3)_3N$	—	—	—	17.6	21
$(C_2H_5)_3N$	38	—	—	—	—
Pyridine	62	—	—	—	—
$(CH_3)_3P$	—	18.9	—	16.47	10
$(CH_3)_3As$	—	—	—	—	10
$(CH_3)_2S$	—	—	6.1	—	—
$(C_2H_5)_2S$	—	—	6.0	—	—
$(CH_2)_4S$	—	—	4.0	—	—

ion were due to CTTS transitions and not due to intramolecular charge transfer of the contact ion pair type, Griffiths and Symons (107) and Singh and Rao (230) examined the electronic spectra of tetra-n-alkylammonium iodides in various solvents. The electronic spectra of tetra-n-heptylammonium iodide and potassium iodide are compared in Fig. 6. Both these iodides give rise to two charge-transfer bands corresponding to $^2P_{3/2}$ and $^2P_{1/2}$ upper states of the iodine atom; the data on the charge-transfer transitions of the two iodides in acetonitrile solvent are given in Tables IX and X.

The CTTS transitions of Br^- and Cl^- ions are at much higher energies than those of the I^- ions. Thus, the wavelengths of absorption of the CTTS transitions of I^-, Br^-, and Cl^- in aqueous solution are respectively 195 and 225 mμ; 187 and 197 mμ; and 175 mμ (133).

Solvent effects on the CTTS transitions of halide ions have been studied extensively, and some data (79) are tabulated in Table XI. It is found that in hydroxylic solvents the halide ions hydrogen-bond fairly strongly, causing appreciable solvent blue shifts of the CTTS transitions (79, 230).

Iodide ion in carbon tetrachloride solvent gives rise to a band at 290 mμ, which has been ascribed to the formation of an $I^- \cdots CCl_4$ complex (32). This equilibrium has been investigated in detail by Dwivedi and Rao (79) in methylene chloride and acetonitrile solvents, employing electronic spectroscopy. The equilibrium constants (at 25°C) and enthalpies of formation are 5.9 liters/mole and -2.5 kcal/mole in CH_2Cl_2 solvent, while the values in

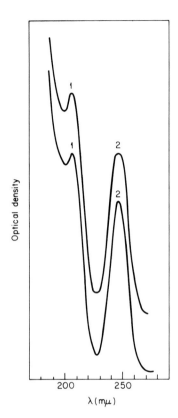

FIG. 6. Iodide-ion (CTTS) transitions in acetonitrile solution: (1) tetra-*n*-hepty-lammonium iodide; (2) potassium iodide.

CH_3CN solvent are 6.5 liters/mole and -9.5 kcal/mole respectively. The results have been discussed in terms of the differences in the ion-pair equilibria in the two solvents. Dwivedi and Rao (79) have also found evidence for the interaction of Br^- with CCl_4; apparently Cl^- does not interact appreciably with CCl_4. It should be noted at this stage that CCl_4 acts as an acceptor with many aromatic donors as well (27). Interaction of several inorganic ions with O_2 (146a) and I_2 (167a) have been examined recently. Blandamer and Fox (33a) have reviewed the literature on CTTS transitions.

F. Solvent Effects on Charge-Transfer Spectra

The effect of solvents on the equilibrium constants of formation in several donor–acceptor systems has been investigated in the literature. For example, the stability of the naphthalene–I_2 complex varies in the order *n*-heptane > cyclohexane > CCl_4 > *n*-hexane > chloroform (69). In most weak complexes

TABLE IX

CTTS TRANSITIONS OF IODIDE ION

	$\nu_{(2P_{1/2}\leftarrow 1S)}$ (cm^{-1})	$\Delta\nu_{1/2}$ (cm^{-1})	$\epsilon_{max} \times 10^{-4}$	f^a
KI	40650	2491	1.7	0.18 (3.1)
$(n\text{-}C_7H_{15})_4N^+I^-$	40570	2546	1.65	0.19 (3.2)

TABLE X

CTTS TRANSITIONS OF IODIDE ION

	$\nu_{(2P_{3/2}\leftarrow 1S)}$ (cm^{-1})	$\Delta\nu_{1/2}$ (cm^{-1})	$\epsilon_{max} \times 10^{-4}$	f^a	$E_{2P_{3/2}-2P_{1/2}}$
KI	48190	3364	2.1	0.31 (3.7)	7540
$(n\text{-}C_7H_{15})_4N^+I^-$	48310	3776	1.8	0.22 (3.6)	7740

^{a} The f (oscillator strength) value in crystalline alkali iodides is found to be 0.25 and the f value of KI in water is 0.25. Apparently, the transitions are not as favored in CH_3CN solvent where the transition energy E_T is also lower than in alcohols or in water. The quantities given in parentheses are the transition dipole moments.

TABLE XI

SOLVENT EFFECTS ON THE CTTS TRANSITIONS OF HALIDE IONS

	λ_{max} (mμ)		
Solvent	I^-	$Br^{-\ a}$	$Cl^{-\ a}$
Carbon tetrachloride	290^b	260^b	246
Chloroform	243	242	245
Dichloromethane	246	220	—
Acetonitrile	246, 207^c	220	197
t-Butanol	225	202	186
i-Propanol	219	~198	—
Ethanol	218	~198	—
Methanol	219	~198	—
Water	226	~198	184

^{a} The second CTTS transition is at very short wavelengths.

^{b} Possibly due to the halide–CCl_4 complex.

^{c} Two CTTS transitions are not seen in other solvents because of solvent cut-off.

it appears that equilibrium constants are affected appreciably, while the enthalpies may or may not show marked variation with solvent (249). Some data on solvent effects on the pyridine–I_2 systems (21) are given in Table XII. The effect of solvent on charge-transfer systems has been explained by Mulliken (174) in terms of dipole–dipole, van der Waals and other nonspecific interactions between solute and solvent molecules. Mulliken's explanation seems to suggest that there should be a variation of enthalpy with the solvent.

Kosower (139) has found that the position of the charge-transfer band in pyridinium iodides is extremely sensitive to solvents. 1-Methyl-pyridinium iodide, for example, shows bands at 3796 and 2945 Å in chloroform solution.

TABLE XII

SOLVENT EFFECTS ON THE PYRIDINE–IODINE CHARGE-TRANSFER COMPLEX (21)

Solvent	D	K (liters/mole, 25°C)	$\epsilon_{max}{}^a$ (liters/ mole-cm)	$\lambda_{max}{}^a$ (mμ)	$\Delta\nu_{1/2}{}^a$ (cm^{-1})	f^a
Chloroform	4.81	63 [b]	1669	401.5	4900	0.035
Carbon tetra-chloride	2.24	111	1389	418.5	4400	0.026
Cyclohexane	2.02	126	1387	424.5	4400	0.026
n-Heptane	1.92	140	1386	424.5	4300	0.026
n-Hexane	1.87	127	1450	426.5	4400	0.028

[a] Of the blue-shifted iodine band.
[b] The $\Delta H°$ in chloroform was found to be -8.4 kcal/mole, compared to -7.8 kcal/mole in heptane.

The difference in these two transition energies are the same in other solvents such as acetonitrile or acetone. It may be noted that this energy difference is the same as that found in the CTTS transition in iodide ion, indicating thereby that the charge-transfer transition of pyridinium iodides also involves the photoexcitation of iodide ion (230). On the basis of charge-transfer energies of 1-ethyl-4-carbomethoxy pyridinium iodide in various solvents, Kosower has proposed a solvent parameter Z that is an empirical measure of solvent polarity. The Z parameter is being widely used in correlations of solvent effects on spectral and reactivity data.

G. Dipole Moments of Charge-Transfer Complexes

Since the formation of donor–acceptor complexes is accompanied by electron transfer, one would expect a change in the electrical properties of the system on complexation. In a complex of a nonpolar donor with a nonpolar acceptor, the dipole moment of the no-bond structure will be negligibly small (nearly zero). However, there will be a dipole moment associated with the dative structure, and it will be directed from the donor to the acceptor (174). Since the actual ground-state wavefunction of the complex is an admixture of both the no-bond and the dative-bond wave functions, this implies that there will be a dipole moment in the complex, the appropriate magnitude of which will depend on the dipole moment of the dative-bond structure.

Dipole moments of a variety of halogen complexes of aromatic hydrocarbons (40), amines (137), carbonyl and thiocarbonyl compounds, and ethers and thioethers (25) have been reported in the literature. The dipole moment of the complex in the ground state μ_N, which is experimentally determined, is given by

$$\mu_N = \int \psi_N M \psi_N \, d\tau \tag{12}$$

$$\mu_N = a^2 \mu_0 + b^2 \mu_1 + 2ab\mu_{01} \tag{13}$$

In Eq. (13), μ_0 is the dipole moment of the no-bond structure (the vector sum of the component moments of A and D) and $\mu_1 (= er_{DA})$, the dipole moment of the charge-transfer state, where e is the electric charge and r_{DA} is the equilibrium distance between D^+ and A^-. The dipole moment of the complex in the ground state, μ_N, can be represented by

$$\mu_N = a^2 \mu_0 + b^2 \mu_1 + abS(\mu_1 - \mu_0) + 2abS\mu_0 \tag{14}$$

When the dipole moment of the no-bond structure μ_0 is small compared with that of the charge-transfer structure μ_1, μ_N can be approximated to

$$\mu_N \approx \mu_1(b^2 + abS) \tag{15}$$

where S is the overlap integral between ψ_0 and ψ_1. Further, the normalization conditions must be satisfied. Equation (15) is valid in weak complexes such as aromatic compounds–iodine. By making use of the various relations among the energies of the ground- and excited-state configurations, the heat of formation, and the $h\nu_{CT}$, one can calculate the percent charge transfer, $100(b^2 + abS)$, as well as the various charge-transfer parameters such as R_E, R_N, W_0, and W_1. Such calculations have been carried out in the literature for iodine complexes (25, 40, 137); some of the important results are shown in Table XIII. We feel that the percent charge transfer estimated in this manner provides a reliable measure of donor basicity and acceptor acidity. In the case of weak charge-transfer complexes such as π-donor–I_2 systems the value of b^2/a^2 can be estimated by the simple relation of Ketelaar (134):

$$\Delta H^\circ / h\nu_{CT} \approx b^2/a^2 \tag{16}$$

TABLE XIII

DIPOLE MOMENTS OF IODINE COMPLEXES[a] (*25, 40, 137*)

Donor	Dipole moment, D	%CT
Benzene	0.72	3
Dioxan	1.30	6.5
Dimethyl sulfide	3.62	12
Tetramethyl urea	5.2	9
Tetramethyl thiourea	7.52	18
Pyridine	4.5	25
Isopropylamine	6.2	36

[a] Solvent, CCl_4 ; temperature, 30°C.

H. Transformation of Outer Charge-Transfer Complexes to Inner Complexes

Mulliken (*174*) pointed out that a donor–acceptor pair can form either the associative outer complex or the dissociative inner complex, depending on the distance of approach between the donor and acceptor and the relative magnitude of the no-bond and dative wave functions. It was also suggested that the formation of the inner complex from the outer complex should be strongly dependent on environmental conditions. If the environmental influence is sufficiently great, the inner complex may become the stable form of the donor–acceptor pair, and with little or no environmental influence, the outer complex may represent the stable form. Thus, for a given donor–acceptor pair under a particular set of environmental conditions, one usually observes either an inner complex or an outer complex. The outer complex once formed can transform to the inner complex under favorable conditions. For example, the increased electrical conductance of iodine in pyridine was explained by Mulliken and Reid (*215*) on the basis of an equilibrium involving the outer and inner complexes:

$$C_5H_5N{\cdot}I_2 \rightleftharpoons C_5H_5N^+I + I^-$$
(outer complex)   (inner complex)

An examination of the literature shows that in many donor–acceptor systems with halogen acceptors, the formation of the trihalide ion is often observed; this can only result through the formation of inner complexes from the initial outer complexes. Recently, Bhat and Rao (*28*) have studied the transformation of outer charge-transfer complexes to the inner complexes in a number of systems by measuring the time dependence of the CT band or the I_3^- band. The time dependence of the charge-transfer bands is considered to be due to the transformation of the initially formed 1 : 1 outer complex into the inner complex; this is followed by the fast reaction of the inner complex with iodine to form I_3^-. The systems studied are triphenylarsine + iodine, triphenylstibine + iodine, dimethyl sulfoxide + iodine, acetone + iodine, and aliphatic nitriles + iodine. In all these systems the activation energy for the transformation of the outer complex to inner complexes is very small (<10 kcal/mole); the activation energy decreases with increasing dielectric constant of the medium. A stable outer complex is generally associated with a larger activation energy for the transformation than a weaker outer complex (22). Thus in triphenyl phosphine + halogen systems the outer complex is not very stable, and such systems readily transform (activation energy being negligibly small) to halophosphine halides, $PR_3X^+X^-$, which are inner complexes (22).

In the case of aliphatic nitriles, the effects of substituents is seen on the thermodynamics of formation of outer complexes as well as in the transformation energetics. Thus the enthalpy, equilibrium constant of formation, and the rate of transformation vary in the order t-Bu > Me > CCl_3 (28). The entropies of activation for the transformation of outer complexes are found to be large and negative; this is expected for reactions involving ionization of neutral molecules. An analysis of the kinetic data in different solvents for the transformations of several n-donor + iodine systems reveals that the electrostatic contribution to the entropy of activation is negligible.

The transformation of outer complexes to inner complexes has recently been studied in triethylamine + iodine system at high donor concentrations by Bist and Person (31). The following reactions have been visualized to occur:

$$Et_3N + I_2 \rightleftharpoons Et_3N \cdot I_2 \qquad \text{very rapid} \qquad (17)$$

$$Et_3N \cdot I_2 \rightarrow B^+ + I^- \qquad \text{rapid} \qquad (18)$$

$$B^+(+Et_3N) \rightarrow X^+ + I^- \qquad \text{slow} \qquad (19)$$

B^+ was believed to be $(Et_3N-I)^+$ (or the diamine). X^+, which does not contain iodine atoms, was suspected to be $(Et_3NH)^+$, or perhaps

$$Et_3N^+ - CH - NEt_2$$
$$\underset{CH_3}{\overset{\textstyle |}{}}$$

by analogy with the results found for the γ–picoline system by Haque and Wood (113).

I. Geometry of Complexes

X-ray crystallographic studies of solid complexes of halogens with π and n donors have provided valuable information on the geometry of the complexes (5, 40, 116, 179). These geometries, however, do not reflect the geometries in solution. This is because systems such as acetone–Br_2 and benzene–I_2 in solid state are of the ∞ : ∞ type, while they are of the 1 : 1 type in solution phase. Hassel and co-workers (*5, 116*) have determined the structure of the solid complexes of benzene and bromine and found the orientation to be **I** rather than **II**.

(I)   (II)

A number of solid adducts of halogens with *n* donors, including various amines, ethers, sulfides, and selenides have been studied by x-ray crystallography (*5, 116*), and it is found generally that the donor atom is collinear with the halogen–halogen bond. The donor–halogen distance in *n* donors is found to depend on the strength of interaction. The *n*-donor strength on the basis of these distances is shown to be in the order $N > Se > S > O$. The complex of bromine with acetone has the structure where each halogen molecule is equivalently linked to two separate donor molecules and the donor oxygen shares two lone pairs with two bromine molecules.

Orientation of donors and acceptors in molecular complexes can be fruitfully investigated by studying the dichroism with respect to the absorption of visible or ultraviolet light or infrared radiation (*5, 40*). Such studies have been carried out with complexes of aromatics with quinones, picric acid, and *sym*-trinitrobenzene and the results are in general agreement with the known crystallographic arrangement of these crystals.

J. Infrared Spectra of Molecular Complexes

The interaction of aromatic hydrocarbons with halogens enhances the intensity of certain infrared absorption bands of the aromatics. Thus, the 850 and 992 cm^{-1} bands of benzene are enhanced in the presence of bromine and iodine, presumably because of changes in the symmetric ring modes (*87*). Sulfur dioxide complexes of benzene and mesitylene, however, do not show major changes in the spectra.

Halogen–halogen stretching vibration frequencies have been reported in a number of charge-transfer complexes. These frequencies are generally shifted downward, and the bands are intensified on complex formation. For example, the Cl—Cl stretching vibration frequency at 557 cm^{-1} in the gas phase is found at 530, 527, and 524 cm^{-1} in benzene, toluene, and xylene solutions respectively (*36*). Similar changes have been found with iodine and bromine. The observation of halogen–halogen stretching vibrations in the vibrational spectra would, offhand, mean that the symmetry of the halogen molecule is perturbed by polarization. One would therefore assign these bands to orientational isomers in which the two halogen atoms are not equivalently located with respect to the aromatic nucleus (88, 177). Mulliken (173) had suggested earlier that the halogen molecule possibly lies parallel to the plane of the benzene ring with its center on the sixfold symmetry axis of the ring. Ferguson and Matsen (89), however, point out that changes in the vertical electron affinity of the halogen occurring during the molecular vibration may be sufficient to account for the observation of halogen–halogen stretching vibration bands, even if a dipole moment parallel to the halogen axis in the complex is not required. Person and co-workers (199) have recently pointed out that the infrared spectrum of solid bromine–benzene complex does not support the presence of a center of symmetry in the complex, as suggested by the x-ray structure determination.

Freidrich and Person (96) have examined the theory of molecular vibrational transition intensities in donor–acceptor complexes with special reference to the intensities of halogen–halogen vibrations and have correlated the coefficients of dative and no-bond wave functions with infrared frequency shifts. The intensification of halogen–halogen stretching vibration is supposed to arise from the electronic orientation during the vibration. Freidrich and Person therefore argue that no conclusion can be arrived at about the geometry of halogen–benzene complexes from a study of the infrared spectra.

Maki and Plyler (159) have reported a band at 184 cm^{-1} in the pyridine–iodine system; they also observed a shift of the 601 cm^{-1} band of pyridine to 618 cm^{-1} in the complex. Lake and Thompson (143) assigned the 180 cm^{-1} band as the modified stretching vibration of the iodine molecule and a new band around 80 cm^{-1} to a stretching of the intermolecular bond. Yarwood and Person (267) have studied the complex of iodine with pyridine and other n donors and have discussed the position and intensities of the I—I stretching bands in terms of the theory of Freidrich and Person (96). In the bromine complexes of substituted pyridines, bands around 110 cm^{-1} and 240 cm^{-1} have been assigned respectively to the vibrations of the intermolecular bond and the modified bromine molecule (144). A band around 187 cm^{-1} has been assigned to the Br$_3$$^{-}$ ion.

Ginn and co-workers (101) have examined the interaction of Br$_2$ and BrCl with pyridine and assigned bands to Py$_2$Br^{+}, Br$_3$$^{-}$, and BrCl$_2$$^{-}$ in the spectra. The interaction of ICl with pyridine shifts the stretching vibration frequency of ICl from 375 cm^{-1} to 355 cm^{-1}; this system has been examined recently by Yarwood and Person (268), who have discussed the intensities and force constants of the N—I stretching vibration. Yarwood (269) has reported integrated intensities for the vibrational fundamental bands of pyridine and pyridine–ICl complex and has interpreted the large intensity changes in terms of the Mulliken charge-transfer theory. The interaction of ICN with benzene, pyridines, and other nitrogen-containing compounds causes pronounced changes in the infrared spectra; the I—C stretching of ICN is shifted by donors (*200, 272*).

Interaction of iodine with ketones, amides, and ureas causes an appreciable decrease in the C=O stretching frequency (*53, 213, 224, 266*); the same is true of C=S frequency in simple thiocarbonyl compounds (*23*). In thioamides and thiourea derivatives the mixed vibration band having a large contribution from the C=S stretching mode is shifted to lower frequency (*24, 103*).

In the sulfur dioxide adducts of aromatic amines, lowering of the C—N stretching frequency has been reported, indicating that SO_2 is directly coordinated with the amino nitrogen atom (*51*). The interaction of olefins, dienes, and acetylenes with metals such as silver and platinum causes marked changes in the carbon–carbon stretching region. In silver complexes (*251*) the infrared spectra have no absorption in the normal C=C stretching region (1550–1650 cm^{-1}). This is a strong evidence for the coordination to C=C bond. In norbornadiene complexes of platinum and palladium the C=C bands are not present in the 1550–1650 cm^{-1} region, indicating coordination with both the double bonds (*2*). The bands of cyclooctadiene at 1600 and 710 cm^{-1} (C=C and =CH vibrations respectively) are replaced by bands at 1612 and 746 cm^{-1} in the cuprous chloride complex (*253*). In the complexes of HCl with aromatic donors, the H—Cl stretching frequency is lowered appreciably; the magnitude of the shifts are linearly dependent on the ionization potentials of donors (*61*).

Interaction of Lewis acids such as Friedel–Crafts catalysts with ketones decreases the C=O stretching frequency. For example, the C=O band in acetone–BF_3 system (*55*) occur around 1635 cm^{-1}. Similarly other halides, such as $AlCl_3$, $TiCl_4$, and $SbCl_5$ also markedly affect the C=O frequency of esters and ketones in donor–acceptor complexes, which indicates that the interaction is in the direction of the carbonyl oxygen (*5, 149*). The interaction of Friedel–Crafts catalysts with acyl and aryl halides is accompanied by a different kind of change in the infrared spectra. The 1 : 1 complex of $AlCl_3$ with acetyl chloride (*240*) in solid state shows no band in the characteristic C=O and C—Cl stretching regions but gives an intense band around 2300 cm^{-1}. This has been taken as evidence for the increased C=O bond order resulting from the formation of the acylium ion. In liquid state, however, this complex gives rise to a band at 1635 cm^{-1} as well as a band around 2300 cm^{-1}, indicating the presence of both the oxocarbonium ion and the donor–acceptor complex (*62*). Adducts of $SbCl_5$ and $TiCl_4$ with acid halides also give rise to mixtures of oxocarbonium ions and the donor–acceptor complexes (*190*).

Solid 1 : 1 adducts of acid fluorides with BF_3, PF_5, and SbF_5 show similar strong absorption in the 2300 cm^{-1} region (61). The metal–F stretching vibrations are also affected in these adducts.

Unlike simple ketones, nitriles show an increase in the C≡N frequency on the formation of Lewis acid–base complexes (44). In a series of $SnCl_4$ adducts of substituted benzonitriles, however, the change in frequency decreases with increasing stability of the complex. For the interaction of a given nitrile with a series of Lewis acids, the C≡N frequency increases with the increasing stability of the complex. $AlCl_3$ is shown by infrared studies to interact with nitro compounds at the oxygen of the nitro group (241). Interaction of dioxan with $AgClO_4$, $SbCl_3$, and $HgCl_2$ are found to give rise to a new band or cause disappearance of some bands of the parent molecule and appearance of new bands (65). The P=O and Se=O frequencies in $POCl_3$ and $SeOCl_2$ are found to be greatly influenced by mixing with $TiCl_4$ and $SnCl_4$ (227). A number of references to the literature on the infrared spectra of inorganic adducts of oxo compounds may be found in the monograph of Lindquist (149).

K. NMR Spectra

NMR spectroscopy has been applied with limited success to the investigation of the structure of molecular complexes. Fratiello (94) has studied the NMR spectra of solutions of iodine in several organic solvents and found that the chemical shifts of solvent molecules reflect the electron-density changes to some extent, if the strength of the interaction is appreciably large. Thus benzene, xylene, and nitro compounds do not show displacement of chemical shift on addition of iodine. NMR spectra have been used to determine equilibrium constants of tetracyanoquinoline–aromatic systems (112); the chemical shifts of the protons are generally shifted to high fields in the complexes. As the complex becomes stronger, the upfield shifts are decreased.

NMR spectra of the adducts of acyl fluoride with metallic fluorides (Lewis acids) clearly show the presence of oxocarbonium ions (190). The NMR spectra of the complexes of olefins and aromatics with Ag^+ and platinum salts have been studied (207, 225). In general the proton chemical shifts are not different from those of free hydrocarbons. This has been taken to indicate that there is not much change in the double bond on complexation. In the Ag^+–benzene complex, only a single signal is shown by aromatic protons, possibly because of rapid exchange. Foster and Fyfe (92a) have recently reviewed NMR studies of donor–acceptor complexes.

III. Radical Ions and Ion Pairs

Organic molecules accept an extra electron to form radical anions or give up an electron to form radical cations. Many of the radical ions are relatively stable in their ground states in solution phase and can be readily examined by spectroscopic methods. Electron-spin resonance (ESR) spectroscopy provides an ideal tool to study the unshared electron in radical ions. Most of the radical ions are colored, exhibiting characteristic absorption bands in the electronic spectra; electronic spectra are generally employed to study not only radical anions and cations but also dianions, which are generally diamagnetic. The spectra of these radical ions and dianions have been interpreted success-fully on the basis of molecular orbital theory. Both ESR and electronic spectroscopy are employed for the characterization of ionic species in solution and to establish the nature of chemical equilibria involving ionic species. Several reviews on various aspects of radical ions including ESR studies have appeared in recent literature (*67, 133, 164, 242*); electronic spectra of radical ions has been reviewed only recently by Rao *et al.* (*212*). At present we shall limit our discussion to radical ions produced by charge-transfer intermediates and to the study of ion pairs in solution by optical spectroscopic methods.

A. *Spectra of Radical Ions Produced from CT Complexes*

If the dative contribution in a donor–acceptor complex is appreciable in the ground state, the magnitude of charge-transfer from the donor to acceptor would be larger leading to radical ions. Such radical ions would be para-magnetic and would show signals in the ESR spectra. The nature of such donor–acceptor complexes in solution depends on solvent polarity. Thus a fairly strong donor such as an aliphatic or heterocyclic amine with a good acceptor such as tetracyanoethylene or iodine may form a weak charge-transfer complex in a nonpolar solvent but form an ion pair in a polar solvent (*212*). These ion pairs may be either intimate ion pairs or dissociated solvent-separated pairs [see Eq. (11)]. The formation of such radical ion pairs in solid state has also been noticed; such solids are generally good organic semicon-ductors. Even though ESR spectroscopy would be ideal to identify the forma-tion of radical ions in such charge-transfer systems, electronic spectroscopy would also provide a useful method for examining the radical ions. These radical ions are generally associated with intense colors, causing characteristic long-wavelength absorption bands in the electronic absorption spectra. A number of ESR studies have been reported on donor–acceptor complexes, particularly in the solid state (*5*). Typical halogen complexes where ESR signals have been found are triphenylamine + iodine, benzidine + iodine, phenothiazine + iodine, phenothiazine + $SbCl_5$, and phenothiazine + $SbCl_4$ (5, 29).

A study of the temperature dependence of the ESR signal intensity in CT complexes provides information on the nature of the unpaired spins in these systems. If the interaction between the ions with unpaired spins, $A^{·-}$ and $D^{·+}$, is negligible, they can align independently in a magnetic field, giving rise to magnetic multiplet levels. If the interaction is appreciable, then the fourfold degenerate magnetic level will be split into singlet and triplet levels; if the singlet level is the ground state, the excited triplet state will be separated by an energy E^{spin}. Complexes between donors of low ionization potential and acceptors of high electron affinity are generally ionic in character, and in such cases the triplet may not be far above the ground state. The triplet state would be sufficiently populated at laboratory temperatures to render the complexes paramagnetic. There are also instances where the triplet state is the ground state. As the temperature is varied, one finds a variation in the ESR signal intensity resulting from changes in the population of the magnetic levels. Thus for a singlet–triplet system we can expect the intensity to increase as the temperature is increased because of the depopulation of the triplet level. For the case of a ground state consisting of one unpaired electron on the donor and one on the acceptor, both acting independently as doublet states, the intensity of the spin-resonance signal would be inversely proportional to temperature (Curie law).

Recently Bhat and Rao (29) have found that benzidine + iodine, p-phenylenediamine + iodine, as well as phenothiazine + $SbCl_4$ show ESR signals that increase in intensity with temperature, as is typical of singlet–triplet systems. The E^{spin} values in these systems have been determined by a study of relative spin concentrations as a function of temperature. In contrast to these systems, phenothiazine + iodine and phenothiazine + $SbCl_5$ show Curie law behavior.

The E^{spin} values in solid charge-transfer complexes have been compared with the activation energies for the conduction of electricity in these complexes (29, 108). Good agreement has been found for perylene + iodine and pyrene + iodine (228), but the agreement is poor for the highly paramagnetic (strongly dative) systems such as amines + iodine or phenothiazine + iodine (29).

There have been several studies on the electronic spectra of radical ions produced through charge-transfer intermediates. Most of these studies are related to acceptors such as TCNE or chloranil (212). Although there is definite evidence for the formation of radical ions in amine–halogen systems, there have not been any studies in the literature on the electronic spectra. Long-wavelength bands extending up to the near-infrared region have been found in phenothiazine $+ I_2$ (795 mμ), phenothiazine $+ SbCl_5$ (1051 mμ), p-phenylenediamine $+ I_2$ (751 mμ), and benzidine $+ I_2$ (1698 mμ) systems; these bands are likely to be due to radical cations rather than to simple charge-transfer complexes (29). Electronic absorption spectra of aromatic cations formed by the interaction of hydrocarbons with a strong Lewis acid such as $SbCl_5$ have been reported by several workers (35, 66, 126, 142, 147). Absorption bands due to radical anions have been found in the interaction of halide ions with acceptors such as tetracyanoethylene and trinitrobenzene (41, 79).

Cations of aromatic compounds such as azines or triphenylamine can be produced by adsorption on surfaces of activated silica and alumina (12), and electronic spectra of such adsorbed cations have been reported (12, 73). Radical anions of polynitro compounds and haloquinones have also been produced by adsorption on surfaces of metal oxides (92, 201). The anions have been identified by ESR and electronic spectra. Triphenylmethane adsorbed on silica–alumina shows evidence for adsorbed species as well as triphenylmethyl radical and the carbonium ion (6).

B. Equilibria Involving Radical Ions and Ion Pairs

Reduction of aromatic hydrocarbons and other organic compounds by alkali metals or amalgams produces radical ions. The solution-phase equilibria between the alkali metal and the hydrocarbon are very much in favor of the radical anions, and it is difficult to determine the equilibrium constants in such systems. It is only in the biphenyl–metal system that the equilibrium constants have been determined in ether solvents by employing the characteristic bands in electronic spectra (216, 226). The equilibria are associated with large negative enthalpy changes; the enthalpy values can be correlated with the electron affinities of the hydrocarbons.

Many radical anions readily form dianions through a disproportionation reaction:

$$2R^{\cdot -} \rightleftharpoons R^{2-} + R \tag{20}$$

The free energy of the disproportionation reaction may be related to the Coulomb repulsion energy, which prevents the addition of an electron to the radical anion. The presence of ionic association in solution markedly affects the free-energy change of the disporportionation reaction; this is because the dianions are more strongly associated with the counter cations than the mono-anions. Electronic spectroscopy has been extensively employed to study the disproportionation of radical anions, particularly to examine the effect of counter cations and solvents on the equilibrium.

Garst and co-workers (98–100) have shown that the disproportionation of triphenylethylene radical anion is considerably affected by the cation and the solvent. Disproportionation is hindered in THF and much more in 1,2-dimethoxyethane, even though they solvate the cation effectively. By a proper choice of solvents and metal ions, the equilibrium constant of disproportionation can be made to vary from 10^{-3} to 10^3, as shown by Zabolotny and Garst (270) in the case of stilbene ions. In solvents such as hexamethylphosphoramide no disproportionation takes place, since the ions are free (93); there is no driving force (entropy gain) in this solvent for the disproportionation reaction. This discussion clearly points out that one has to really understand the nature of the ionic species in solution in order to interpret the various features in the electronic spectra of radical ions.

There is considerable evidence in the literature to show that ion pairs exist in solution in radical ions. The influence of alkali metal ions on the electronic spectra of radical anions was first investigated by Carter and co-workers (54), who found that the absorption maxima of the radical ions shifted to longer wavelengths with increasing radius of the alkali-metal ion. They proposed the relation

$$v_{max} \propto \frac{1}{r_c} + 2 \tag{21}$$

The value of λ_{max} extrapolated to infinite radius of the cation gives the value of the solvated anion. This dependence of λ_{max} on cation radius arises from the presence of contact ion pairs in solution. In contact pairs the two ions are in close proximity and are surrounded by the solvent. McClelland (165, 166) has theoretically treated this perturbing effect of cations on the anion spectra. Hoijtink and co-workers (11) have discussed ion-pair formation in terms of the solvation radius of the cation and temperature. Contact ion pairs are generally favored by the increasing radius of the cation, decreasing solvent dielectric constant, and increasing temperature.

The influence of ionic association on the electronic spectrum of quaterphenyl anions has been studied by Hoijtink and co-workers (49), who have shown the presence of following equilibria in solution. The spectra have been

$$R^{\cdot} + M^+ \rightleftharpoons R^{\cdot}, M^+ \tag{22}$$

$$R^{2-} + M^+ \rightleftharpoons R^{2-}, M^+ \tag{23}$$

$$R^{2-} + 2M^+ \rightleftharpoons R^{2-}, 2M^+ \tag{24}$$

accounted for by assuming that only one type of species is present in the solvent separated ion pair. From these studies Hoijtink and co-workers (49) have established that the spectra of the associated forms are similar to those of the free solvated ions. Their theoretical calculations have also shown that the counter ion of the monoanions are located at positions different from those of the dianions; the distance between the dianion and cation is much shorter than that between the cation and the monoanion.

Szwarc and co-workers (57) find that the sodium salts of aromatic radical ions are associated into contact pairs even in DME, and the proportion of the solvent-separated pair is greater in THF. The thermodynamic data on the biphenyl radical ion in the two solvents (obtained by electronic spectroscopy) clearly demonstrates the different behavior of ion pairs. The solvation of the ions produced by the dissociation of contact pairs leads to large negative values of ΔS (-52 eu); lower negative value (-20 to -3 eu) of ΔS are found for the dissociation of solvated or partially separated pairs.

Smid and co-workers (56, 118) have recently examined the equilibrium between contact and solvent-separated ion pairs of carbanions, particularly of fluorene and its derivatives, by employing spectrophotometric methods. The contact pairs generally absorb at lower wavelengths than the solvent-separated pairs. The fraction of solvent-separated pairs decreases in the order Li > Na > K > Cs. Solvent separation of Li and Na salts occur on addition of small quantities of DMSO or di-i-propylamine. In fluorenyl-lithium the maxima for the contact-ion pair are at 420, 430, and 460 mμ in diethyl ether, while it exists as a solvent-separated pair with λ_{max} at 458, 487, and 522 mμ in oxetane. Apparently the basicity of oxygen in cyclic ethers, as well as steric hindrance, plays an important role in determining the equilibrium between the two types of ion pairs.

In the spectra of most radical ions the solvent-separated pairs absorb at longer wavelengths than the contact pairs. In addition to the association of the anion with the counter cation, it is also known that radical ions become associated into other ion pairs in ethereal solutions. For example, the absorption maxima of lithium diphenyl ketyl in THF is shifted appreciably by addition of LiBr because of the formation of ion-pair agglomerates (208).

Edgell and co-workers (80, 81) have examined the far-infrared and Raman spectra of alkali-metal salts of chromium, manganese, and cobalt carbonyls in THF and DMSO solvents, and have assigned bands to the vibrations of solvated cations in these systems. The ions are found to be "free" in DMSO solutions, while more than one species is present in THF. Popov and co-workers (163) have recently studied a large number of alkali-metal salts in dialkyl sulfoxide solvents. These workers have identified bands in the 500–100 cm^{-1} region characteristic of cations in the solvent cage. We feel such studies of the vibrational spectra of ionic species in solution, particularly in non-aqueous solvents, would be of value in understanding ion-pair and ion–solvent interactions.

IV. Hydrogen-Bonded Systems

A hydrogen bond, X—H $\cdots$ Y, is generally formed between an X—H bond, where X is a considerably electronegative atom and Y has a lone pair of electrons. Hydrogen-bond energies vary anywhere between 1 and 10 kcal/mole. Although the electrostatic model was earlier considered to be quite satisfactory in explaining the properties of the hydrogen bond, quantum-mechanical calculations have shown the importance of other forces, such as the delocalization (charge-transfer) forces, dispersion forces, and exchange repulsion forces. Detailed molecular orbital calculations on entire molecular systems containing hydrogen bonds have been carried out in recent years and the theory of the hydrogen bond has been reviewed by Bratoz (*38*) and Murthy and Rao (*183*). The potential function for the hydrogen bond has been examined by several workers (*183, 202*). The most widely used potential is that proposed by Lippincott and Schroeder (*141*); this potential is very similar to the Morse function for diatomic molecules. There have been improvements of this model to include bent hydrogen bonds as well. Molecular orbital methods have also been employed to derive potential functions for protons in hydrogen-bonded systems

Spectroscopic methods provide a ready means for the identification and characterization of hydrogen bonds (*184, 202*). Infrared and Raman spectroscopy are particularly valuable for the study of the hydrogen bond (*111, 214*), since hydrogen bonding not only alters the vibrational modes of molecules but also gives rise to new vibrational modes. Some of the important vibrational modes associated with the hydrogen bond are the following:

ν_{X-H} or ν_s	$\begin{array}{c}\text{X} \leftarrow \text{H} \rightarrow \cdots \text{Y}\\ \diagup\\ \text{R}\end{array}$	X—H stretch (3500–2500 cm^{-1})
ν_b	$\begin{array}{c}\text{X—H} \cdots \text{Y}\\ \diagup\\ \text{R}\end{array}$	R—X—H in-plane bend (1700–1000 cm^{-1})
ν_t	$\begin{array}{c}\text{X—H}\odot \cdots \text{Y}\\ \diagup\\ \text{R}\end{array}$	R—X—H torsion† (out of plane) (900–300 cm^{-1})
$\nu_{X\cdots Y}$ or ν_σ	$\begin{array}{c}\text{X—H} \cdots \leftarrow \text{Y} \rightarrow\\ \diagup\\ \text{R}\end{array}$	X $\cdots$ Y stretch (250–50 cm^{-1})
ν_β	$\begin{array}{c}\text{X—H} \cdots \text{Y}\\ \diagup\\ \text{R}\end{array}$	X—H $\cdots$ Y bend (<50 cm^{-1})

†The symbol $\odot$ indicates a vibrational movement of the hydrogen atom perpendicular to the RXY plane.

In addition to providing spectroscopic data on hydrogen-bonded systems, infrared and Raman spectra also provide a means of determining equilibrium constants of hydrogen-bond association. By measuring the intensity of ν_{X-H}

as a function of the concentration of X—H or of an added donor, one can evaluate equilibrium constants for 1 : 1 complex formation: equilibrium constants determined at different temperatures readily provide the enthalpy ($\Delta H°$) of formation, which is a measure of hydrogen-bond strength. The frequency shift Δv_{X-H} of the X—H stretching vibration band on hydrogen bonding has considerable significance; Δv_{X-H} has been correlated with various physical and chemical properties. Thus Δv_{X-H} is found to be proportional to the hydrogen-bond distance $R_{X \dots Y}$, the enthalpy ($\Delta H°$), and the free energy (ΔF^0) of formation of hydrogen bonds, as well as the half-bandwidth and intensity of the X—H stretching band. Of these, the relation between Δv_{X-H} and $\Delta H°$ or $R_{X \dots Y}$ is particularly informative. Such correlations are generally applicable to hydrogen bonds of intermediate energies (3–6 kcal/mole), but not to very weak or very strong hydrogen bonds. Very short (or strong) hydrogen bonds give rise to large shifts of v_{X-H}; in some cases the band due to v_{X-H} is not seen in the normal range found in most systems. Bent hydrogen bonds, as in many intramolecular bonded systems, do not obey the Δv_{X-H}–$\Delta H°$ or Δv_{X-H}–$R_{X \dots Y}$ relations. Bands due to the hydrogen-bond stretching vibrations, $v_{X \dots Y}$, normally found in the far-infrared region, are valuable for the characterization of hydrogen bonds, and these bands have been assigned in many systems in the last few years.

NMR studies provide information on (1) chemical shifts of the hydrogen-bonded proton, (2) hydrogen-bond formation equilibria and exchange times, (3) changes in relaxation times, and (4) position of hydrogen atoms in hydrogen-bonded crystals. High-resolution NMR spectroscopic studies of hydrogen bonding have been discussed in the texts by Pople et al. (206) and Emsley et al. (83). Recently Lippert (150) and Laszlo (145) have briefly reviewed NMR studies of hydrogen bonding. Wide-line NMR spectroscopy on hydrogen-bonded solids has provided a powerful tool to establish the structure of these solids.

Electronic spectroscopy has been employed in identifying hydrogen bonds in many organic systems (127, 211). Generally the effects of hydrogen bonding are seen in terms of changes in wavelength and intensity of $n \rightarrow \pi^*$ and $\pi \rightarrow \pi^*$ transitions. Electronic spectroscopy is of limited value in the study of hydrogen bonding in inorganic compounds.

In this section we shall discuss the spectroscopic studies of hydrogen bonds in inorganic systems such as water, water–donor complexes, acids, acid–donor complexes, etc.

A. Water

Water liquid is denser than ice and is supposed to have some crystalline order. The two important models proposed for the structure of water are as follows: (1) Water is essentially homogeneous and is quasicrystalline. Free molecules of water, if any, behave as if present in the matrix of a solid. (2) Water is a mixture of different molecular complexes involving different degrees of hydrogen bonding. A special case of this model is the one where only significant structures are considered. Agreement with thermodynamic properties is found to be good if the proportion of free molecules is anywhere between 25 and 40%. In many ways the first model involving the continuum seems to be more reasonable; however, experimental data, particularly those based on infrared and Raman studies have been cited in support of both models. The structure and properties of water have been reviewed by Eisenberg and Kauzmann (*81a*).

The vibrational spectrum of ice is considerably simpler than that of liquid water because of the presence of long-range order in the former. We shall therefore first examine some of the features in the spectra of ice. The infrared spectra of ice of different forms have been investigated extensively, and the complex nature of ice has made it difficult to assign the observed bands to the various normal modes. The problem is particularly severe because there is splitting and broadening of bands by intermolecular coupling. The infrared spectrum of ice in the range 10,000–30 cm^{-1} is shown in Fig. 7. The four distinct regions in the spectrum are (1) the intramolecular O—H stretching vibrations around 3300 cm^{-1}, (2) the intramolecular bending vibrations around 1600 cm^{-1}, (3) the intermolecular rotational vibrations around 840cm^{-1}, and (4) the intermolecular translational vibrations at 230 and 165 cm^{-1}. The bands are broad and nearly overlap with each other. Orientational disorder in ice plays a significant role in deciding the normal vibrations and their infrared activity. Since orientationally disordered crystals (e.g., ice I) have no unit cell, no selection rules will be valid and all vibrations will be spectroscopically active. Various vibrations will, how-

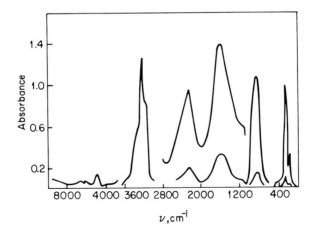

FIG. 7. Infrared spectrum of ice in the range 10,000–30 cm^{-1}. [From Whalley (*263a*).]

ever, have different intensities, which can be understood only by a detailed analysis. In this regard the low-frequency translational band is best understood, and the nature of this band has been discussed at length by Bertie and Whalley (*17*) in terms of lattice modes and acoustic waves. Vitreous ice shows the translational band around 230 cm^{-1}; the band has no fine structure as in ice and is considerably broader. In liquid water this band undergoes further broadening due to further decrease in the short-range order (*17*).

The O—H stretching band is strongly affected by the static field of the crystal as well as the intermolecular coupling of the motion of neighboring molecules. The static field can be estimated in ice and water by recording the spectra of dilute solutions of HDO (*17, 109, 122*). The spectrum of HDO under these conditions will represent the spectrum of a hydrogen-bonded molecule of ice; the uncoupled stretching vibrations of OD and OH occur at 2416 and 3275 cm^{-1}, characteristic of hydrogen bonds of intermediate strength. The O—D stretching vibration band of HDO is a well-resolved narrow band, in contrast to O—H stretching bands of many hydrogen-bonded systems, which are generally broad. The width of the O—H stretching bands in ice is explained to be due to the presence of a variety of O$\cdots$O distances in the crystal. Bertie and Whalley (*17*) estimate that a half-width of a few hundredths of an angstrom in the distribution of O $\cdots$ O distances is sufficient to account for the observed width of the band. Hornig and co-workers (*122*) find that the anharmonicity of the potential function for the O—H stretch increases from -80 cm^{-1} in vapor phase to -120 cm^{-1} in crystal; this small increase indicates that the shape of the potential function remains essentially unchanged in ice.

The coupled O—H or O—D vibrations give rise to a series of closely spaced bands; this is responsible for the large widths of the stretching bands. A detailed discussion of the coupled and uncoupled vibrations in ice has been presented by Whalley (*263a*). The overtones and combination bands of ice are more difficult to understand, since the fundamental bands themselves are so broad. The only useful overtones would be those of the uncoupled O—H and O—D stretching vibrations, but these are not clearly seen. The librational frequency of ice in the neighborhood of 800 cm^{-1} (and 600 cm^{-1} for D_2O) is comparable to the values found in crystalline hydrates. This band does not provide much information of value on the structure of ice.

Dilute solutions of HDO in ice II have been found to show sharp O—H and O—D stretching bands with bandwidths of only 5 cm^{-1} (*17*). These results are in agreement with x-ray studies, which show the presence of distorted oxygen arrangement (where there are no equally favored sites for hydrogen present) in an ordered structure of this high-pressure form. Ices III, V, and VI are known to be tetrahedrally coordinated structures; infrared and Raman spectra of these ices have been reported (*17, 43, 160*). The frequency shifts found in the infrared spectra indicate less favorable hydrogen bonding in the high-pressure polymorphs, but there is little difference among these phases. The fundamental stretching modes in these ices have been assigned the following frequencies: ice I, 3085 cm^{-1}; ices II, III, V, and VI, 3159–3204 cm^{-1}; ice VII, 3348 cm^{-1}.

The infrared spectra of different forms of ice have been examined. Lippincott and co-workers (152) have recorded infrared spectra of water and ice at different temperatures and pressures and have shown that at high pressures the hydrogen bonds responsible for the open structure of ice collapse. In the close-packed structure of dense ice, hydrogen bonding produces a minor effect on the vibrational modes. It has been concluded that weak hydrogen bonds still exist in the dense ice, but with an O—H $\cdots$ O angle considerably different from 180°. Water and heavy water show a band around 170 cm^{-1} in the far-infrared region; the relation $[v_{D_2O}/v_{H_2O}]^2 = [M_{H_2O}/M_{D_2O}] = 0.90$ has been verified for several bands. Low-frequency bands in the spectra have been interpreted as due to vibrations of hydrogen bonds (235).

Wall and Hornig (257) investigated the Raman spectrum of HDO in liquid water and found that the vibrations of HDO molecule were uncoupled because of dilution, but found no distinct O—H and O—D stretching bands at different temperatures. Wall and Hornig therefore concluded that this observation contradicted the existence of two or more distinct types of water molecules in liquid water. Instead, they found a continuous distribution of frequencies with maxima at 2516 cm^{-1} for O—D and 3439 cm^{-1} for O—H stretching vibrations. The corresponding values in ice are 2421 and 3277 cm^{-1}, and in vapor these values are 2727 and 3704 cm^{-1}. Employing the bond-length–frequency relationship, a distribution of O $\cdots$ O bond lengths can be obtained from frequency distribution. Such an analysis gives about 2.85 Å as the most probable value for the O $\cdots$ O bond length; the actual distances vary anywhere between 2.75 and 3.1 Å.

The infrared and Raman spectra of liquid water have been investigated by several workers. A typical spectrum in the O—H stretching region is shown in Fig. 8. Three characteristic frequencies at 3440, 3190, and 3650 cm^{-1} are generally quoted in this region for H_2O; the corresponding bands for D_2O are 2515, 2360, and 2680 cm^{-1}. The spectra in the liquid and solid state are quite similar, but the liquid shows considerably broader bands. The normal vibrations in liquid water are undoubtedly more localized in H_2O than in ice because of greater disorder, but they are still highly coupled. We therefore know less about the stretching vibrations of water compared to those of ice. Comparison with vapor frequencies can be erroneous because intermolecular coupling can shift bands to frequencies higher than those of the uncoupled vibrations. Liquid water shows a number of bands in the overtone region (~ 6500 cm^{-1}), but the origin of these bands is not clearly understood. Many workers interpret them as the overtones of the O—H stretching vibrations of differently hydrogen-bonded species of water. It is difficult to understand

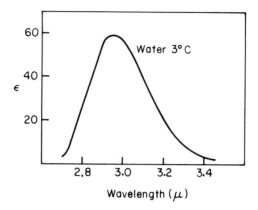

FIG. 8. O—H stretching band of liquid H_2O. [From Whalley (*263a*).]

why such discrete bands should occur in the overtone region and not in the fundamental region. Thus, in spite of numerous investigations, the nature of water and its vibrational spectrum still remain little understood. Recent laser Raman studies of Walrafen (*259, 260*) seem to offer some hope in understanding water structure.

Walrafen (*259, 260*) has provided valuable evidence for the existence of various hydrogen-bonded species in water by carefully analyzing the Raman band contours; the results seem to disprove the continuum model. Hydrogen-bond bending and stretching modes have been assigned by Walrafen to bands at 60 and 165 cm^{-1}. In addition, Walrafen finds three bands at 450, 550, and 720 cm^{-1} due to the librational modes. All these bands show variations in intensity with temperature consistent with the presence of hydrogen-bonded species and free water molecules. Walrafen has treated his Raman intensity data in terms of an equilibrium between free water and a (tetrahedral unit) pentamer and has obtained a value of 2.5 kcal/mole for the hydrogen-bond energy. Stepwise hydrogen-bond breakage has been proposed as the mechanism for the variation of Raman intensity with temperature. Walrafen has also been able to observe a pronounced shoulder at 2645 cm^{-1} on the O—D stretching band of D_2O (solution in H_2O) centered at 2525 cm^{-1}. These two bands also vary in intensity with temperature in a manner consistent with the presence of the equilibrium mentioned earlier.

As pointed out by Schiffer and Hornig (*222a*), the uncoupled OH or OD bands of Walrafen are also much too broad and can only result from a continuum. It appears that a model combining the features of both the mixture and continuum (*225a*) models would be necessary to describe the properties of water, in particular, the presence of a variety of molecular environments as shown by the large width of the uncoupled OH and OD bands.

Pimentel and co-workers (254) studied the infrared spectra of liquid water at low temperatures by the matrix-isolation technique and assigned bands to the O—H stretching vibrations (3600–3000 cm^{-1}) of the monomeric, dimeric, and polymeric species. Buijs and Choppin (46, 47) investigated the spectrum of water in the 1.16–1.25-μ region and suggested that the bands in the overtone region were due to the presence of three types of water molecules, those having zero (1.16 μ; ϵ, 7.56 × 10^3), one (1.20 μ; ϵ, 7.58 × 10^3), and two (1.25 μ; ϵ, 4.17 × 10^3) hydrogen bonds, respectively, to neighboring molecules. By assigning extinction coefficients to each of the species on the basis of the spectra of H$_2$O liquid at different temperatures, as well as of ice and aqueous salt solutions, the fraction of each type of water molecule has been calculated at several temperatures. Like Buijs and Choppin (46), Goldstein and Penner (102) have assigned values to the extinction coefficient of the doubly bonded species using absorbance data of ice and have evaluated the extinction coefficients for the other species from various experimental data. The fractions of hydrogen bonds unbroken in liquid water at various temperatures obtained in these investigations, were in fair agreement with those calculated by Nemethy and Scheraga (187).

Luck (155) has investigated several of the near-infrared bands of water, but his calculations on the relative proportions of different species are not very extensive. Luck (156) suggests that there is a considerable fraction of hydrogen-bonded species in water up to the vicinity of the critical point, basing his conclusions on experiments with water under high pressure. The assignment of the bands by Buijs and Choppin (46) has been challenged by Hornig (121). These criticisms have been answered by Buijs and Choppin (47), but more definitive assignments need further investigations.

Scheraga and co-workers (248) have examined the spectra of liquid H$_2$O, liquid D$_2$O, H$_2$O ice, and D$_2$O ice in the near-infrared region at different temperatures and have calculated the concentrations of unbonded (1.16 μ; ϵ, 19.06 × 10^3), singly hydrogen bonded (1.20 μ; ϵ, 14.03 × 10^3), and doubly hydrogen bonded (1.25 μ; ϵ, 8.97 × 10^3) water molecules in liquid H$_2$O and liquid D$_2$O. The results on H$_2$O and D$_2$O seem to be consistent with their earlier statistical thermodynamic calculations (187). Luck (157) has recently studied the temperature dependence of four overtone bands of water and has interpreted the results in terms of a simple cluster model. Worley and Klotz (265) have studied the near-infrared spectra of H$_2$O–D$_2$O solutions at different temperatures and have estimated ΔH values of -2.4 kcal/mole for the O—H $\cdots$ O bond; they have also examined the structure-breaking effects of various salts and macromolecules.

The proton chemical shift in water varies with temperature, which is undoubtedly related to the changes in the hydrogen-bonded structure. While the chemical shift is generally taken as a direct measure of the number of hydrogen bonds in water, it is also possible that other structural changes (stretching or bending) involving the hydrogen bonds may be responsible for the changes in the shielding parameters. Muller (*170a*) has calculated the shielding constants of monomeric water molecules and hydrogen-bonded clusters by using the equilibrium values for the fraction of hydrogen-bonded species (in a given model) at two temperatures and then comparing the value for the nonbonded water molecule with that for water in an organic solvent. Rüterjans and Scheraga (*220a*) have calculated the chemical shifts for different models by assuming that the temperature coefficient of the chemical shift is a constant determined by the mole fraction of the hydrogen-bonded species. Hindman (*116a*) has used variations of these procedures as well as other models in interpreting proton chemical-shift data of water at different temperatures. His studies show that there are discrepancies among the results from different models. Hindman (*116a*) also considers the possibility of changes other than breaking of hydrogen bonds, such as bending or stretching of hydrogen bonds. In spite of all these NMR studies on water, it appears that the variation of proton chemical shifts is far from being understood. Since the chemical shift is an average property (of all the species present), the problem essentially boils down to interpreting the average chemical shift in terms of the different species present in the equilibrium. Any such interpretation will have severe limitations inherent in the nature of the problem itself.

Recently the infrared and Raman spectra of a new form of water prepared in fused-quartz capillaries has been examined by Lippincott, Bellamy and co-workers (*14*, *153*). The infrared spectrum of this new water, now named "polywater," appears to be unique with the apparent absence of the O—H stretching bands found in normal water. However, new bands are found around $1600-1400$ cm^{-1}. These observations have been interpreted in terms of a strong symmetric O—H $\cdots$ O bonds (aking to HF_2^-). In polywater, oxygen sp^2 orbitals, rather than sp^3 orbitals, seem to be involved in hydrogen bonding (*101a*).

1. HYDRATES

Hydrogen bonding in some of the crystalline hydrates has been examined by neutron diffraction and by NMR spectroscopy. The NMR method makes use of the well-known Pake method (*192*) to determine the interproton distances in single crystals. The best-known example of the utility of this method is the analysis of the structure of the hydrate gypsum. El Saffar (*82*) has reviewed the results obtained on hydrates by NMR spectroscopy.

Data on the infrared spectra of various crystalline hydrates have been reported in the literature, but there has not been much valuable information from this study. Hydrated inorganic salts generally show a broad maximum in the range 3600–3100 cm^{-1} due to the O—H stretching and another band in the range 1670–1600 cm^{-1} due to the bending mode. Wherever water is strongly coordinated to a metal ion, an additional band due to bending motion has been found in the 1000–800 cm^{-1} region. Hydration markedly affects the spectra of many of the oxyanions, such as SO_4^{2-}, PO_4^{3-}, and NO_3^{-}, since the symmetry of these oxyanions would be different in hydrated crystals compared to anhydrous salts.

2. AQUEOUS SOLUTIONS

Fritzsche (*97*) measured the frequency shifts occurring with the infrared bands of dioxane when water is added and concluded that dioxane–water interaction was the cause of the shifts. Recent studies (*95*) on dioxane–water and pyridine–water mixtures have shown that the high-frequency shifts of the O—H stretching band in either solvent are due to a reduction of hydrogen bonding present in pure water. The low frequency shift of the C—O stretching mode of dioxane in aqueous mixtures indicates that oxygen is the hydrogen-bonding site. High-frequency displacements observed for the C—H stretching modes in both dioxane and pyridine in their aqueous mixtures cannot be interpreted unambiguously. Raman studies on dimethylsulfoxide–water mixtures (*148*) have shown that the intensities of Raman bands decrease continuously with increasing water content of the mixture. The S=O stretching frequency of dimethylsulfoxide is shifted to lower frequencies with increasing water content. The association of tributyl phosphine oxide with water has been examined recently (*218*).

Evidence for hydrogen bonding between nitrate ion and water was found by Marcus and Fresco (*161*) by an examination of the infrared spectra of sodium nitrate in aqueous and nitric acid solutions; the infrared-inactive vibration at 1050 cm^{-1} becomes very intense in solution. The shape of the O—H stretching bond in the Raman spectra (*263*) of water is markedly affected by the presence of electrolytes such as KF, KCl, KBr, KI, and LiBr. All the salts except KF increase the stretching frequencies but do not affect the bending frequency. The influence of different anions is in the order F^- < Cl^- < Br^- < I^-. Part of the change in the spectra on the addition of electrolytes has been attributed to the combined effect of the electrostatic interaction between the ions and solvent dipoles and a decrease in hydrogen bonding of the solvent. Raman spectra of solutions of several metal halides by Wall and Hornig (*258*) have shown small anion effects, but no cation effects; these workers find no indication for ion–water association leading to large changes in frequencies or bandwidths of the water stretching bands. The effects of temperature and electrolytes (lithium, sodium, potassium, and ammonium salts) on the Raman spectra of water has been studied by Walrafen (*259*).

All electrolytes produce intensity decrease of the hydrogen-bond stretching vibration (~ 175 cm^{-1}); no marked cationic effects are observed. Hydrogen-bond bending (~ 60 cm^{-1}) and stretching bands decrease in intensity with increase in temperature.

Barrow and co-workers (*169*) have examined the spectra of water with bases and anions in inert solvents such as carbon tetrachloride. They find appreciable shifts of O—H stretching frequency of monomeric water on the addition of anions, the shifts varying in the order I$^-$ < Br$^-$ < Cl$^-$ $\simeq$ F$^-$. This is in contrast to the order found by Schleyer, West and co-workers (*223, 261*) for covalent organic halides where the $\Delta\nu_{OH}$ varies in the order I > Br > Cl > F; the enthalpy values for the interaction, however, vary in the order F > Cl > Br > I, which is contrary to the Badger–Bauer rule. Dwivedi and Rao (*79*) have recently examined the interaction of halide ions with monomeric water and found them to vary in the order Cl$^-$ > Br$^-$ > I$^-$. The results of Barrow and co-workers (*169*) and Dwivedi and Rao (*79*) in CCl$_4$ solution are shown in Table XIV. It may be noted that the $\Delta\nu_{O-H}$ for the interaction of methanol with quarternary ammonium halides varies in the order Cl$^-$ > F$^-$ > Br$^-$ > I$^-$ (*4*).

Holmes and co-workers (*119*) have investigated the behavior of water protons in a series of organic solvents by NMR and infrared spectroscopy. The chemical shifts of the water proton and the infrared frequency shifts in various solvents have been correlated with each other and with hydrogen-bond energies. A good linear relationship has also been obtained between $\Delta\nu_{O-H}$ and ΔH°. The thermodynamics of hydrogen bonding between water and the donors dimethylacetamide, acetone, and tetrahydrofuran in cyclohexane has been studied by Takahashi and Li (*243*). Muller and Simon (*170*) have studied the interaction of water protons with dioxane in carbon tetrachloride by employing NMR spectroscopy and evaluated the equilibrium constants and enthalpies for the 1 : 1 and 1 : 2 complexes.

TABLE XIV

HYDROGEN BONDING WITH HALIDE IONS

Donor	$\Delta\nu_{O-H}$ (cm^{-1})
Bu$_4$N$^+$F$^-$	372
Bu$_4$N$^+$Cl$^-$	370
Bu$_4$N$^+$Br$^-$	300
(n-C$_7$H$_{15}$)$_4$N$^+$Cl$^-$	355
(n-C$_7$H$_{15}$)$_4$N$^+$Br$^-$	290
(n-C$_7$H$_{15}$)$_4$N$^+$I$^-$	240

The positions of the absorption bands in the electronic spectra of many organic compounds, particularly those containing atoms with lone-pair electrons, are markedly shifted in aqueous solutions compared to those in nonpolar solvents. A major portion of the solvent effects is undoubtedly due to hydrogen bonding. The solvent blue shifts of $n \rightarrow \pi^*$ transitions of organic molecules are much larger in aqueous solutions than in other polar and hydroxylic solvents (211). The separation of the hydrogen-bonding contribution from other effects on the electronic absorption spectra is, however, difficult.

B. Acids and Related Compounds

Hydrogen bonding in some of the simple hydrogen halides has been studied. The hydrogen-bond energies in polymers of HF in the gas phase have been determined by infrared spectroscopy. The energy per bond is about 6.5 kcal/mole (202). The hydrogen-bond energy in dimers of HCl has been found to be 2.14 kcal/mole by gas-phase infrared spectroscopy (210). Hydrogen bonding in HCN causes a lowering of C—H stretching frequency from 3312 cm^{-1} in the vapor phase to 3132 cm^{-1} in the crystal. The bending frequency increases from 712 to 830 cm^{-1} (117). The spectrum of HCN dimer in vapor phase has been examined by Sheppard and co-workers (129) and the hydrogen-bond energy found to be 5.7 kcal/mole with an N $\cdots$ C distance of 3.1 Å. Hydrogen bonding in hydrazoic acid has been studied by Pimentel and co-workers (74, 203) by infrared spectroscopy; the hydrogen-bond energy is about 5 kcal/mole. The association of thiocyanic acid and its deutero derivatives have been reported (13); association seems to be negligible up to $10^{-2} M$.

Hydrogen sulfide shows a band at 2611 cm^{-1} due to the S—H stretching vibration in the gas phase; the band is found at 2574 cm^{-1} in the liquid state. Three bands at 2546, 2554, and 2521 cm^{-1} are seen in the solid state. These results indicate specific hydrogen-bonding interaction in the condensed state. Accordingly, hydrogen sulfide has different crystal structures, a cubic phase at high temperatures and two low-temperature tetragonal phases.

Ammonia is known to associate in the gas phase (203), forming relatively strong N—H $\cdots$ N bonds (4 kcal/mole). The infrared spectrum of ammonia suspended in solid nitrogen has been examined by Pimentel and co-workers (203, 204). The hydrogen bonding of hydrazine has recently been studied by Durig and co-workers (78). The hydrogen bonding in hydroxylamine has also been examined by infrared spectroscopy (188).

Infrared spectra of the extracts of nitric acid from aqueous solution in aprotic solvents shows the presence of dimeric form as the predominant species (114). The structure of the dimer has been suggested to be as shown:

$$\begin{array}{ccc} O & & H \cdots O \\ \diagdown & & \diagup & \diagdown \\ N-OH\cdots O & & N-OH \\ \diagup & \diagdown & \diagup \\ O & H\cdots O \end{array}$$

In concentrated solutions, larger aggregates seem to be present. Hydrogen bonding in polycrystalline boric acid (125) has been studied by NMR spectroscopy, and the second moment derived from these measurements seems to be in agreement with the structure proposed by Zachariasen (271).

The hydronium ion, H_3O^+, has been investigated by a number of methods, including x-ray crystallography, and NMR and infrared spectroscopy. On the basis of NMR studies of acid hydrates, Richards and Smith (217) concluded that the hydrate contains H_3O^+ ion. These workers proposed the structure to be that of a flat pyramid. However, further infrared studies by Bethell and Sheppard (19), Ferriso and Hornig (91), and Taylor and Vidale (245) made possible the assignment of the fundamental modes of H_3O^+, and there is little doubt that the ion has a flat pyramidal structure. Busing and Hornig (50) have studied the effect of dissolved KBr, KOH, and KCl on the Raman spectrum of water and identified bands due to H_3O^+, free, and bonded OH^- peaks.

Hydrogen bonding in chromous acid (220, 233), $HCrO_2$, shows some unexpected features in the infrared spectrum. The comparison of the spectrum of $HCrO_2$ with $DCrO_2$ revealed that the differences are not mainly because of mass change. The O—H stretching band of $HCrO_2$ is at 1640 cm^{-1}, while the O—D stretching is not found in the expected 1200–1300 cm^{-1} region. Instead, strong bands are found around 1750 cm^{-1} in $DCrO_2$. While the O—H stretching band is a singlet, the O—D band is a doublet. There are two lattice modes in the 300–700 cm^{-1} region in $HCrO_2$, whereas there are six bands in the same region for $DCrO_2$ (220). These results could only be interpreted in terms of two different potentials for $HCrO_2$ and $DCrO_2$; apparently there is a high barrier in $DCrO_2$, whereas in $HCrO_2$ the barrier is negligible. On the basis of two different potential functions, Snyder and Ibers (233) have been able to explain all the observed features. The infrared spectra of $HCoO_2$ and $DCoO_2$ are quite analogous to those of corresponding chromous acids (68).

Sheppard and co-workers (39) have recognized a well-defined trio of broad bands in the regions 2800–2400 (A), 2350–1900 (B), and 1720–1600 cm^{-1} (C) in systems containing the structural grouping X(=O)OH, where X = P, As, S, Se, and less commonly C. These bands are associated with strongly hydrogen-bonded C—H groups. Bands (A) and (B) correspond to the O—H stretching and (C) to bending vibrations. Hydrogen bonding studies on trifluoromethyl phosphinic acid (48) have indicated that the monomer is mostly CF_3HPOOH rather than $CF_3P(OH)_2$. The presence of bands corresponding to P—H and P=O stretching vibrations as well as the O—H $\cdots$ O vibrations are strongly in favor of the phosphinic acid dimer. Infrared spectra of solid and liquid phosphorous acids show that compounds containing the POOH grouping associate (*246*). Alkyl phosphonic and alkyl phosphothionic acids also undergo self-association (*247*). Raman and infrared investigation of selenic and methylselenic acids show that methylselenic acid is strongly associated (*191*).

Ferraro and Peppard (*90*) have studied hydrogen bonding in several organophosphorus acids by NMR spectroscopy and have estimated the energy to be around 6.0 kcal/mole, which indicates that the bond is a little stronger than that in acetic acid. An examination of the infrared spectra of organic acids containing P(=O)OH or P(=O)(OH)$_2$ grouping show that these acids have a strong tendency to form aggregates containing O—H $\cdots$ O bonds (*15*). Wolf and co-workers (*264*) have found that the P—H stretching band is shifted to lower frequencies (~ 17 cm^{-1}) because of association in compounds of the general formula

$$
\begin{array}{ccc}
\text{O} & & \text{S} \\
\| & & \| \\
(X)(Y)P—H & \text{or} & (X)(Y)P—H
\end{array}
$$

and these authors suggest that the PH proton can form weak hydrogen bonds. Trithiocarbonic acid and related thioacids show hydrogen bonding in the liquid phase and in concentrated solutions (*250*). Evidence for hydrogen bonding has been found in phosphorotetrathioic acids, CH$_3$SP(=S)(SR)SH, with $\Delta\nu_{SH}$ of 140 cm^{-1} (*3*).

1. INTERACTION OF ACIDS WITH DONORS

There have been several studies in recent years on the interaction of hydrogen halides with various donors. Thus Millen and co-workers (*9, 18*) have studied the interaction of HCl and HF with ether and have found evidence for hydrogen bonding in terms of broadening of the structure of the X—H vibration band. These workers have also studied the interaction of HF with a few ketones, methanol, methyl sulfide, acetonitrile, and alkyl halides (*10*). In all these complexes the hydrogen-bond force constant is found to be of the order 0.1 mdyn/Å. Arefev and Malyshev (*7*) find that the hydrogen-bond energies of hydrogen halides with acetone, dioxane, and ethyl ether vary in the order HCl > HBr > HI. The interaction of HCl with dimethyl ether has been examined by Bernstein and co-workers (*105*) by employing NMR spectroscopy, who find the hydrogen-bond energy to be 7.1 kcal/mole. A deshielding of the HCl proton by 3 ppm has been noticed. Vapor-phase association of HCl with acetone has been studied by infrared spectroscopy (*60*), and the equilibrium constant for the formation of the 1 : 1 complex is estimated to be 4.01×10^{-4} mm^{-1}; the C=O stretching frequency of acetone decreases from 1740 to 1725 cm^{-1} because of hydrogen bonding.

Interaction of HCN with ammonia has been studied by vapor-phase infrared spectroscopy by Sheppard and co-workers (129). The band contour of the C—H stretching vibration has the shape expected of a symmetric top molecule. The C $\cdots$ N distance has been estimated to have a lower limit of 3.0 Å. The interaction of nitric acid with ether has been investigated (52, 168). The structure of the O—H band in this complex shows evidence for hydrogen bonding. A band at 175 cm^{-1} has been attributed to the O $\cdots$ O stretching vibration.

The infrared spectra of adducts of phosphine-, arsine-, and amine- oxides and sulfoxides with HNO$_3$, HCl, and HBr as well as their deuterated derivatives have been examined by Hadzi (110); the hydrogen-bond frequencies are found in the region 180–240 cm^{-1}. In amineoxide hydrohalides there seems to be proton transfer to the basic oxygen, a phenomenon not found in other systems studied by Hadzi. From a study of the infrared spectra of hydrohalides of pyridine in various solvents it was concluded that they exist as hydrogen-bonded ion pairs (140).

C. Inorganic Salts

The tetrahedral ammonium ion generally exhibits a strong band in the N—H stretching region (3340–3030 cm^{-1}) and another band in the region 1480–1390 cm^{-1}. In many ammonium salts a number of bands appear in the N—H stretching region, depending on the extent of hydrogen bonding of the NH$_4^+$ ion. In ammonium fluoride the N—H symmetric stretching frequency is at 2870 cm^{-1}, about 450 cm^{-1} lower than in crystals where there is only weak hydrogen bonding (205). Ammonium chloride, ammonium bromide, and ammonium iodide possess similar structures and undergo phase transformations. The infrared spectra of these halides have been studied by Hornig and co-workers (205, 256) to provide evidence for hydrogen bonding between NH$_4^+$ and halide ions.

Sandorfy and co-workers (42, 59, 221) and Waddington and co-workers (189) have investigated hydrogen bonding in various amine hydrohalides and ammonium salts containing different anions. In the solid state, the hydrogen bonds are of the type N$^+$—H $\cdots$ X$^-$. The broad bands around 3000 cm^{-1} in primary amine salts are shifted to lower frequencies in secondary and tertiary derivatives. The shift to high frequencies is in the order HCl, HBr, HI. The absence of band shifting in aqueous solutions has been interpreted as due to the formation of N$^+$—H—O type bonds. The NH frequencies in aromatic amine hydrochlorides are lower than in aliphatic amine hydrochlorides. Fluoroborates and tetraphenylborates are not hydrogen bonded (*189*).

Weak hydrogen bonding is known to be present in alkali-metal hydroxides. Thus the O—H stretching frequency in KOH is at 3600 cm^{-1}, compared to 3650 cm^{-1} in Ca(OH)$_2$. Evidence for hydrogen bonding has been found in crystalline lithium hydroxide (*236*).

Bifluoride ion (*63, 135*), which possesses a strong, short, and symmetric hydrogen bond, exhibits a very low asymmetric stretching frequency at 1460 cm^{-1} and a bending frequency at 1225 cm^{-1}. On the basis of infrared spectrum of tetramethylammonium bichloride, the bichloride ion has been shown to have a linear symmetrical hydrogen bond (*255*). Evans and Lo (*85*) have found two types of bichloride salts: type I, such as cesium, tetramethylammonium, tetrabutylammonium, and hexadecyltrimethylammonium bichlorides showing strong bands in the region 1520–1670 and 1200 cm^{-1}; and type II, such as tetraethylammonium, tetrapropylammonium, and tetrapentylammonium bichlorides in the region 600–1300 cm^{-1}. Evans and Lo have concluded that type-II salts have the linear symmetrical bond, while type-I salts may deviate from linear structure. The comparative stabilities of HI$_2^-$, HBr$_2^-$, HCl$_2^-$, and HF$_2^-$ have been examined (*115*) by studying the reaction R$_4$NX(s) + HX(g) $\rightleftharpoons$ R$_4$NXHX(s); the respective hydrogen-bond energies are, -12.4, -12.8, -14.2 and -37 kcal/mole.

Blinc *et al.* (*34*) have discussed the nature of potential functions of hydrogen bonds as determined by infrared and NMR studies. They classify the double minimum potentials of hydrogen bonds into two categories, one with low barriers and the other with high barriers. The different types of hydrogen bonds have been distinguished from infrared frequencies as well as the temperature dependence of tunneling frequency determined by NMR studies. The tunneling frequency is temperature dependent if the potential is of a single minimum type or a double minimum type with low barrier. In particular, Hadzi and co-workers (*34*) note two ranges for the infrared O—H stretching frequency in systems with short hydrogen bonds. In one group, which is supposed to have the symmetric single minimum potential, the average frequency is at 1694 cm^{-1} (the O—D frequency being at 1367 cm^{-1}). Compounds with relatively short hydrogen bonds and low barriers show an O—H frequency in the range 2190–2510 cm^{-1}. When the O $\cdots$ O distances are fairly long (above 2.7 Å), the barrier heights become much larger and the frequency ranges generally found are 2257–2943 and 2350–3200 cm^{-1}. These frequency ranges and their relation to the O $\cdots$ O distance are grossly approximate and at best serve as rough guidelines.

Hadzi and co-workers (*34, 34a*) and also several other authors have carried out detailed studies of many systems containing short hydrogen bonds, such as the acid salts. Potassium hydrogen maleate shows an O—H stretching frequency below 1650 cm^{-1}; the hydrogen bond in maleate ion is short (2.3 Å) and symmetric. The infrared assignments of potassium bicarbonate have been made on the basis of normal vibration analysis (*186*). The bicarbonate ion forms dimers where the three different C—O distances have different force constants. The hydrogen bond itself is rather strong and the O ⋯ O distance is 2.61 Å. The hydrogen atom in such a symmetric hydrogen bond would essentially move as a harmonic oscillator, and the barrier in the double minimum potential, if any, would be very small. Such conclusions have found some support from neutron diffraction studies as well. Somorjai and Hornig (*234*) have indicated that the double minimum potentials may be asymmetric, and such asymmetry cna be examined by band intensities as well as frequencies of all levels up to and above the barrier.

Phase transitions in many ferroelectric materials are associated with the reorientation of hydrogen bonds. Detailed studies have been carried out on several hydrogen-bonded ferroelectric crystals, the most important of them being Rochelle salt, $NaK(C_4H_4O_6)\cdot 4H_2O$, colemanite, $CaB_3O_4(OH)_3\cdot H_2O$, ammonium sulfate, potassium ferrocyanide trihydrate, and potassium dihydrogen phosphate. NMR spectroscopy has been particularly useful to study the ferroelectric transitions in these crystals; the line widths and shapes are studied as a function of temperature through the transition to provide information on the hydrogen bonding and orientation of groups. Infrared and Raman spectroscopy are also useful in studying hindered motion, rotational modes, and hydrogen bonding in these crystals. Optical spectroscopy does not often provide useful information on the structural changes in these hydrogen-bonded crystals responsible for ferroelectricity. Thus Raman spectra indicate no major change in the strength of the hydrogen bonds in KH_2PO_4 at the ferroelectric transition.

D. *Hydrogen Bonding of Halides and Other Ions with Hydroxylic Compounds*

There is a considerable infrared evidence for hydrogen bonding between covalent organic halogen derivatives and hydroxylic compounds (*128, 223, 261*). The effect of addition of electrolytes such as metal halides and perchlorates on the infrared and Raman spectra of methanol has been examined by several workers, who find definite evidence for specific hydrogen-bonding interaction between the ions and the O—H bond of the alcohol. Allerhand and Schleyer (*4*) found that the O—H stretching frequency of methanol is appreciably shifted by quaternary ammonium halides, the Δv_{O-H} varying in the order Cl$^-$ > F$^-$ > Br$^-$ > I$^-$. Variations in the cation had little effect in the case of I$^-$ and Br$^-$. Studies on the effect of anions such as ClO$_4{}^-$ and I$^-$ on the O—H stretching band of methanol show that ClO$_4{}^-$ is ineffective, but

I^- causes the appearance of new bands due to the O—H $\cdots I^-$ hydrogen bond. Hydrogen bonding of anions can be related to their size, radii, and polarizability. Generally speaking, anions where the negative charge is highly localized would form a better hydrogen bond. The perchlorate ion, which is noted for its small polarizability, appears to be a poor hydrogen-bond acceptor. The frequency shifts as well as the band intensities of the bonded O—H peaks have been studied (*158*) in several tetraalkylammonium salts–hydroxylic compound systems and some of the data in CCl_4 solution are given in Table XV (*79, 158, 230*).

The effect of electrolytes on the vibrational spectra of hydrogen-bonded compounds such as methanol, *t*-butanol, and N-methylacetamide in benzene solution has been examined by Bufalini and Stern (*45*), who have interpreted the results in terms of hydrogen-bond formation by the anion of the electrolyte. The hydrogen-bond energy for the system methanol–tetrabutylammonium bromide is found to be 6.7 kcal/mole. Singh and Rao (*230*) and Dwivedi and Rao (*79*) have studied the thermodynamics of various alcohols with quaternary ammonium halides; the 1 : 1 equilibrium constants are quite large

TABLE XV

INTERACTION OF HALIDES WITH HYDROXYLIC COMPOUNDS

Donor	Acceptor	$\Delta\nu_{O-H}$ (cm^{-1})
Tetrabutylammonium bromide	*t*-Butanol	275
	i-Propanol	275
	Ethanol	285
	2,2,2-Trichloroethanol	330
	Phenol	405
	p-Cresol	402
	o-Cresol	400
	Mesitol	365
	2,4-Dichlorophenol	450
	2,4-Dibromophenol	420
	2,4,6-Trichlorophenol	535
	2,4,6-Tribromophenol	505
	Dibutylamine	130
	N-Methyl aniline	170
	Diphenylamine	210
	Indole	285
Tetrabutylammonium iodide	*p*-Cresol	340
Benzyltributylammonium chloride	*p*-Cresol	445
Tetra-*n*-heptylammonium chloride	Methanol	371
Tetra-*n*-heptylammonium bromide	Methanol	304
Tetra-*n*-heptylammonium iodide	Methanol	255
Tetrabutylammonium perchlorate	*p*-Cresol	180

(42–55 liters/mole at 25°C) and the enthalpies of formation are of the order 5–8 kcal/mole. Evidence for the formation of a hydrogen bond between the iodide and an alcohol is also found in the electronic absorption spectra (CTTS transitions) of the ternary system $I^- - CH_3OH - CCl_4$, where isobestic points have been observed (*33, 79, 230*).

V. Scales of Basicity and Acidity

It was mentioned earlier (in Section II) that the magnitude of charge transfer in the ground state of donor–acceptor complexes provides a quantitative measure of the donor basicities or acceptor acidities. Similarly we would expect the properties of hydrogen bonds, $X—H \cdots Y$, being primarily determined by the electronegativities of atoms X and Y, to directly reflect the acidity of the X—H bond or the basicity of donor site Y. Many workers have employed thermodynamic and spectroscopic data for the interaction of donors and acceptors to evaluate the acidities of acceptors and basicities of donors. Recently Pearson (*194*) has proposed empirical scales for the softness of acids and bases on the basis of equilibrium and kinetic data. Although such measures of acidities and basicities are likely to be qualitative and empirical, they may still be of value in the prediction of properties and behavior of inorganic chemical systems. We shall briefly examine the scales for donor basicities and acceptor acidities employed by various workers.

A. Basicities of Donors and Acidities of Acceptors from Hydrogen-Bonding Studies

Generally speaking, the change in the free energy or the enthalpy associated with hydrogen-bonding equilibria between donors and acceptors depends to some extent on the acidity of the acceptor and the basicity of the donor. Thus Singh et al. (*229*) found that the equilibrium constants and Δv_{O-H} increased with the acidities of alcohols in benzophenone–alcohol systems. Studies of the interaction of a hydroxylic compound with a series of donors show that in many systems $\Delta F°$ or $\Delta H°$ is proportional to Δv_{O-H}. For example, the following relation has been proposed between $\Delta H°$ and Δv_{O-H} for phenol–donor systems, when $\Delta H°$ is between 3 and 6 kcal/mole:

$$-\Delta H° \text{ (kcal/mole)} = 0.012\Delta v_{O-H} + 1.87 \qquad (25)$$

This relation has been verified by determining ΔH independently by calorimetric measurements (*84*). On the basis of proportionality of Δv_{O-H} to ΔH, a few workers have estimated donor basicities or acceptor acidities on the basis of $\Delta \nu_{O-H}$ values. Thus West and co-workers (262a) have estimated basicities of olefins and acetylenes on the basis of their interaction with phenol. Basicities of alkyl ethers, alkoxysilanes, and siloxanes have been determined from the extents of hydrogen bonding with various acceptors such as phenol and pyrrole (120, 262). The basicity of the oxygen atom was found to decrease in the order, COC, COSi, and SiOSi; here frequency shifts are taken as measures of the relative Lewis basicities. The donor properties of triphenyl derivatives of Vb elements as well as of pyridine derivatives have been discussed in terms of $\Delta \nu_{O-H}$ values with phenol (22).

Drago and co-workers (162) have measured the donor strengths of trialkyl phenoxy compounds of Group-IV elements in terms of $\Delta \nu_{O-H}$ with phenol as acceptor. Typical data obtained for the interaction of phenol with derivatives of a family of elements in CCl_4 solution are shown in Table XVI. It has been noted that the trends in $\Delta \nu_{O-H}$ values with triphenylamine, triphenylphosphine, and triphenylarsine are quite different from those found for the interaction of these donors with iodine. The equilibrium constant for the interaction of phenol with these donors, however, varies in the order triphenylphosphine > triphenylamine $\simeq$ triphenylarsine; this trend is somewhat similar to that found for the interaction with iodine, where triphenylamine interacts strongly to form the inner complex while the other two donors form relatively weak charge-transfer complexes. With various Lewis acids the order of basicity is found to be triphenylamine > triphenylphosphine > triphenylarsine > triphenylstibine.

The acidities of the compounds $(C_6H_5)_3MOH$ (M = C, Si, and Ge) have been examined by investigating the change in the O—H stretching frequency upon hydrogen bonding with the Lewis bases diethyl ether and tetramethylene sulfoxide in CCl_4 solution (Table XVII). On the basis of these results, it was concluded that triphenylsilanol was a stronger acid than the corresponding carbinol or the corresponding germanium derivatives. The donor and acceptor strengths of compounds of a family of elements determined in

TABLE XVI

$\Delta \nu_{OH}$ AS A MEASURE OF DONOR STRENGTH

Compound	$\Delta \nu_{O-H}$ (cm^{-1})
$(CH_3)_3SiOC_6H_5$	177
$(CH_3)_3SnOC_6H_5$	352
$(C_2H_5)_3SnOC_6H_5$	371
$(C_2H_5)_3PbOC_6H_5$	410
Triphenylamine	460
Triphenylphosphine	430
Triphenylarsine	460

TABLE XVII

$\Delta\nu_{OH}$ AS A MEASURE OF ACCEPTOR ACIDITY

	$\Delta\nu_{O-H}$ (cm^{-1})		
Compound	TMSO	Et$_2$O	Et$_2$O[a]
$(C_6H_5)_3COH$	248	174	176
$(C_6H_5)_3SiOH$	440	316	317
$(C_6H_5)_3GeOH$	295	198	196

[a] Data from (262b).

this manner are greatly influenced by π bonding effects, and there appears to be no correlation between the trends in strengths obtained by hydrogen bonding and the electronegativities of central elements determined by NMR chemical-shift data (75). It may also be remarked at this stage that the proportionality between ΔH° and $\Delta\nu_{O-H}$ is not strictly valid in all situations, particularly with data for the interaction of a series of unrelated acceptors with a donor (184).

By examining the NMR spectra of silicochloroform in a number of solvents, Huggins and Carpenter (124) have concluded that this compound is a considerably weaker acid than chloroform. This observation has been taken to account for the existence of d_π—p_π bonding between silicon and chlorine. Singh et al. (229) have studied the hydrogen bonding of methanol and t-butanol with several donors by the NMR method; the association shifts decrease in the order tri-n-butylamine > acetone > acetonitrile > benzene. With acetone as the donor the association shifts with different alcohols vary in the order trifluoroethanol > t-butanol > aniline > thiophenol > chloroform > bromoform > iodoform. Triphenylconbinol shows a smaller association shift with acetone than tributylsilanol, probably because the latter is a stronger acid (229). Association shifts of chloroform, bromoform, and iodoform in various donor solvents show that the proton-donating ability varies in the order chloroform > bromoform > iodoform (193). Creswell and Allred (64) have obtained thermodynamic data for the association of tetrahydrofuran with fluoroform, chloroform, bromoform, and iodoform; the enthalpies vary in the order $CHCl_3 > CHBr_3 = CHF_3 > CHI_3$. Eyman and Drago (86) have proposed a correlation between the phenol OH chemical shift and the ΔH of hydrogen bonding based on their data for thirty phenol–base systems; $\delta_{obsd} = 0.748\Delta H - 4.68$. Here again one should be careful in deriving information on donor or acceptor strengths from such correlations or association-shift data.

B. Soft and Hard Acids and Bases

Pearson (*194, 195*) has introduced the concept of hard and soft acids and bases, which is of value in understanding a wide variety of chemical phenomena. Hard and soft acids and bases are defined as follows:

Hard acid. The acceptor atom is of high positive charge and small size, and does not have easily excitable electrons.

Soft acid. The acceptor atom is of low positive charge and large size, and has several easily excitable electrons.

Hard base. The donor atom is of low polarizability and high electronegativity, is hard to oxidize, and is associated with empty orbitals of high energy and hence inaccessible.

Soft base. The donor atom is of high polarizability and low electronegativity, is easily oxidized, and is associated with empty low-lying orbitals.

A classification of Lewis acids and bases into "hard" and "soft" classes is shown in Table XVIII. The borderline cases have also been listed. It is clear from the table that the acceptors in hydrogen-bonding equilibria are hard acids and the acceptors in charge-transfer interactions are soft acids. Pearson (*194–196*) has made an impressive correlation of a large number of facts in inorganic and organic chemistry by proposing a general principle that hard acids prefer to coordinate with hard bases and soft acids prefer soft bases. This principle is by no means a theory in itself. We shall briefly discuss the classification of acids and bases and examine the validity of the classification to hydrogen-bonded and charge-transfer systems. Critical reviews of Pearson's classification of bases and acids have appeared in the literature (*123*).

According to Pearson's classification a typical hydrogen-bond acceptor, phenol, is a hard acid; iodine, an acceptor in charge-transfer complexes, is a soft acid. It is to be expected that the interaction with phenol will be effective with hard donor atoms such as fluorine, oxygen, and nitrogen, and not with soft donor atoms such as iodine, sulfur, and phosphorus. This is in agreement with the experimental data of Schleyer, West and co-workers (*223, 261*), who find that the enthalpy for the interaction of phenol with alkyl halides varies in the order $F > Cl > Br > I$. The relative differences of the enthalpy of interaction of oxygen and sulfur donors with the acids phenol and iodine have been explained by Drago (*76*) on a similar basis. Selenides form stronger complexes with iodine than with phenol. The triiodide ion I_3^- is more stable than I_2F^- because the iodide ion is a soft base towards iodine. Similar considerations can be employed to explain the stronger interaction of NH_3 and primary amines with phenol than with iodine and the opposite behavior of a tertiary amine with these acids (*76*). It can be seen from Table VI that thiocarbonyl compounds are better donors for the interaction with iodine than with phenol.

TABLE XVIII

PEARSON'S CLASSIFICATION OF LEWIS ACIDS AND BASES

Acids		Bases	
Hard	Soft	Hard	Soft
H^+, Li^+, Na^+, K^+	Cu^+, Ag^+, Au^+, Tl^+, Hg^+, Cs^+	H_2O, OH^-, F^-, $CH_3CO_2^-$	R_2S, RSH, RS^-, I^-, SCN^-
Be^{2+}, Mg^{2+}, Ca^{2+}, Sr^{2+}, Sn^{2+}	Pd^{2+}, Cd^{2+}, Pt^{2+}, Hg^{2+}, CH_3Hg^+	PO_4^{3-}, SO_4^{2-}, CO_3^{2-}, ClO_4^-	$S_2O_3^{2-}$, Br^-, R_3P, R_3As, $(RO)_3P$
Al^{3+}, Sc^{3+}, Ga^{3+}, In^{3+}, La^{3+}	Tl^{3+}, $Tl(CH_3)_3$, BH_3	NO_3^-, ROH, RO^-, R_2O	CN^-, RNC, CO, C_2H_4
Cr^{3+}, Co^{3+}, Fe^{3+}, As^{3+}, In^{3+}	RS^+, RSe^+, RTe^+	NH_3, RNH_2, N_2H_4	C_6H_6, H^-, R^-
Si^{4+}, Ti^{4+}, Zr^{4+}, Th^{4+}, Pu^{4+}, VO^{2+}	I^+, Br^+, HO^+, RO^+		
UO_2^{2+}, $(CH_3)_2Sn^{2+}$	I_2, Br_2, ICN, etc.		
$BeMe_3$, BF_3, BCl_3, $B(OR)_3$	Trinitrobenzene, etc.		
$Al(CH_3)_3$, $Ga(CH_3)_3$, $In(CH_3)_3$	Chloranil, quinones, etc.		
RPO_2^+, $ROPO_2^+$	Tetracyanoethylene, etc.		
RSO_2^+, $ROSO_2^+$, SO_3	O, Cl, Br, I, R_3C (?)		
I^{7+}, I^{5+}, Cl^{7+}	M^0 (metal atoms)		
R_3C^+, RCO^+, CO_2, NC^+	Bulk metals		
HX (hydrogen-bonding molecules)			

Borderline		Borderline	
Fe^{2+}, Co^{2+}, Ni^{2+}, Cu^{2+}, Zn^{2+}, Pb^{2+}, $B(CH_3)_3$, SO_2, NO^+		$C_6H_5NH_2$, C_5H_5N, N_3^-, Cl^-, NO_2^-, SO_3^{2-}	

Boron trifluoride (BF_3) is a hard acid, and borane (BH_3) is a soft acid. The boron atom in both these compounds is in the trivalent. The soft hydride ions lose a portion of the negative charge to the boron atom. The effective charge of the boron atom in borane is thus much less than $+3$, and so it can accept another soft base such as carbon monoxide to form a stable complex BH_3CO. The analogous BF_3CO is not known. The soft H^- groups have made it possible for boron to also hold the soft CO group. This flocking together of either soft or hard ligands around a central metal atom, leading to stabilization, has been called symbiosis by Jorgensen (*130*).

Aromatic hydrocarbons and olefins are soft bases. The charge-transfer complexes of aromatic donors with several soft acceptors are known. Stable complexes with metal ions are best formed if the metal ion is soft. The typical soft acids such as Ag^+, Pt^{2+}, and Pd^{2+} have long been known to form stable complexes with these bases.

The empirical criteria that led Pearson to classify acids and bases as hard and soft further suggest that the relative softness or hardness of a species is related to the nature of bonds it tends to form. The softer an acceptor and donor, the more covalent the bond formed between them; the harder they are, the more electrostatic the bond, Recently Drago and Wayland (*76, 77*) have proposed a double-scale equation to correlate enthalpies of adduct formation in the gas phase and poor solvating solvents:

$$-\Delta H^\circ = E_A E_B + C_A C_B \qquad (26)$$

The subscripts A and B indicate acceptor and donor, while E and C are two empirically derived parameters assigned to each. The parameters E and C are interpreted as being related to the electrostatic- and covalent-bond-forming abilities of the acids and bases, normalized to $E_A = C_A = 1$ for iodine. The product $E_A E_B$ provides the electrostatic contribution and $C_A C_B$ the covalent contribution to the enthalpy. It has been found that most hydrogen-bonding acids have a large electrostatic parameter (E_A) and a smaller covalent parameter (C_A), compared to iodine. The E and C parameters have been found to be consistent with the hard–soft classification. Phenol has a larger covalent parameter than aliphatic alcohols. Both the E and C parameters for ICl (dipole moment $1.2D$) are larger than those for I_2 (zero dipole moment). On the basis of the E and C parameters, ICl would be expected to be stronger acceptor than I_2 toward donors; experimental results agree with this prediction.

Pearson and Mawby (*197*) have introduced a softness parameter σ_P, obtained as a relative difference of coordination-bond energies of F^- and I^-, the hardest and softest ligands respectively. This parameter has shown a good trend in the degree of softness, but only if the comparison is restricted to acceptors of the same charge.

Klopman (*136*) has employed polyelectronic perturbation theory to this problem. He proposes two softness parameters, one for the acceptor (σ_K) and another for the donor (σ_L). These parameters have been derived from the energies of frontier orbitals of an acceptor or donor and their respective desolvation energies. The softness sequence found for acids is on the whole reasonable, although Na^+ occupies a questionable position between Ni^{2+} and Cu^{2+}.

Ahrland (*1*) has proposed another softness parameter (σ_A) for acceptors. According to Ahrland, the larger the difference between the total ionization potential for the formation of the gaseous ion and the dehydration energy, the softer is the ion. A unified comparison can be made for ions of different charges if the difference is divided by the charge of the ion. Ahrland has compared his σ_A values with Klopman's σ_K and the σ_P values of Pearson and Mawby and found the sequence to be the same. A similar softness parameter for donors could not be obtained because of the unavailability of good de-solvation-energy and electron-affinity values.

Recently Lohmann (*154*) has found that the relative difference of the free energies of solvation of single ions in water (hard solvent) and acetonitrile (soft solvent) is a softness parameter. This approach is very similar to that of Pearson and Mawby (*197*).

References

1. Ahrland, S., *Chem. Phys. Letters* **2**, 303 (1968).
2. Alexander, R. E., *J. Am. Chem. Soc.* **82**, 535 (1960).
3. Alford, D. O., Menefee, A., and Scott, C. B., *Chem. Ind. (London)* 514 (1959).
4. Allerhand, A., and Schleyer, P. V. R., *J. Am. Chem. Soc.* **85**, 1233 (1963).
5. Andrews, L. J., and Keefer, R. M., "Molecular Complexes in Organic Chemistry," Holden Day, San Francisco, 1964.
6. Arai, H., Saito, Y., and Yoneda, Y., *Bull. Chem. Soc. Japan* **40**, 312 (1967).
7. Arefev, I. M., and Malyshev, V. I., *Opt. i Spektroskopiya* **13**, 206 (1962).
8. Arnett, E. M., *in* "Progress in Physical Organic Chemistry" (S. Cohen, A. Streitwieser, and R. W. Taft, eds.), Vol. I. Wiley (Interscience), New York, 1963.
9. Arnold, J., and Millen, D. J., *J. Chem. Soc.* 503 (1965).
10. Arnold, J., and Millen, D. J., *J. Chem. Soc.* 510 (1965).
11. Aten, A. C., Dieleman, J., and Hoijtink, G. J., *Discussions Faraday Soc.* **29**, 182 (1960).
12. Barachevskii, V. A., Kholmogorov, V. E., Kotov, E. E., and Terenin, A. N. *Dokl. Akad. Nauk SSSR* **147**, 1108 (1962); *Chem. Abstr.* **58**, 12094h (1963).
13. Barakat, T. M., Legee, N., and Pullin, A. D. E., *Trans. Faraday Soc.* **59**, 1764 (1963).
14. Bellamy, L. J., Osborn, A. R., Lippincott, E. R., and Bandey, A. R., *Chem. Ind. (London)* 686 (1969).
15. Bellamy, L. J., and Beecher, L., *J. Chem. Soc.* 1701 (1952); 728 (1953).

16. Benesi, H. A., and Hildebrand, J. H., *J. Am. Chem. Soc.* **71**, 2703 (1949).
17. Bertie, J. E., and Whalley, E., *J. Chem. Phys.* **40**, 1637, 1646 (1964); **46**, 1271 (1967).
18. Bertie, J. E., and Millen, D. J., *J. Chem. Soc.* 497 (1965).
19. Bethell, D. E., and Sheppard, N., *J. Chem. Phys.* **21**, 1421 (1953).
20. Betts, J., and Robb, J. C., *Trans. Faraday Soc.* **64**, 2402 (1968).
21. Bhaskar, K. R., and Singh, S., *Spectrochim. Acta* **23A**, 1155 (1967).
22. Bhaskar, K. R., Bhat, S. N., Singh, S., and Rao, C. N. R., *J. Inorg. Nucl. Chem.* **28**, 1915 (1966).
23. Bhaskar, K. R., Bhat, S. N., Murthy, A. S. N., and Rao, C. N. R., *Trans. Faraday Soc.* **62**, 788 (1966).
24. Bhaskar, K. R., Gosavi, R. K., and Rao, C. N. R., *Trans. Faraday Soc.* **62**, 29 (1966).
25. Bhat, S. N., and Rao, C. N. R., *J. Am. Chem. Soc.* **90**, 6008 (1968).
26. Bhat, S. N., and Rao, C. N. R., *J. Chem. Phys.* **47**, 1863 (1967).
27. Bhat, S. N., Bhaskar, K. R., and Rao, C. N. R., *Proc. Indian Acad. Sci.* **66**, 97 (1967).
28. Bhat, S. N., and Rao, C. N. R., *J. Am. Chem. Soc.* **88**, 3216 (1966).
29. Bhat, S. N., and Rao, C. N. R., *Can. J. Chem.* **47**, 3899 (1969).
30. Bist, H. D., and Person, W. B., *J. Phys. Chem.* **71**, 2750 (1967).
31. Bist, H. D., and Person, W. B., *J. Phys. Chem.* **73**, 482 (1969).
32. Blandamer, M. J., Gough, T. E., and Symons, M. C. R., *Trans. Faraday Soc.* **62**, 286, 301 (1966).
33. Blandamer, M. J., Gough, T. E., and Symons, M. C. R., *Trans. Faraday Soc.* **60**, 488 (1964).
33a. Blandamer, M. J. and Fox, M. F., *Chem. Revs.* **70**, 59 (1970).
34. Blinc, R., Hadzi, D., and Novak, A., *Z. Elektrochem.* **64**, 567 (1960).
34a. Blinc, R., and Hadzi, D., *Spectrochim. Acta* **16**, 852 (1960).
35. Blomgren, G. E., and Kommandeur, J., *J. Chem. Phys.* **45**, 1636 (1961).
36. Borod'ko, T. G., and Syrkin, Y. K., *Zh. Strukt. Khim.* **2**, 480 (1961).
37. Brasch, J. W., Mikawa, Y., and Jakobsen, R. J., *Appl. Spectry. Rev.* **1**, 187 (1968).
38. Bratoz, S., *in* "Advances in Quantum Chemistry" (P. O. Lowdin, ed.), Vol. 3. Academic Press, New York, 1967.
39. Braunholtz, J. T., Hall, G. E., Mann, F. G., and Sheppard, N., *J. Chem. Soc.* 868 (1959).
40. Briegleb, G., "Elektronen-Donator-Acceptor Komplexe." Springer-Verlag, Berlin, 1961.
41. Briegleb, G., Liptay, W., and Fick, R., *Z. Elektrochem.* **66**, 859 (1962).
42. Brissette, C., and Sandorfy, C., *Can. J. Chem.* **38**, 34 (1960).
43. Brown, A. J., and Whalley, E., *J. Chem. Phys.* **45**, 4360 (1966).
44. Brown, T. L., and Kubota, M., *J. Am. Chem. Soc.* **83**, 4175 (1961).
45. Bufalini, J., and Stern, K. H., *J. Am. Chem. Soc.* **83**, 4362 (1961).
46. Buijs, K., and Choppin, G. R., *J. Chem. Phys.* **39**, 2035 (1963).
47. Buijs, K., and Choppin, G. R., *J. Chem. Phys.* **40**, 3120 (1964).
48. Burg, A. B., and Griffiths, J. E., *J. Am. Chem. Soc.* **83**, 4333 (1961).
49. Buschow, K. H. J., Dielman, J., and Hoijtink, G. J., *J. Chem. Phys.* **42**, 1993 (1965).
50. Busing, W. R., and Hornig, D. F., *J. Phys. Chem.* **65**, 284 (1961).
51. Byrd, W., *Inorg. Chem.* **1**, 762 (1962).
52. Carlson, G. L., Witkowski, R. E., and Fateley, W. G., *Nature* **211**, 1289 (1966).
53. Carlson, R. L., and Drago, R. S., *J. Am. Chem. Soc.* **84**, 2320 (1962).
54. Carter, H. V., McClelland, B. J., and Warhurst, E., *Trans. Faraday Soc.* **56**, 455 (1960).
55. Chalandon, P., and Susz, B. P., *Helv. Chim. Acta* **41**, 697 (1958).
56. Chan, L. L., and Smid, J., *J. Am. Chem. Soc.* **90**, 4654 (1968).

57. Chang, P., Slates, R. V., and Szwarc, M., *J. Phys. Chem.* **70**, 3180 (1966).
58. Chaudhuri, J. N., and Basu, S., *Trans. Faraday Soc.* **55**, 898 (1959).
59. Chenon, B., and Sandorfy, C., *Can. J. Chem.* **36**, 1181 (1958).
60. Christian, S. D., Tucker, E. E., and Affsprung, H. E., *Spectrochim. Acta* **23A**, 1185 (1967).
61. Cook, D., *J. Chem. Phys.* **25**, 788 (1956).
62. Cook, D., *Can. J. Chem.* **37**, 48 (1959).
63. Cote, G. L., and Thompson, H. W., *Proc. Roy. Soc. (London)* **A210**, 206 (1951).
64. Creswell, C. J., and Allred, A. L., *J. Am. Chem. Soc.* **85**, 1723 (1963).
65. Daasch, L. W., *Spectrochim. Acta* 726 (1959).
66. Das, M., and Basu, S., *Spectrochim. Acta* **17**, 897 (1961).
67. de Boer, E., *Advan. Organometal. Chem.* **2**, 115 (1964).
68. Delaplane, R. G., Ibers, J. A., Ferraro, J. R., and Rush, J. R., *J. Chem. Phys.* **50**, 1920 (1969).
69. de Maine, P. A. D., and Peone, J., *J. Mol. Spectry.* **4**, 262 (1960).
70. Dewar, M. J. S., and Lepley, A. R., *J. Am. Chem. Soc.* **83**, 4560 (1961).
71. Dewar, M. J. S., and Rogers, H., *J. Am. Chem. Soc.* **84**, 395 (1962).
72. Dilke, M. H., Eley, D. D., and Sheppard, M. G., *Trans. Faraday Soc.* **46**, 261 (1950).
73. Dollish, F. R., and Keithhall, W., *J. Phys. Chem.* **69**, 2127 (1965).
74. Dows, D. A., and Pimentel, G. C., *J. Chem. Phys.* **23**, 1258 (1955).
75. Drago, R. S., *Record Chem. Progr.* **26**, 157 (1965).
76. Drago, R. S., *Chem. Brit.* **3**, 516 (1967).
77. Drago, R. S., and Wayland, B. B., *J. Am. Chem. Soc.* **87**, 3571 (1965).
78. Durig, J. R., Bush, S. F., and Mercer, E. E., *J. Chem. Phys.* **44**, 4238 (1966).
79. Dwivedi, P. C., and Rao, C. N. R., *Spectrochim. Acta* (In Press).
80. Edgell, W. F., Watts, A. T., Lyford, J., and Risen, W. M., *J. Am. Chem. Soc.* **88**, 1815 (1966).
81. Edgell, W. F., Pauuwe, N., and Lyford, J., Abstracts of 7th Nat. Meeting of Soc. of Appl. Spectry., 1968.
81a. Eisenberg, D., and Kauzmann, W., "The Structure and Properties of Water." Oxford Univ. Press (Clarendon), London and New York, 1969.
82. El Saffar, Z. M., *J. Chem. Phys.* **45**, 4643 (1966).
83. Emsley, J. W., Feeney, J., and Sutcliffe, L. H., "High Resolution Nuclear Magnetic Resonance Spectroscopy," Vols. I, II. Pergamon Press, Oxford, 1966.
84. Epley, T. D., and Drago, R. S., *J. Am. Chem. Soc.* **89**, 5770 (1967).
85. Evans, J. C., and Lo, G. Y.-S., *J. Phys. Chem.* **70**, 11, 20 (1966).
86. Eyman, D. P., and Drago, R. S., *J. Am. Chem. Soc.* **88**, 1617 (1966).
87. Ferguson, E. E., and Chang, I. Y., *J. Chem. Phys.* **34**, 628 (1961).
88. Ferguson, E. E., *J. Chem. Phys.* **25**, 577 (1956).
89. Ferguson, E. E., and Matsen, F. A., *J. Chem. Phys.* **29**, 105 (1958).
90. Ferraro, J. R., and Peppard, D. F., *J. Phys. Chem.* **67**, 2639 (1963).
91. Ferriso, C. C., and Hornig, D. F., *J. Chem. Phys.* **23**, 1464 (1955).
92. Flockhart, B. D., Leith, I. R., and Pink, R. C., *J. Catal.* **9**, 45 (1967); *Chem. Abstr.* **68**, 6723 (1968).
92a. Foster, R., and Fyfe, F. *in* "Progress in NMR Spectroscopy," Vol. 4 (J. W. Emsley, J. Feeney, and L. H. Sutcliffe, eds.). Pergamon Press, Oxford, 1969.
93. Franta, E., Chaudhuri, J., Cserhegyi, A., Jagur-Grodzinski, J., and Szwarc, M., *J. Am. Chem. Soc.* **89**, 7129 (1967).
94. Fratiello, A., *J. Chem. Phys.* **41**, 2204 (1964).
95. Fratiello, A., and Luongo, J. P., *J. Am. Chem. Soc.* **85**, 3072 (1963).

96. Freidrich, H. B., and Person, W. B., *J. Chem. Phys.* **44**, 2161 (1966).
97. Fritzsche, H., *Spectrochim. Acta* **17**, 352 (1961).
98. Garst, J. F., and Zabolotny, E. R., *J. Am. Chem. Soc.* **87**, 495 (1965).
99. Garst, J. F., Zabolotny, E. R., and Cole, R. S., *J. Am. Chem. Soc.* **86**, 2257 (1964).
100. Garst, J. F., and Cole, R. S., *J. Am. Chem. Soc.* **84**, 4353 (1962).
101. Ginn, S. G. W., Haque, I., and Wood, J. L., *Spectrochim. Acta* **24A**, 1531 (1968).
101a. Goel, A., Murthy, A. S. N., and Rao, C. N. R., *Chem. Commun.*, 423 (1970).
102. Goldstein, R., and Penner, S. S., *J. Quant. Spectry. Radiative Transfer* **4**, 441 (1964).
103. Gosavi, R. K., Agarwala, U., and Rao, C. N. R., *J. Am. Chem. Soc.* **89**, 235 (1967).
104. Gover, T. A., and Porter, G., *Proc. Roy. Soc. (London)* **A262**, 476 (1961).
105. Govil, G., Clague, A. D. H., and Bernstein, H. J., *J. Chem. Phys.* **49**, 2821 (1968).
106. Griffiths, T. R., and Symons, M. C. R., *Trans. Faraday Soc.* **56**, 1125 (1960).
107. Griffiths, T. R., and Symons, M. C. R., *Mol. Phys.* **3**, 90 (1960).
108. Gutman, F., and Lyons, L. E., "Organic Semiconductors." Wiley, New York, 1967.
109. Haas, C., and Hornig, D. F., *J. Chem. Phys.* **32**, 1763 (1960).
110. Hadzi, D., *J. Chem. Soc.* 5128 (1962).
111. Hallam, H. E., *in* "Infrared Spectroscopy and Molecular Structure" (M. Davies, ed.), Elsevier, Amsterdam, 1963.
112. Hanna, M. W., and Ashbough, A. L., *J. Phys. Chem.* **68**, 811 (1964).
113. Haque, I., and Wood, J. L., *Spectrochim. Acta* **23A**, 2523 (1967).
114. Hardy, C. J., Greenfield, B. F., and Scargill, D., *J. Chem. Soc.* 90 (1961).
115. Harrell, S. A., and McDaniel, D. H., *J. Am. Chem. Soc.* **86**, 4497 (1964).
116. Hassel, O., and Romming, C., *Quart. Rev. (London)* **16**, 1 (1962).
116a. Hindman, J. C., *in* "Developments in Applied Spectroscopy," Vol. 6. Plenum Press, New York, 1968.
117. Hoffman, R. E., and Hornig, D. F., *J. Chem. Phys.* **17**, 1163 (1949).
118. Hogen-Esch, T. E., and Smid, J., *J. Am. Chem. Soc.* **88**, 307, 318 (1966).
119. Holmes, J. R., Kivelson, D., and Drinkard, W. C., *J. Am. Chem. Soc.* **84**, 4677 (1962).
120. Horak, M., Bazant, V., and Chavalovsky, V., *Collection Czech. Chem. Commun. Suppl.* **25**, 2822 (1960).
121. Hornig, D. F., *J. Chem. Phys.* **40**, 3119 (1964).
122. Hornig, D. F., White, H. F., and Reding, F. P., *Spectrochim. Acta* **12**, 338 (1958).
123. Hudson, R. F., Williams, R. J. P., and Hale, J. D., Jorgensen, C. K., and Ahrland, S., *in* "Structure and Bonding," Vol. I. Springer-Verlag, New York, 1967.
124. Huggins, C. M., and Carpenter, D. R., *J. Phys. Chem.* **63**, 238 (1959).
125. Ibers, J. A., and Holm, C. H., *J. Phys. Soc. Japan* **16**, 839 (1961).
125a. Ichiba, S., Mishima, M., Sakai, H., and Negita, H., *Bull. Chem. Soc. Japan*, **41**, 49 (1968).
126. Ij Alabersberg, W., Hoijtink, G. J., Mackor, E. L., and Weijland, W. P., *J. Chem. Soc.* 3049, 3055 (1959).
127. Jaffe, H. H., and Orchin, M., *in* "Theory of Applications of Ultraviolet Spectroscopy." Wiley, New York, 1962.
128. Jones, D. A. K., and Watkinson, J. G., *J. Chem. Soc.* 2366, 2371 (1964).
129. Jones, W. J., Seel, R. M., and Sheppard, N., *Spectrochim. Acta* **25A**, 385 (1969).
130. Jorgensen, C. K., *Inorg. Chem.* **3**, 1201 (1964).
131. Jortner, J., and Sokolov, U., *Nature* **190**, 1003 (1961).
132. Julien, L. M., and Person, W. B., *J. Phys. Chem.* **72**, 3059 (1968).
133. Kaiser, E. T., and Kevan, L., *in* "Radical Ions." Wiley (Interscience), New York, 1968.
134. Ketelaar, J. A. A., *J. Phys. Radium* **15**, 197 (1954).

135. Ketelaar, J. A. A., and Bedder, W., *J. Chem. Phys.* **19**, 654 (1951).
136. Klopman, G., *J. Am. Chem. Soc.* **90**, 223 (1968).
137. Kobinata, S., and Nagakura, S., *J. Am. Chem. Soc.* **88**, 3905 (1966).
138. Kosower, E. M., *in* "Progress in Physical Organic Chemistry" (S. Cohen, A. Steitwieser, and R. W. Taft, eds.). Wiley (Interscience), New York, 1963.
139. Kosower, E. M., *J. Am. Chem. Soc.* **77**, 3883 (1955); **78**, 5700 (1956); **80**, 3253 (1958).
140. Kotowycz, G., Schaefer, T., and Bock, E., *Can. J. Chem.* **42**, 2541 (1964).
141. Kroll, M., *J. Am. Chem. Soc.* **90**, 1097 (1968).
142. Kuroda, H., Sakurai, T., and Akamatu, H., *Bull. Chem. Soc. Japan* **39**, 1893 (1966).
143. Lake, R. F., and Thompson, H. W., *Proc. Roy. Soc. (London)* **A297**, 440 (1967).
144. Lake, R. F., and Thompson, H. W., *Spectrochim. Acta* **24A**, 1321 (1968).
145. Laszlo, P., *in* "Progress in NMR Spectroscopy" (J. W. Emsley, J. Feeney, and L. H. Sutcliffe, eds.), Vol. 3. Pergamon Press, Oxford, 1967.
146. Lepley, A. R., *J. Am. Chem. Soc.* **84**, 3577 (1962).
146a.Levanon, H., and Navon, G., *J. Phys. Chem.* **73**, 1861 (1969).
147. Lewis, I. C., and Singer, L. S., *J. Chem. Phys.* **43**, 2712 (1965).
148. Lindberg, J. J., and Majani, C., *Acta Chem. Scand.* **17**, 1477 (1963).
149. Lindqvist, I., "Inorganic Adduct Molecules of Oxo Compounds." Academic Press, New York, 1963.
150. Lippert, E., *Ber. Bunsenges Physik. Chem.* **67**, 267 (1963).
151. Lippincott, E. R., and Schroeder, R., *J. Chem. Phys.* **23**, 1099 (1955); Schroeder, R., and Lippincott, E. R., *J. Phys. Chem.* **61**, 921 (1957).
152. Lippincott, E. R., Weir, C. R., and Valkenburg, A. V., *J. Chem. Phys.* **32**, 612 (1960).
153. Lippincott, E. R., Stromberg, R. R., Grant, W. H., and Cessac, G. E., *Science* **164**, 1482 (1969).
154. Lohmann, F., *Chem. Phys. Letters* **2**, 659 (1968).
155. Luck, W., *Z. Elektrochem.* **67**, 186 (1963).
156. Luck, W., *Z. Elektrochem.* **68**, 895 (1964).
157. Luck, W., *Ber. Bunsenges Physik. Chem.* **69**, 627 (1965).
158. Lund, H., *Acta Chem. Scand.* **12**, 298 (1958).
159. Maki, A. G., and Plyler, E. K., *J. Phys. Chem.* **66**, 766 (1962).
160. Marckmann, J. P., and Whalley, E., *J. Chem. Phys.* **41**, 1450 (1964).
161. Marcus, R. A., and Fresco, J. M., *J. Chem. Phys.* **27**, 564 (1957).
162. Matwiyoff, N. A., and Drago, R. S., *J. Organomet. Chem.* **3**, 393 (1965).
163. Maxey, B. W., and Popov, A. I., *J. Am. Chem. Soc.* **91**, 20 (1969).
164. McClelland, B. J., *Chem. Rev.* **64**, 301 (1964).
165. McClelland, B. J., *Trans. Faraday Soc.* **57**, 2073 (1961).
166. McClelland, B. J., *Trans. Faraday Soc.* **58**, 1458 (1962).
167. McConnell, H., Ham, J. S., and Platt, J. R., *J. Chem. Phys.* **21**, 66 (1953).
167a.Meyerstein, D., and Terenin, A., *Trans. Faraday Soc.* **59**, 1114 (1963).
168. Millen, D. J., and Samsanov, O. A., *J. Chem. Soc.* 3085 (1965).
169. Mohr, S. C., Wilk, W. D., and Barrow, G. M. *J. Am. Chem. Soc.* **87**, 3048 (1965).
170. Muller, N., and Simon, P., *J. Phys. Chem.* **71**, 568 (1967).
170a.Muller, N., *J. Chem. Phys.* **43**, 2555 (1965).
171. Mulliken, R. S., *J. Am. Chem. Soc.* **72**, 600 (1950).
172. Mulliken, R. S., *J. Chem. Phys.* **19**, 514 (1951).
173. Mulliken, R. S., *J. Am. Chem. Soc.* **74**, 811 (1952).
174. Mulliken, R. S., *J. Phys. Chem.* **56**, 801 (1952).
175. Mulliken, R. S., "Symposium on Molecular Physics," p. 45. Nikko, Japan, 1953.
176. Mulliken, R. S., *J. Chim. Phys.* **51**, 341 (1954).

177. Mulliken, R. S., *J. Chem. Phys.* **23**, 397 (1955).
178. Mulliken, R. S., *Rec. Trav. Chim.* **75**, 845 (1956).
179. Mulliken, R. S., and Person, W. B., *Ann. Rev. Phys. Chem.* **13**, 107 (1962).
179a.Mulliken, R. S., and Person, W. B., "Molecular Complexes," Wiley (Interscience), New York, 1969.
180. Mulliken, R. S., *J. Chim. Phys.* 20 (1963).
181. Mulliken, R. S., *J. Am. Chem. Soc.* **91**, 1237 (1969).
182. Murrell, J. N., *Quart. Rev. (London)* **15**, 191 (1961).
182a.Murrell, J. N., *Chem. in Britain* **5**, 107 (1969).
183. Murthy, A. S. N., and Rao, C. N. R., *J. Mol. Struct.* (In Press).
184. Murthy, A. S. N., and Rao, C. N. R., *Appl. Spectry. Rev.* **2**, 69 (1968).
185. Nagakura, S., *J. Chim. Phys.* **61**, 217 (1964).
186. Nakamoto, K., Sarma, Y. A., and Ogoshi, H., *J. Chem. Phys.* **43**, 1177 (1965).
187. Nemethy, G., and Scheraga, H. A., *J. Chem. Phys.* **36**, 3382 (1962).
188. Nightingale, R. E., and Wagner, E. L., *J. Chem. Phys.* **22**, 203 (1954).
189. Nuttall, R. H., Sharp, D. W. A., and Waddington, T. C., *J. Chem. Soc.* 4965 (1960).
190. Olah, G. A., Kuhn, S. J., Tolygyesi, W. S., and Baker, E. B., *J. Am. Chem. Soc.* **84**, 2733 (1962).
191. Paetzold, R., Schumann, H. D., and Simar, A., *Z. Anorg. Allgem. Chem.* **305**, 88 (1960).
192. Pake, G. E., *J. Chem. Phys.* **16**, 327 (1964).
193. Paterson, W. G., and Cameron, D. M., *Can. J. Chem.* **41**, 198 (1963).
194. Pearson, R. G., *J. Am. Chem. Soc.* **85**, 3533 (1963).
195. Pearson, R. G., *Chem. Brit.* **3**, 103 (1967).
196. Pearson, R. G., and Songstad, J., *J. Am. Chem. Soc.* **89**, 1827 (1967).
197. Pearson, R. G., and Mawby, R. J., *in* "Halogen Chemistry" (V. Gutmann, ed.), Vol. 3. Academic Press, New York, 1967.
198. Person, W. B., *J. Am. Chem. Soc.* **87**, 167 (1965).
199. Person, W. B., Cook, C. F., and Freidrich, H. B., *J. Chem. Phys.* **46**, 2521 (1967).
200. Person, W. B., Humphrey, R. E., and Popov, A. I., *J. Am. Chem. Soc.* **81**, 273 (1959).
201. Pimenov, Y. D., Kholmogorov, V. E., and Terenin, A. N., *Dokl. Akad. Nauk SSSR* **163**, 935 (1965); *Chem. Abstr.* **63**, 14081C (1965).
202. Pimentel, G. C., and McClellan, A. L., "The Hydrogen Bond." Freeman, San Francisco, 1960.
203. Pimentel, G. C., Charles, S. W., and Rosengren, K., *J. Chem. Phys.* **44**, 3029 (1966).
204. Pimentel, G. C., Bulanin, M. O., and Van Thiel, M., *J. Chem. Phys.* **36**, 500 (1962).
205. Plumb, R. C., and Hornig, D. F., *J. Chem. Phys.* **21**, 366 (1953); **23**, 947 (1955).
206. Pople, J. A., Bernstein, H. J., and Schneider, W. G., *in* "High Resolution Nuclear Magnetic Resonance." McGraw Hill, New York, 1959.
207. Powell, D. B., and Sheppard, N., *J. Chem. Soc.* 2519 (1960); *Spectrochim. Acta* **13**, 69 (1958).
208. Powell, D. G., and Warhurst, E., *Trans. Faraday Soc.* **58**, 952 (1962).
209. Puranik, P. G., and Kumar, V., *Proc. Indian Acad. Sci.* **A58**, 29, 327 (1963).
210. Rank, D. H., Sitaram, P., Glickman, W. A., and Wiggins, T. A., *J. Chem. Phys.* **39**, 2673 (1963).
211. Rao, C. N. R., *in* "Ultraviolet and Visible Spectroscopy—Chemical Applications," 2nd ed. Butterworths, London and Washington, D. C., and Plenum Press, New York, 1967.
212. Rao, C. N. R., Kalyanaraman, V., and George, M., *Appl. Spectry. Rev.* **3**, 153 (1970).
213. Rao, C. N. R., Chaturvedi, G. C., and Gosavi, R. K., *J. Mol. Spectry.* **28**, 526 (1968).

213a.Rao, C. N. R., Chaturvedi, G. C., and Bhat, S. N., *J. Mol. Spectry.* **33**, 554 (1970).
214. Rao, C. N. R., *in* "Chemical Applications of Infrared Spectroscopy." Academic Press, New York, 1963.
215. Reid, C., and Mulliken, R. S., *J. Am. Chem. Soc.* **76**, 3869 (1954).
216. Renbaum, A., Eisenberg, A., Haack, R., and Landel, R. F., *J. Am. Chem. Soc.* **89** 1062 (1967).
217. Richards, R. E., and Smith, J. A. S., *Trans. Faraday Soc.* **47**, 1261 (1951).
218. Roland, G., and Duyckaerts, G., *Spectrochim. Acta* **22**, 793 (1966).
219. Rose, N. J., and Drago, R. S., *J. Am. Chem. Soc.* **81**, 6138 (1959).
220. Rush, J. J., and Ferraro, J. R., *J. Chem. Phys.* **44**, 2496 (1966).
220a.Rüterjans, H. H., and Scheraga, H. A., *J. Chem. Phys.* **45**, 3296 (1966).
221. Sauvageau, P., and Sandorfy, C., *Can. J. Chem.* **38**, 1901 (1960).
222. Schenk, G. H., *Record Chem. Progr.* **28**, 135 (1967).
222a.Schiffer, J., and Hornig, D. F., *J. Chem. Phys.* **49**, 4150 (1968).
223. Schleyer, P. V. R., and West, R., *J. Am. Chem. Soc.* **81**, 3164 (1959).
224. Schmulback, C. D., and Drago, R. S., *J. Am. Chem. Soc.* **82**, 4484 (1960).
225. Schug, J. C., and Martin, A. R., *J. Phys. Chem.* **66**, 1554 (1962).
225a.Senior, W. A., and Verrall, R. E., *J. Phys. Chem.* **73**, 4242 (1969).
226. Shatenstein, A. I., Petrov, E. S., and Yakovlava, E. A., *J. Polymer Sci.* **16**, 1729 (1967).
227. Shelden, J. C., and Tyree, S. Y., *J. Am. Chem. Soc.* **81**, 2290 (1959).
228. Singer, L. S., and Kommandeur, J., *J. Chem. Phys.* **34**, 133 (1961).
229. Singh, S., Murthy, A. S. N., and Rao, C. N. R., *Trans. Faraday Soc.* **62**, 1056 (1966).
230. Singh, S., and Rao, C. N. R., *Trans. Faraday Soc.* **62**, 3310 (1966).
231. Smith, M., and Symons, M. C. R., *Trans. Faraday Soc.* **54**, 346 (1958).
232. Smith, M., and Symons, M. C. R., *Discussions Faraday Soc.* **24**, 206 (1957).
233. Snyder, R. G., and Ibers, J. A., *J. Chem. Phys.* **36**, 1356 (1962).
234. Somorjai, R. L., and Hornig, D. F., *J. Chem. Phys.* **36**, 1980 (1962).
235. Stanevich, A. E., and Yavoslaosku, N. G., *Dokl. Akad. Nauk SSSR* **137**, 60 (1961).
236. Stekhanov, A. I., and Popova, E. A., *Opt. i Spektroskopiya* **11**, 203 (1961).
237. Stone, F. G. A., *Chem. Rev.* **58**, 101 (1958).
238. Strong, R. L., Rand, S. J., and Britt, J. A., *J. Am. Chem. Soc.* **82**, 5053 (1960).
239. Strong, R. L., and Perano, J., *J. Am. Chem. Soc.* **89**, 2535 (1967).
240. Susz, B. P., and Wuhrman, J. J., *Helv. Chim. Acta* **40**, 971 (1957).
241. Susz, B. P., and Chalandon, P., *Helv. Chim. Acta* **41**, 1332 (1958).
242. Szwarc, M., *in* "Progress in Physical Organic Chemistry" (S. Cohen, A. Streitwieser, and R. W. Taft, eds.), Vol. 6. Wiley (Interscience), New York, 1963.
243. Takahashi, F., and Li, N. C., *J. Am. Chem. Soc.* **88**, 1117 (1966).
244. Tamres, M., and Goodenow, J. M., *J. Phys. Chem.* **71**, 1982 (1967).
245. Taylor, R. C., and Vidale, G. L., *J. Am. Chem. Soc.* **78**, 5999 (1956).
246. Thomas, L. C., and Chittenden, R. A., *J. Opt. Soc. Am.* **52**, 829 (1962).
247. Thomas, L. C., Chittenden, R. A., and Hartley, H. E. R., *Nature* **192**, 1283 (1961).
248. Thomas, M. R., Scheraga, H. A., and Schrier, E. E., *J. Phys. Chem.* **69**, 3722 (1965).
249. Thompson, C. C., and de Maine, P. A. D., *J. Am. Chem. Soc.* **85**, 3096 (1963); *J. Phys. Chem.* **69**, 2766 (1965).
250. Tice, P. A., and Powell, D. B., *Spectrochim. Acta* **21**, 835 (1965).
251. Traynham, J. G., *J. Org. Chem.* **26**, 4694 (1961).
252. Tsubomura, H., and Mulliken, R. S., *J. Am. Chem. Soc.* **82**, 5966 (1960).
253. Van den Hende, J. H., and Baird, W. C., *J. Am. Chem. Soc.* **85**, 1009 (1963).
254. Van Thiel, M., Becker, E. D., and Pimentel, G. C., *J. Chem. Phys.* **27**, 486 (1957).
255. Waddington, T. C., *J. Chem. Soc.* 1708 (1958).

256. Wagner, E. L., and Hornig, D. F., *J. Chem. Phys.* **18**, 305 (1950).
257. Wall, T. T., and Hornig, D. F., *J. Chem. Phys.* **43**, 2079 (1965).
258. Wall, T. T., and Hornig, D. F., *J. Chem. Phys.* **47**, 784 (1967).
259. Walrafen, G. E., *J. Chem. Phys.* **40**, 3249 (1964); **44**, 1546 (1966); **48**, 244 (1968); **50**, 560, 568 (1969).
260. Walrafen, G. E., *in* "Equilibria and Reaction Kinetics in Hydrogen Bonded Solvent Systems" (A. K. Covington, ed.). Taylor and Francis, London, 1968.
261. West, R., Powell, D. L., Whatley, L. S., Lee, M. K. T., and Schleyer, P. V. R., *J. Am. Chem. Soc.* **84**, 3221 (1962).
262. West, R., Whatley, L. S., and Lake, K. J., *J. Am. Chem. Soc.* **83**, 761 (1961).
262a. West, R., and Kraihanzel, C. S., *J. Am. Chem. Soc.* **83**, 765 (1961).
262b. West, R., Baney, R. H., and Powell, D. L., *J. Am. Chem. Soc.* **82**, 6269 (1960).
263. Weston, Jr., R. E., *Spectrochim. Acta* **18**, 1257 (1962).
263a. Whalley, E., *in* "Developments in Applied Spectroscopy," Vol. 6. Plenum Press, New York, 1968.
264. Wolf, R., Houalla, D., and Mathis, F., *Spectrochim. Acta* **A23**, 1641 (1967).
265. Worley, J. D., and Klotz, I. M., *J. Chem. Phys.* **45**, 2868 (1966).
266. Yamada, H., and Kojima, K., *J. Am. Chem. Soc.* **82**, 1543 (1960).
267. Yarwood, J., and Person, W. B., *J. Am. Chem. Soc.* **90**, 594 (1968).
268. Yarwood, J., and Person, W. B., *J. Am. Chem. Soc.* **90**, 3930 (1968).
269. Yarwood, J., *Trans. Faraday Soc.* **65**, 934 (1969).
270. Zabolotny, E. R., and Garst, J. F., *J. Am. Chem. Soc.* **86**, 1645 (1964).
271. Zachariasen, W. H., *Acta Cryst.* **7**, 305 (1954).
272. Zingaro, R. A., and Witmer, W. B., *J. Phys. Chem.* **64**, 1705 (1960).

MASS SPECTROSCOPY

K. G. Das

NATIONAL CHEMICAL LABORATORY
POONA, INDIA

I. Introduction

After the discovery of "canal rays" (*43*) in 1886, the fundamental principles of mass spectrometry were established by Wien (*95*) and Thomson (*91*). Instruments to analyze positive rays were later built by Aston (*2*), Dempster (*34*), and others. Since then there has been a rapid increase in the number and types of mass spectrometers.

Mass spectrometry in the early days was mainly used for isotope analysis, but the range of applications has broadened so much that the technique is now used in nearly all branches of physics and chemistry. In the field of inorganic chemistry this technique is mainly used in organometallic chemistry, as an analytical tool, and in high-temperature chemistry of inorganic compounds. Several books (*1, 44, 61, 81*) and reviews (*13, 20, 64*) have appeared that deal with the various applications of inorganic mass spectrometry.

One of the sources of information in this growing subject is the *Mass Spectrometry Bulletin* published by the Mass Spectrometry Data Centre, A.W.R.E., Aldermaston, Berkshire, England. There are two new journals on mass spectrometry: *International Journal of Mass Spectrometry and Ion Optics* and *Organic Mass Spectrometry*.

A mass spectrometer is generally a positive-ion analyser. The ions are produced by electron bombardment of the sample vapor. They are sorted out and recorded on the basis of their mass/charge ratio (*m/e*). The mass spectrum of a compound is the distribution of the *m/e* of the ionized products formed by electron impact on molecules in the gas phase.

The ion abundances reflected in the spectrum include species formed by several competing fragmentation paths. Even in simple molecules the number of such species is so great and the interrelationships are so obscure that it is very difficult to disentangle satisfactorily the products of all the competing and consecutive processes. To define the reaction paths, four kinds of mass-spectral data have been widely used: (1) correlation of spectra with molecular structures (2) spectra of labeled compounds, (3) metastable peaks, and (4) appearance potentials.

II. Instrumentation

A. Ion Production

In a mass spectrometer ions are produced by any one of the following techniques: (1) electron bombardment of the vapor; (2) thermal evaporation of an inorganic compound from a heated metal surface; (3) irradiation with ultraviolet light of sufficiently short wavelength; (4) creation of an electrical

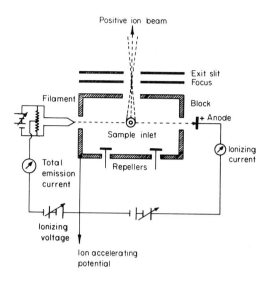

FIG. 1. Schematic diagram of an electron bombardment ion source.

spark between two electrodes of a metal or semiconductor; and (5) desorption of substances from a surface, as ions, by application of a strong electric field.

A diagrammatic sketch of a simple electron bombardment ion source is shown in Fig. 1. The electron source is a heated tungsten or rhenium wire. Sample vapor is introduced into the ionization chamber perpendicular to both the electron beam and the final ion beam. In high-temperature work, crucibles are placed a few millimeters from the electron beam and the sample is distilled into the electron beam as a well-collimated molecular beam.

B. Ion Analysis

The positive ions produced are accelerated by a large potential difference whereby they acquire large kinetic energy. They are separated by deflection in a magnetic field or by utilization of differences in time of flight (in an electric field) down an evacuated tube or by use of various selectors employing radio-frequency voltages.

In double-focusing mass spectrometers, which are used for high-resolution work, the separation of the ions is effected in two stages. The electrostatic and the magnetic analyzers are used to achieve velocity and direction focusing respectively. High-resolution mass spectrometers are available commercially that give resolutions of the order of 30,000 or more.

C. Ion Detection

The ion collection and recording are done electrically or photographically. In the former method the separated ions are collected and the ion currents amplified and recorded on a strip-chart recorder, an oscillograph, a digitizer, or a tape recorder. In photographic recording the emulsion of a photographic plate becomes exposed at the point of contact of each separated ion beam.

D. Ion Optics

Schematic diagrams of the Nier–Johnson and Mattauch–Herzog ion optics are given in Figs. 2 and 3 respectively. In the Nier–Johnson ion optics, where the ions make a 180° deflection, the magnetic field is scanned and the resolved ion beam is recorded electrically. The Mattauch–Herzog geometry ensures that the analyzed ions are focussed at the same time on the plane of the photoplate. The recording of the spectrum is generally done photographically, but it is also possible to record the spectrum electrically by magnetic scanning. In photographic recording the density of each line is proportional to the abundance of the corresponding ion. The precise mass of an ion is obtained by accurately measuring its position on the photoplate with respect to a standard ion.

After acceleration by the voltage V an ion with unit charge e has a potential energy eV. This is equal to its kinetic energy $\frac{1}{2}(mv^2)$, where m is its mass and v its velocity:

$$eV = \tfrac{1}{2}(mv^2) \tag{1}$$

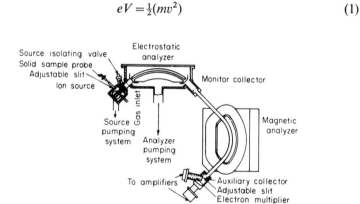

FIG. 2. Schematic diagram of the MS 902 Nier–Johnson geometry. [After G. W. A. Milne, *Quart. Rev.* **22**, 79 (1968), by permission of the author, the Chemical Society, and A.E.I.]

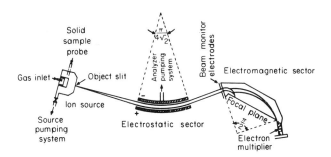

F<small>IG</small>. 3. Schematic diagram of the CEC 21-110B Mattauch–Herzog geometry. [After G. W. A. Milne, *Quart. Rev.* **22**, 80 (1968), by permission of the author, the Chemical Society, and CEC.]

The centripetal force exerted by the magnetic field on the ion, *Hev*, is balanced by the centrifugal force mv^2/R (R is the radius of the ion path and H is the magnetic field):

$$Hev = \frac{mv^2}{R} \qquad \text{or} \qquad R = \frac{mv}{eH} \qquad (2)$$

Elimination of v between Eqs. (1) and (2) results in the fundamental equation for a single focusing mass spectrometer:

$$\frac{m}{e} = \frac{H^2R^2}{2V}$$

E. Resolving Power

The resolving power of a mass spectrometer is its ability to separate ions of different mass-to-charge ratio. To separate two ions of masses M_1 and M_2 the resolving power necessary is defined by the expression $M_1/\Delta m$ ($\Delta m = M_2 - M_1$). The following factors determine resolving power: (1) variation of the kinetic energy of the ions of the same m/e value, (2) the divergence of the beam due to space-charge effects, (3) the radius of the magnet, and (4) the width of the source and collector slits. For accurate mass measurements a resolution of 10,000 is the minimum requirement.

The sensitivity of the instrument is a measure of its ability to detect ions of a particular component at an arbitrary m/e value. It can be expressed in many ways. In gas analysis it is often measured as amperes of ion current per torricelli of pressure of the gas in the ion source. Sensitivity and resolving power are approximately inversely proportional to each other.

III. Mass Spectrum

Mass spectra are usually published in the form of bar graphs prepared by replotting the data obtained from the original spectrum. The relative abundance of each peak is expressed as a percentage of the base peak (most intense peak) or as a fraction of the total ion current. The latter is an expression of the percent each peak contributes to the total ionization (e.g., $\sum_{40}$ means that all peaks above m/e 40 are considered in calculating the total ionization).

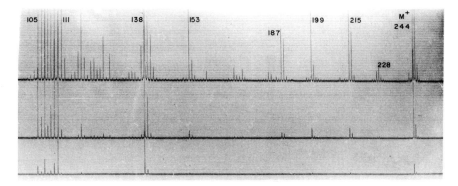

Fig. 4. Spectrum recorded on an oscillograph.

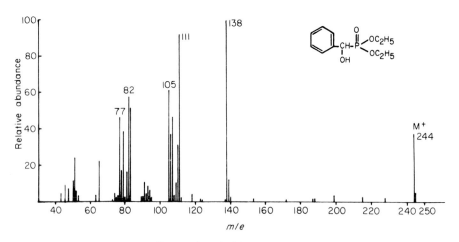

Fig. 5. A typical bar spectrum.

A. Sample Handling

Organometallic compounds are usually introduced through the direct inlet system. A few micrograms are volatilized at 10^{-6} mm pressure at the lowest temperature possible. In high-temperature work different techniques are used to volatilize elements and inorganic compounds. Both the Knudsen and Langmuir vapor-pressure methods provide suitable source of vapor species to the mass spectrometer.

B. Electron-Impact-Induced Processes

The detector records the ionic species formed by ionization processes and by the various reactions of these species. With moderate-energy electrons the following reactions are possible:†

$$P + e \xrightarrow{(7-15eV)} P^{+\cdot} + 2e \qquad \text{(Simple ionization)}$$
$$P + e \longrightarrow P^{n+} + (n+1)e \qquad \text{(Multiple ionization)}$$
$$P + e \longrightarrow P^{-} \qquad \text{(Electron capture)}$$

As the energy of the electron beam is increased, the ion abundance increases. If the energy is increased further, fragmentation of the molecule takes place. One or more of the following processes occur by electron impact: (1) ionization, (2) fragmentation, (3) pair production, (4) resonance capture, and (5) ion–molecule reaction. The last process gives rise to peaks with mass greater than the molecular weight. It is dependent on the sample pressure and is often detected from the variations in the relative intensities of the concerned peaks at different pressures.

C. Types of Ions

The following types of ions have been reported: (1) parent-molecule ion, (2) fragment ion, (3) rearrangement ion, (4) metastable ion, and (5) multiply charged ion.

Rearrangement ions are formed by the migration of hydrogen atom or functional groups such as alkyl, aryl, etc. as observed in organic compounds (*12*).

A metastable ion is one that decomposes after acceleration during its passage from the ion source to the detector. In a double-focusing instrument it is formed in the field-free region between the magnetic and electrostatic

---

† Throughout this review P is used to represent the molecular ion and M to denote metal atom.

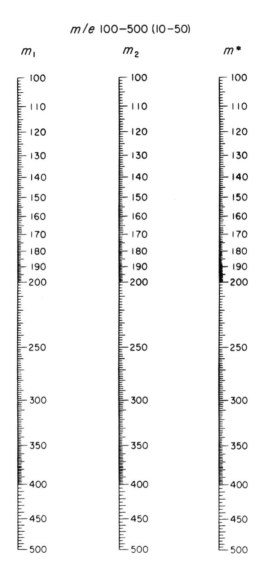

Fig. 6. Metastable-ion nomogram. [From F. W. McLafferty, "Interpretation of Mass Spectra." Benjamin, New York, 1966.]

analyzers. A diffuse peak called a metastable peak is usually observed at nonintegral mass due to the product ion from such a decomposition. The following mathematical formula relates the ions $m_1{}^+$, $m_2{}^+$ and m^*:

$$m_1{}^+ \rightarrow m_2{}^+ + \text{neutral fragment}$$

$$m^* = \frac{m_2{}^2}{m_1}$$

Beynon (8) has devised nomograms to correlate the parent, daughter, and metastable peaks. One example is shown in Fig. 6. These peaks are not always symmetrical. The isotope abundance pattern is evident in the metastable peak for ions containing polyisotopic elements. The pattern is distorted if the neutral fragment contains a polyatomic element (Fig. 7).

McLafferty *et al.* (85) have made use of metastable-ion characteristics to gain insight into the structures of many intermediate ions formed during the fragmentation of some organic compounds. Recently Williams and co-workers (26) have pointed out the pitfalls in such an approach.

Multiply charged ions result from the loss of more than one electron from the molecule. By suitably reversing the fields in the mass spectrometer it is possible to study negative ions in the same manner as positive ions. In negative-ion mass spectrometry the molecular ion peaks are very intense and fragmentation modes are insignificant processes.

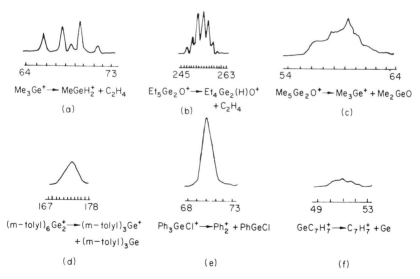

FIG. 7. Metastable peaks. [After F. Glockling and J. R. C. Light, *J. Chem. Soc.* A, 718 (1968) by permission of the authors and the Chemical Society.]

IV. Applications in Organometallic Chemistry

Systematic investigation into the behavior of organometallic compounds under electron impact was initiated after the advent of high-resolution mass spectrometers. Sufficient information is now available in this field to show the significant differences from organic molecules. In organometallic chemistry the three main uses of mass spectrometry are (1) molecular-weight and molecular-composition determination, (2) correlation of fragmentation modes and structure, and (3) determination of the appearance and ionization potentials and bond energies.

A. Molecular Formula

The peak at the highest m/e value is considered as the molecular ion $(P^{+\cdot})$. Generally this is true, but there are many instances where the parent ion is not observed because of thermal decomposition or fragmentation. In such cases it is sometimes possible to arrive at the molecular weight from a detailed study of the rest of the spectrum or with the help of computer techniques or by using a field-emission source (7). Mass markers such as perfluoroalkyl phosphonitriles are used to determine the exact integral mass. Peaks with m/e values above the mass of the parent ion have been observed in the spectra of nickel (36) and other metal chelates (37, 66). This is similar to the observations of Biemann et al. (90) in alkaloids. It is necessary to be cautious in determining molecular weights by mass spectrometry when other potential chelating sites are available for ion–molecule reactions to take place.

The high-resolution peak-matching technique (77) is of very great importance in computing exact empirical formulas from accurate molecular weights. The molecular formulas of several new compounds derive support from their mass-spectral data.

It is easy to identify metal-containing ions from the isotope patterns (21). With many polyatomic elements in a molecule the calculation of abundance patterns and mass combinations is carried out by computer methods (18). In arriving at the relative ion abundances allowance must be made for the ^{13}C content of each ion.

A rough indication of the stability of a fragment ion is its relative intensity. It is often rather difficult to assign specific structures for ions. Sometimes one has to invoke valency change in the metal atom. MacDonald and Shannon (66) have drawn a distinction between odd- and even-electron molecular ions and fragment ions by assuming valency changes of the metal in these ions—e.g., **I** shows an odd-electron ion, **II** an even-electron ion.

$$\left[\begin{array}{c} R \\ \diagdown \\ \diagup \\ R \end{array} M^{II} \right]^{+} \qquad\qquad \left[\begin{array}{c} R \\ \diagdown \\ \diagup \\ R \end{array} M^{III} \right]^{+}$$

(I)   (II)

Because of the empirical approach to mass spectrometry it is meaningful to arrive at definite conclusions on structural assignments only after comparative studies on model compounds.

B. General Fragmentation Modes

Few correlations on the influence of the metal atom on the fragmentation of the ligand have emerged from the information available in literature. It is to be expected that in the next few years the situation will be different, and it may be possible to predict fragmentation of organometallic compounds with greater confidence. Most of the observed differences in the behavior of organometallic and organic compounds arise from the low M—C and M—H bond strengths compared with those of C—C and C—H.

Because of the less electronegative character of metals compared with carbon and hydrogen, when a positive ion decomposes, the charge prefers to stay with the metal-containing moiety. In general a high proportion of the ion current is usually carried by metal-containing fragments. However, elimination of the metal-containing moiety as a neutral molecule or a radical has also been reported in a few cases. The relatively poor thermal stability of organometallic compounds is well known (75). Hence the ion abundance reflected in the mass spectrum can depend on the experimental conditions under which the spectrum is recorded.

One of the major problems in mass spectrometry is the location of the exact site of primary ionization process. The concept of charge localization has been successfully applied by McLafferty et al. (70) and others for satisfactory interpretation of fragmentation reactions. Biemann et al. (69) have shown that this assumption is unwarranted. " The situation may more closely resemble a dynamic distribution of charge, which for a period required for electron impact induced fragmentation is statistically maximized at the site of lowest ionisation potential, i.e. the average charge density decreases with increasing relative ionisation potential of the site." It is often not clear from the ion abundances whether in the initial ionization step a bonding or nonbonding electron is lost. So far no general rules have emerged out that correlate the molecular ion abundance and structure.

One of the well-established general fragmentation reactions is the decomposition of the molecular ion by radical elimination. This leads to the formation of stable even-electron ions. The subsequent decomposition process is often dominated by the elimination of neutral molecules to form further even-electron ions:

$$P^{+\cdot} \xrightarrow{\text{$-$Radical}} F_1^{+} \xrightarrow[\text{molecule}]{\text{$-$Neutral}} F_2^{+}$$

Almost all types of organometallic compounds undergo hydrogen transfer rearrangement processes resulting in a variety of metal hydrides of varying abundance:

$$Et_3M^{+} \xrightarrow{-C_2H_4} Et_2MH^{+} \xrightarrow{-C_2H_4} EtMH_2^{+} \xrightarrow{-C_2H_4} MH_3^{+}$$

The relative intensity of the metal ion M^{+} seems to depend on the ionization potential and also on the bond energies of M—C, M—H, and M—M bonds.

Rearrangement ions formed by the electron-impact-induced migration of functional groups such as alkyl, aryl, etc. are often of high abundance in the spectra of many organometallic compounds.

The following discussion is mainly concerned with the behavior under electron impact of main-group metal compounds, organotransition-metal complexes, and a few other metal–organic complexes.

C. Main-Group Metal Compounds

The metal alkyls of Li (*48*), Be, Mg (*20*), Zn, Hg, Al (*99*), Si (*5*), Ge (*79*), Sn (*101*), Pb (*79*), As, Sb, and Bi (*62*) have been subjected to electron-bombard-ment studies. Evidence for the polymeric nature of some of these compounds in the vapor phase has been obtained. Thus ethyl–lithium vapor was found to consist of tetramer and hexamer units, both of which decompose on ioniza-tion. Only under very carefully controlled spectrometer conditions is it possible to see appreciable amounts of the polymeric structures. At high source temperatures dissociation or thermal degradation can result.

The characteristic fragmentation mode of beryllium alkyls is alkane elimination, leading to the formation of stable even-electron ions (*20*):

$$Et_4Be_2^{+\cdot} \xrightarrow{-Et} Et_3Be_2^{+} \xrightarrow{-C_2H_4} Et_2Be_2H^{+} \xrightarrow{-C_2H_4} EtBe_2H_2^{+}$$

If the ionization potential of the hydrocarbon radical is smaller than that of beryllium, the metal elimination is a preferred fragmentation process. In zinc and mercury alkyls the major fragmentation reactions are simple cleavage processes. Metal hydride ions are not very significant in lead alkyls.

In the alkyls of Si, Ge, Sn, and Pb the molecular ions are minor peaks. A significant fragmentation reaction is radical elimination from the molec-ular ion and from odd-electron fragment ions.

In compounds of the type R_3MR' the relative probability of elimination of the various radicals depends on bond strengths and relative stabilities of the radicals and ions produced.

Metal–metal bonded compounds of the type X_3M—$M'Y_3$ (M, M' = Si, Ge, Sn) were reported (40) to undergo extensive rearrangement reactions in which exchange or transfer of organic groups X and Y takes place across the metal–metal bond. Similar reorganization processes were reported in some transition-metal complexes—e.g., Ph_3Sn-$Fe(CO)_2\pi$-C_5H_5 shows ions in which C_5H_5 is directly bonded to Sn and Ph to Fe (65). In order to explain these processes satisfactorily it is necessary to assume that the molecular ion forms a five-coordinate intermediate with d-orbital participation. This also explains why the hydrocarbon Ph_3C—CMe_3 does not undergo similar rearrangement processes.

Four different types of metastable ions were observed for the following transitions (see Fig. 7):

$$Me_5Ge_2O^+ \rightarrow Me_3Ge^+ + Me_2GeO$$
$$(m\text{-tolyl})_6Ge_2{}^+ \rightarrow (m\text{-tolyl})_3Ge^+ + (m\text{-tolyl})_3Ge$$
$$Ph_3GeCl^{+\cdot} \rightarrow Ph_2{}^+ + PhGeCl$$
$$GeC_7H_7{}^+ \rightarrow C_7H_7{}^+ + Ge$$

In the spectra of R_3M—$M'Ph_3$ compounds, the $Ph_3M'^+$ ion is more abundant irrespective of the nature of the metal. This has been rationalized by assuming that the ionization potential of Ph_3M' radical is lower than that of R_3M, that charge delocalization over the phenyl group stabilizes $Ph_3M'^+$ ion, and that fewer low-energy decomposition processes are available for $Ph_3M'^+$ than for R_3M^+.

Metastable-ion-supported alkene elimination is observed for all σ-bonded metal alkyls. Some of the reactions leading to the elimination of neutral molecules from even electron ions are as follows:

$$Ph_3M^+ \rightarrow PhM^+ + Ph_2$$
$$Ph_2MCl^+ \rightarrow PhM^+ + PhCl$$
$$Ph_2MH^+ \rightarrow PhM^+ + C_6H_6$$

A few odd-electron ions also show the loss of neutral molecules:

$$Ph_4M^{+\cdot} \rightarrow Ph_2M^{+\cdot} + Ph_2$$
$$Et_3GeH^{+\cdot} \rightarrow Et_2Ge^{+\cdot} + C_2H_6$$
$$MeSiH^{+\cdot} \rightarrow Si^{+\cdot} + CH_4$$
$$Ph_3MX^{+\cdot} \rightarrow Ph_2^{+\cdot} + PhMX$$

The trialkyl compounds of Group III have been studied intensively in order to characterize electron-deficient bridge bonds and the exchange phenomenon associated with them. Evidence for the dimeric vinyl bridge structure (**III**) in trivinyl gallium was obtained from the NMR spectrum (93).

(III)

D. Organotransition-Metal Compounds

A sizable amount of information is now available on the mass spectrometric behavior of organotransition-metal compounds. These compounds are conveniently classified on the basis of the type of ligand present: (1) transition-metal carbonyls, (2) metal hydride compounds, (3) metal carbonyl halides, and (4) complexes containing π-bonded ligands.

1. TRANSITION-METAL CARBONYLS

The mononuclear carbonyls of iron, nickel, vanadium, manganese, chromium, tungsten, cobalt, and molybdenum (9, 75, 97) have been found to show significant molecular ion peaks. They lose carbon monoxide successively until the metal ion is formed:†

$$M(CO)_6^{+} \xrightarrow[*]{-CO} M(CO)_5^{+} \xrightarrow[*]{-CO} M(CO)_4^{+} \xrightarrow{\text{etc.}} M^{+}$$

From the appearance-potential measurements it has been suggested that a nonbonding electron is lost in the electron-impact-induced ionization process. Metal ions are most abundant in their spectra. Doubly charged molecular and fragment ions also break down by stepwise loss of carbon monoxide:

$$M(CO)_n^{2+} \rightarrow M(CO)_{n-1}^{2+} + CO$$

Cleavage of the carbon–oxygen bond to form metal carbides occurs in heavier metal carbonyls, and these ions in turn fragment by loss of carbon monoxide (98):

$$M(CO)_nC^{+} \rightarrow M(CO)_{n-1}C^{+} + CO$$

The mass spectra of homonuclear carbonyls of manganese, rhenium, osmium, cobalt, rhodium, and iridium and heteronuclear carbonyls $MnRe(CO)_{10}$, $Ru_2Os(CO)_{12}$, $RuOs_2(CO)_{12}$, and $Mn_2Fe(CO)_{14}$ have been studied (56, 84). These polynuclear metal carbonyls also show progressive loss

† An asterisk indicates that a metastable peak is observed.

of carbon monoxide from the parent ion. However, cleavage of the metal cluster depends on the nature of the metal. Binuclear carbonyls break into charged and neutral mononuclear fragments.

From the differences in the fragmentation of bi- and polynuclear carbonyls it is possible to obtain evidence on the extent of metal–metal bonding. In metal–metal bonded carbonyls a high proportion of the ion current is carried by ions containing metal–metal bonds. The doubly charged dimetallic ions are also abundant in their spectra. Compounds with bridging carbonyl groups, on the other hand, show extensive cleavage to mononuclear ions. Thus $Mn_2(CO)_{10}$ shows consecutive loss of ten carbonyl groups to give Mn_2 ion. Cleavage into monomeric units is a minor fragmentation process. The dimeric nature of $Fe_2(CO)_9$ is evident from its molecular weight. The appearance of $Fe(CO)_5{}^+$ is in agreement with the presence of bridging carbonyl groups. Mononuclear ions of the type $[Fe(CO)_n]$ ($n = 0$–5) are abundant, suggesting that cleavage of the bridging carbonyl bond occurs in preference to the cleavage of terminal carbonyl bond. Structure **IV** shows $Mn_2(CO)_{10}$, **V** shows $Fe_2(CO)_9$.

(IV)   (V)

From a study of the spectra of cyclic and noncyclic trinuclear metal carbonyls of iron, ruthenium, and osmium and tetranuclear carbonyls of cobalt, rhodium, and iridium, Lewis and associates (64) have shown that (1) the parent molecular ions are observed in all cases, (2) stepwise loss of carbon monoxide occurs to give the polynuclear metal ion, (3) carbon monoxide transfer does not occur, (4) ions of the type $[M(CO)_xC]^+$ are common for heavier transition metals, and (5) no ions of the type $[M(CO)_xO]^+$ are observed.

2. METAL-HYDRIDE COMPOUNDS

One of the main difficulties in the correct identification of metal hydrides is the determination of the number of hydrogen atoms present per molecule. Mass spectrometry has provided satisfactory solutions to these problems in many cases (57)—e.g., $H_3Mn_3(CO)_{12}$, $H_3Re_2(CO)_{12}$, and $H_4Ru_4(CO)_{12}$. Since most metals are polyisotopic, it is easy to identify metal-containing fragments from the isotopic clusters, without recourse to accurate mass measurement.

Only a limited number of transition-metal hydrides have been examined by this technique. Parent molecular ions are observed in all mononuclear hydrides. The parent and fragment ions retain the hydrogen atoms. The carbonyl hydrides $HMn(CO)_5$, $HRe(CO)_5$, and $HCo(CO)_4$ show competitive loss of hydrogen and carbon monoxide:

$$[HMn(CO)_5] \xrightarrow{-CO} [HMn(CO)_4]^+ \xrightarrow[\text{(stepwise)}]{-4CO} [MnH]^+$$

$$\searrow{-H} \qquad\qquad {-H}\nearrow$$

$$[Mn(CO)_5]^+ \xrightarrow[\text{(stepwise)}]{-5CO} [Mn]^+$$

The weakly acidic cyclopentadienyl complexes $HMo(CO)_3(C_5H_5)$ and $HW(CO)_3(C_5H_5)$ show successive loss of carbon monoxide and hydrogen radicals. Polynuclear carbonyl hydrides lose carbon monoxide in preference to hydrogen. The parent molecular ions have been observed in the spectra of $H_3Mn_3(CO)_{12}$ and $H_3Re_2(CO)_{12}$.

The spectrum of $H_7B_2Mn_3(CO)_{10}$ shows stepwise loss of ten carbonyl groups. All the hydrogen atoms are retained until four carbonyl groups are lost. It is reasonable to assume that the bridging hydrogen atoms are more strongly bonded than the terminal hydrogens (87). The structures of several new compounds have been confirmed from the mass-spectral evidence—e.g., $HMCo_3(CO)_{12}$ (M = Fe, Ru), $H_4Os_4(CO)_{12}$, $HRu_4(CO)_{13}$, $H_2Os_4(CO)_{13}$, $HRe_3(CO)_{14}$, $HMnRe_2(CO)_{14}$, and $HMnRe_2(CO)_{14}$.

3. METAL CARBONYL HALIDES

Bridging and terminal metal-carbonyl-halide spectra were also studied. Those containing terminal-bonded halogen include the mononuclear carbonyl halides of manganese, rhenium [$M(CO)_5X$ (X = Cl, Br, or I)], iron, and ruthenium [$M'(CO)_4X_2$ (X = Cl, Br, or I)]. The bridging carbonyl halides consist of the dimeric compounds $M_2(CO)_8X_2$ (M = Mn or Re; X = Cl, Br, or I) and $Rh_2(CO)_4X_2$ (X = Cl, Br, or I). Parent ions are generally observed in their spectra. They show considerable differences in their fragmentation modes. Mononuclear complexes decompose by the loss of CO or X radicals with equal ease. The intermediate ions also show loss of halogen:

$$[M(CO)_5X]^+ \xrightarrow{-CO} [M(CO)_4X]^+ \xrightarrow[\text{(stepwise)}]{-4CO} (MX)^+$$

$$\searrow{-X} \qquad\qquad {-X}\swarrow$$

$$[M(CO)_5]^+ \xrightarrow[\text{(stepwise)}]{-5CO} [M]^+$$

The binuclear bridged complexes lose carbon monoxide stepwise to give the M_2X_2 ion. The loss of halogen is observed only from the breakdown of this ion:

$$[M_2(CO)_nX_2]^+ \xrightarrow[\text{(stepwise)}]{-nCO} [M_2X_2]^+ \xrightarrow{-X} [M_2X]^+ \text{ etc.}$$

The diagnostic value of mass spectrometry in distinguishing between these carbonyl halides is thus evident. Both the hydrides and halides show significant difference in the case of elimination of terminal and bridging ligands under electron impact.

The spectra of nitrosyl halide complexes $M_2(NO)_4X_2$ indicate that the primary fragmentation involves the successive loss of nitrosyl groups before cleavage of the M_2X_2 ion:

$$[M_2(NO)_4X_2]^+ \xrightarrow[\text{(stepwise)}]{-NO} [M_2(NO)_nX_2]^+ + (3-n)NO$$

where $n = 0-3$.

In the fragmentation of nitrosyl carbonyl complexes the loss of nitric oxide competes with carbon monoxide loss. The loss of nitric oxide involves the transfer of three electrons, while removal of carbon monoxide results from a two-electron transfer. The dimeric nature of metal nitrosyl thiol complexes has been ascertained from their mass spectra. The metal–sulphur ring structure appears to be very stable under electron impact. Carbonyl-sulphur complexes $Fe_3(CO)_9S_2$, $Fe_2(CO)_6S_2R_2$, $Mn_2(CO)_8S_2R_2$, and $Re_2(CO)_8S_2Ph_2$ lose carbon monoxide, leaving the ion in which sulphur is attached to the metal.

4. π-BONDED COMPLEXES

Many complexes containing π-bonded ligands (alkyl, cyclobutadiene cyclopentadiene, arene, olefin, and acetylene) have been extensively studied (13, 71, 73) (Fe, Co, Ni, V, Cr, Ru, Re, Os). Metal–ring-bond cleavage processes lead to the most abundant ions in their spectra:

$$(C_5H_5)_2M^{+\cdot} \begin{cases} (C_5H_5)M^+ \longrightarrow M^+ \\ M^+ + C_{10}H_{10} \end{cases}$$

The cyclopentadienyl group fragments by the loss of acetylene. Insignificant ions also result from the elimination of H, CH_3, and other hydrocarbons:

$$(C_5H_5)_2M^+ \rightarrow C_8H_8M^+ + C_2H_2$$

Ion–molecule reactions were reported to occur in some bis-cyclopenta-dienyls. The fragmentation of ferrocene and nickelocene have been described. Ferrocene shows ions originating from $(C_5H_5)_3Fe_2{}^+$. In ring-substituted ferrocenes the abundance of metal-containing fragments is less (*80*). Bis-π-alkyl complexes of nickel fragment by elimination of ethylene and alkyl radicals. The diagnostic value of mass spectrometry to characterize complex cyclopentadienyl complexes has been now well established. The field-ioniza-tion technique is used to reduce fragmentation reactions in π-alkyl complexes so that the molecular ion is observed in greater abundance.

E. Organophosphorus Compounds

Many phosphorus compounds have been studied mass spectrometrically. Kiser and associates (*94*) reported the fragmentation and appearance poten-tials of phosphine and diphosphine. Their interpretation of the fragmenta-tion reactions was supported by deuterium-labeling experiments. From the appearance-potential measurements the formation of $PH_4{}^+$ ion in the mass spectrum of phosphine has been shown to be due to ion–molecule reaction (*46*): $PH_3{}^+ + PH_3 \rightarrow PH_4{}^+ + PH_2$. Williams and co-workers (*96*) have evaluated deuterium–hydrogen scrambling in deuterium-labeled benzene ring from the electron-impact studies on triphenylphosphine, triphenylphosphine oxide, etc. The mass spectrum of triphosphine was reported recently (*39*). The behavior of many phosphine–metal carbonyl complexes (*10*) and fluoro-alkyl phosphines (*19*) under electron impact has been studied.

The organophosphates are known to behave anomalously in the mass spectrometer (*76*). Rearrangement ions resulting from the transfer of one or two hydrogen atoms from the substituent to the electrophilic phosphate moiety have been observed in the spectra of trialkyl and alkyl/arylphosphates. Similar atomic migrations were reported in the spectra of a series of alkyl phosphites (*47*). It has been concluded from these studies that phosphorus(V) is a stable valency state of the element and the quadricovalent state is not achieved by electron bombardment.

The fragmentation of the esters of phosphorus and phosphonic acids (*74*) is dominated by hydrogen-transfer rearrangement processes as well as simple bond-cleavage reactions. Pentavalent phosphorus ions are formed by the transfer of hydrogen to both phosphorus and oxygen atoms. Some of the general fragmentation modes of phosphonic acid esters are shown in **VI**.

An interesting rearrangement process characteristic of α-hydroxy phos-phonic esters (**VII**) (*31*) is the formation of the (M—O) ion. This type of fragmentation mode has been reported in oximes (*42*).

Some of the characteristic fragmentation modes leading to bond-forming

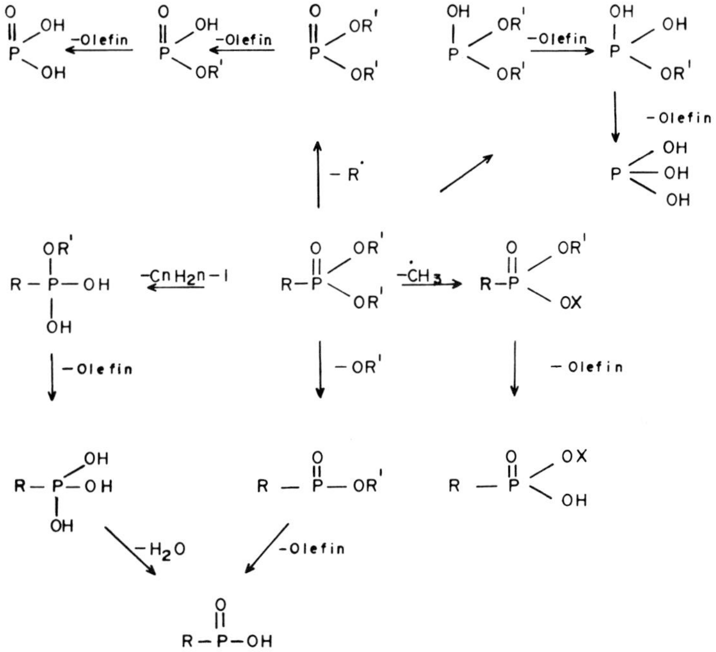

(VI)

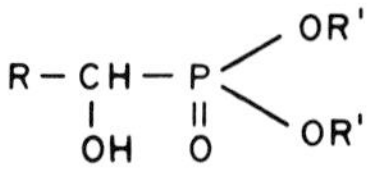

(VII)

reactions of compounds containing the P=S bond are (27) (1) skeletal rearrangement of the type **VIII**, (2) migration of aryl groups to phosphorus, (3) formation of the (P—SH) ion from the transfer of a hydrogen atom from the alkyl or aryl group, and (4) formation of phenol ion in phenoxy derivatives (**IX**).

The spectrum of p-nitrophenyl phosphor-amidothioate shows that P—O bond fission is competing with P—N bond cleavage (**X**).

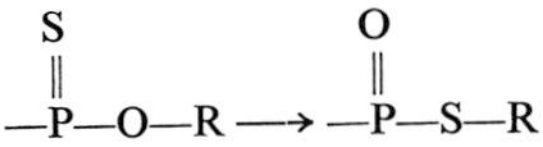

$$\overset{S}{\underset{\|}{}}\;P{-}O{-}R \longrightarrow \overset{O}{\underset{\|}{}}\;P{-}S{-}R$$

(VIII)

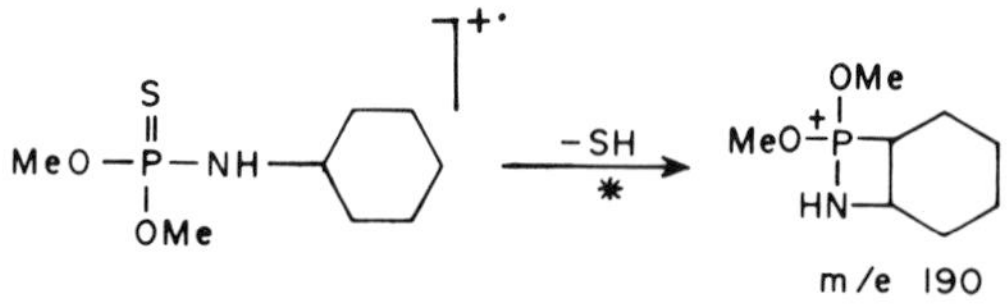

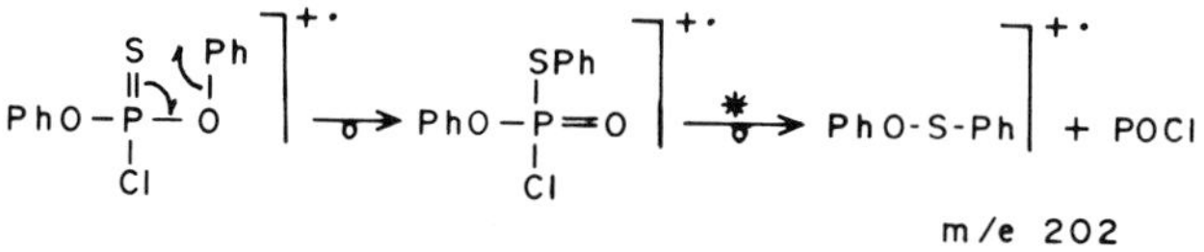

m/e 190

(IX)

m/e 202

MeO—P
OPhNO$_2$ m/e 232

m/e 180

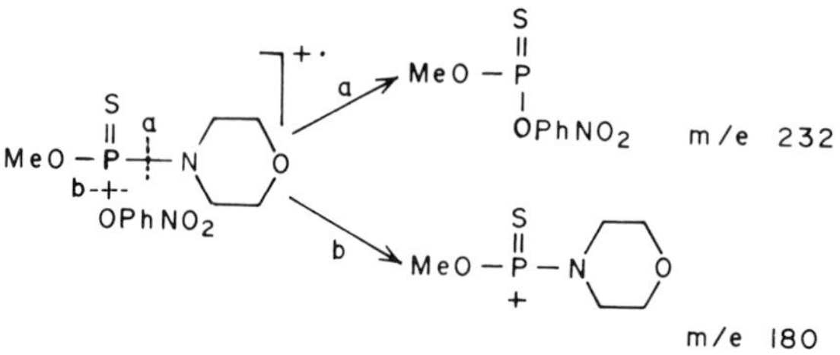

(X)

Isomerization of phosphorothioates to the S-*p*-nitrophenyl phosphor-amidate has been reported to take place thermally (*50*) and by electron impact (*58*).

The molecular ion is the base peak in the spectrum of diphenyl phos-phorochloridothioate. The fragmentation reactions of this compound include the loss of SH, POCl, and PO$_2$Cl. Many rearrangement processes operating in these molecules involve the nucleophilic migration of aryl groups to posi-tive centers, which leads to the elimination of neutral molecules (**XI**).

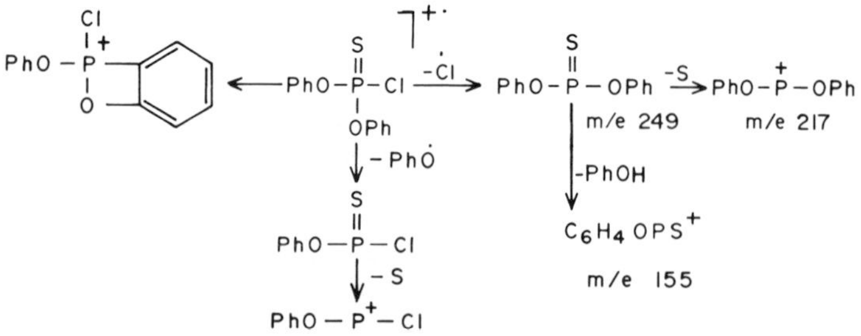

m/e 249   m/e 217

C$_6$H$_4$OPS$^+$

m/e 155

(XI)

The mass-spectral behavior of some pentavalent organophosphorus pesticide esters has been described (30). They include phosphorodithioates, phosphorothionates, phosphorothiolates, and phosphates (**XII**).

(**XII**)

The elemental compositions of the major fragments were established from accurate mass measurements. In most cases the base peaks are formed by simple cleavage and rearrangement processes. The major rearrangement ions are formed by migration of hydrogen atom from the alkyl ester group to the thiophosphite oxygen skeleton and migration of hydrogen from the Z moiety to the phosphorus–oxygen skeleton or to the thiophosphite–oxygen skeleton. Migration to the Z moiety of hydrogen from the ester group bonded to phosphorus, and alkyl migration to the Z moiety were also observed. The base peak has been found to be not characteristic of the different groups or compounds within a group. Cleavage $\beta-$ to Z moiety produces relatively intense peaks in a given group of compounds. Some typical rearrangement processes are shown in **XIII**.

The mass spectra of fluorophosphonitriles (11), chlorophosphonitriles (83), and bromophosphonitriles (28) have been reported in the literature. In the mass-spectral fragmentation of chlorophosphonitriles the relative ion stabilities have been explained in terms of rehybridization of the σ bond and alteration of the π molecular orbital. The primary ionization process has been assumed to be the loss of an electron from a bonding π molecular orbital of $P_3N_3Cl_6$. This is followed by elimination of a chlorine atom accompanied by a σ-bond rehybridization and some p_π–p_π overlap in the ring. The same reasoning has been used to explain the ion abundances in the spectra of similar type of compounds. A metastable ion-supported process of ring contraction leading to the elimination of PN moiety has been observed:

$$P_nN_nCl_{2n} + e \rightarrow P_nN_nCl^+_{2n-1} + Cl^- + e$$
$$P_4N_4Cl_5^+ \rightarrow P_3N_3Cl_5^+ + PN$$
$$P_4N_4Cl_5^{2+} \rightarrow P_3N_3Cl_5^{2+} + PN$$

Ring contraction by loss of $PNCl_2$ has also been postulated:

$$P_4N_4Cl_7^+ \rightarrow P_3N_3Cl_5^+ + PNCl_2$$

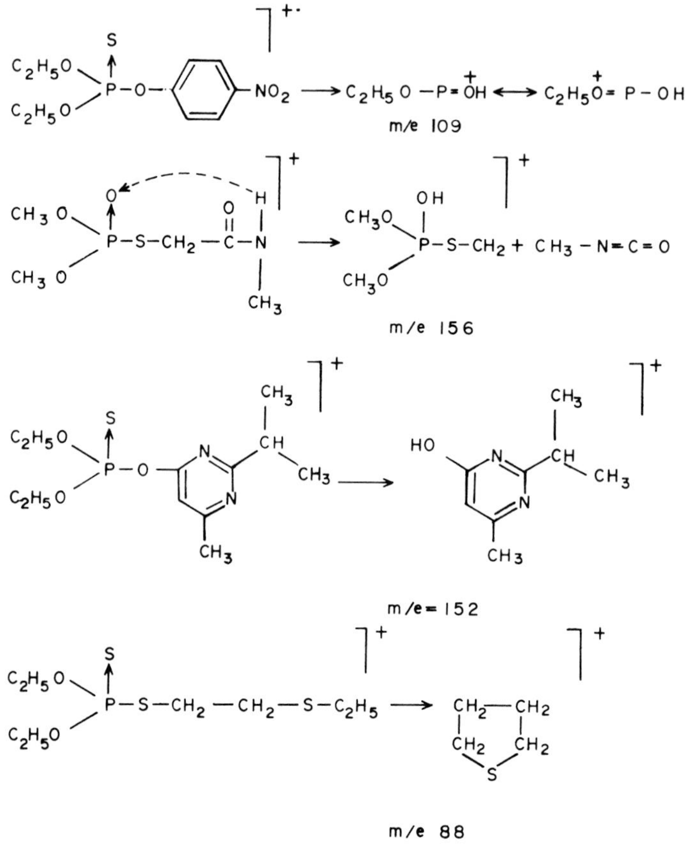

(XIII)

The ions $P_4^{+\cdot}$, $P_3N^{+\cdot}$, and $P_2N_2^{+\cdot}$ reported represent a group of related ions resulting from the replacement of the phosphorus atoms by nitrogen atoms in the well-known P_4 tetrahedral molecule.

Coxon and co-workers (29) have described the behavior of nongeminally substituted cyclic chlorobromotriphosphonitriles, $P_3N_3Cl_xBr_{6-x}$. Cyclic and linear ions are formed as in $P_3N_3Br_6$ and $P_3N_3Cl_6$. Of the total ion current $\sim 72\%$ is carried by cyclic ions and $\sim 17\%$ by linear ions. The decrease in ionization potential resulting from the introduction of bromine atoms has been explained as due to the increased localization of electron density on the ring nitrogen atom. The following scheme shows some fragmentation paths operating in these compounds.

$$P_3N_3Cl_3Br^{+\cdot} \xrightarrow{\;*\;} P_3N_3Cl_3Br_2^+ \xrightarrow{\;*\;} P_3N_3Cl_3Br^{+\cdot} \xrightarrow{\;*\;} P_3N_3Cl_3^+$$

| (3.9%) | (100%) | (1.9%) | (12.2%) |

$$P_3N_3Cl_2Br_3^+ \longrightarrow P_3N_3Cl_2Br^{+\cdot} \longrightarrow P_3N_3Cl_2Br^{+\cdot} \longrightarrow P_3N_3Cl_2^{+\cdot}$$

| (1.5%) | (0.5%) | (7.7%) | (3.6%) |

$$P_3N_3ClBr_2^+ \longrightarrow P_3N_3ClBr^{+\cdot} \longrightarrow P_3N_3Cl^+$$

| (0.4%) | (0.9%) | (3.5%) |

F. Silicon Compounds

The mass spectral fragmentation modes of only very few silicon compounds have been investigated in detail. The Russian group has mainly studied the behavior of many silahydrocarbons (22–24) siloxanes (35, 72, 102), etc. They have reported that in silahydrocarbons containing two silicon atoms separated by a methylene group, the stability of the molecular ion falls with increase in the chain length. Branching in the bridge leads to slight reduction in the stability as compared with the unbranched bridge with the same number of carbon atoms. The

$$\diagdown \diagup$$
$$—Si—CH_2—Si—$$
$$\diagup \diagdown$$

group has been shown to be more stable than

$$\diagdown \diagup$$
$$—C—CH_2—C—$$
$$\diagup \diagdown$$

group.

The mass spectra of only a few siloxanes (35, 71, 102) have been published. The decomposition of siloxanes under electron impact, like that of other organosilicon compounds, goes extremely selectively. The characteristic features are (1) extremely small P^+, (2) intense P-15 ion, (3) multiply charged ion, (4) elimination of hydrocarbon groups from the organic periphery of the molecule, and (5) abundant metastable peaks.

$$(CH_3)_3SiOH^+ \xrightarrow{-CH_3} C_2H_6SiOH^+—$$
$$(P+ = 0.4\%) \qquad\qquad (100\%)$$

$$\xrightarrow{-CH_2} CH_5SiO^+ \xrightarrow{-H_2O} CH_3Si^+$$
$$(2\%) \qquad\qquad (7\%)$$

$$\xrightarrow{-CH_4} CH_3SiO^+$$
$$(6\%)$$

$$\xrightarrow{-C_2H_4} H_3SiO^+ \xrightarrow{-H_2O} HSi^+$$
$$(12\%) \qquad\qquad (8\%)$$

$$\xrightarrow{-C_2H_6} HSiO^+$$
$$(20\%)$$

It has been shown from the mass spectra of alkyl silanes (3, 59) that change from a tertiary carbon atom to a tertiary silicon atom results in an increase in the stability of the molecule to electron impact by a factor of 2–3. However, replacement of a quaternary carbon atom by a quaternary silicon atom leads to an increase in stability by a factor of 100. A probable scheme for the fragmentation of alkyl silanes is as follows:

$$[C_5SiH_{14}]^{+\cdot} \begin{cases} \xrightarrow{-\cdot CH_3} [C_4SiH_{11}]^+ \xrightarrow{-C_2H_4} [C_2SiH_7]^+ \xrightarrow{-C_2H_4} [SiH_3]^+ \xrightarrow{-2H} [SiH]^+ \\ \quad\quad\quad (22\%) \quad\quad\quad\quad (49\%) \quad\quad\quad\quad (2.6\%) \quad\quad (3.7\%\cdot) \\ \xrightarrow{-\cdot C_2H_5} [C_3SiH_9]^+ \xrightarrow{-C_2H_4} [CSiH_5]^+ \xrightarrow{-2H} [CSiH_3]^+ \\ \quad\quad\quad (100\%) \quad\quad\quad (8.6\%) \quad\quad\quad (10\%) \end{cases}$$

The typical fragmentation modes of a cyclic hydrocarbon containing a quaternary silicon is as follows:

$$[C_6SiH_{14}]^+ \begin{cases} \xrightarrow{-H} [C_6SiH_{13}]^+ \xrightarrow{-C_2H_4} [C_4SiH_9]^+ \xrightarrow{-C_2H_4} [C_2SiH_5]^+ \\ \quad\quad (27\%) \quad\quad\quad (29\%) \quad\quad\quad (6\%) \\ \xrightarrow{-CH_3} [C_5SiH_{11}]^+ \xrightarrow{-C_2H_4} [C_3SiH_7]^+ \xrightarrow{-C_2H_4} [CSiH_3]^+ \\ \quad\quad (26\%) \quad\quad\quad (44\%) \quad\quad\quad\quad (34\%) \\ \quad\quad\quad\quad\quad\quad\quad\quad\quad\quad\quad\quad\quad\quad\quad\quad -CH_3 \uparrow \\ \xrightarrow{-C_2H_4} [C_4SiH_{10}]^+ \xrightarrow{\quad\quad -C_2H_4 \quad\quad} [C_2SiH_6]^+ \\ \quad\quad (100\%) \quad\quad\quad\quad\quad\quad\quad\quad\quad (61\%) \\ \xrightarrow{-C_3H_5} [C_3SiH_9]^+ \xrightarrow{-C_2H_4} [CSiH_5]^+ \\ \quad\quad (11\%) \quad\quad\quad (28\%) \\ \xrightarrow{-C_3H_6} [C_3SiH_8]^+ \xrightarrow{-C_2H_4} [CSiH_4]^+ \\ \quad\quad (27\%) \quad\quad\quad (9\%) \\ \xrightarrow{-C_4H_7} [C_2SiH_7]^+ \\ \quad\quad (23\%) \end{cases}$$

The mass spectra of some silazanes and chloro and methoxy silazanes have been presented (86). Metastable supported fragmentation modes leading to the expulsion of NH_3, CH_4, HCl, and CH_2O were observed. 1,3-Dimethoxy tetra methyl disilazane, for example, undergoes the following fragmentation sequence:

$$\begin{bmatrix} CH_3 & CH_3 \\ | & | \\ CH_3O-Si-NH-Si-OCH_3 \\ | & | \\ CH_3 & CH_3 \end{bmatrix}^{+\cdot} \xrightarrow{-CH_3} \begin{array}{c} CH_3 \\ | \\ CH_3O-Si-NH^+=Si-OCH_3 \\ | \quad\quad\quad | \\ CH_3 \quad\quad CH_3 \end{array}$$

$$\downarrow -CH_2O$$

$$\begin{array}{c} CH_3 \\ | \\ H-Si-NH^+=Si-H \\ | \quad\quad | \\ CH_3 \quad CH_3 \end{array} \xleftarrow{-CH_2O} \begin{array}{c} CH_3 \\ | \\ H-Si-NH^+=Si-OCH_3 \\ | \quad\quad\quad | \\ CH_3 \quad\quad CH_3 \end{array}$$

The existence of the radical anion of dodecamethyl cyclohexasilane in the gas phase has been shown by Gohlke (*41*). The formation of this anion involves electron capture. The delocalization of the added electron (probably in the 3*d* orbitals of adjacent silicon atoms) imparts an unusual degree of stability to the radical ion.

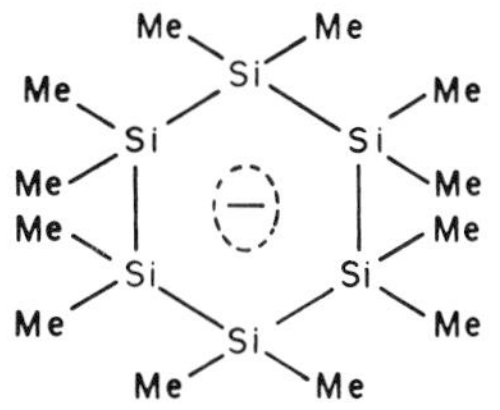

(**XIV**)

The mass spectra of tetraaza-3,6-disilacyclohexanes and silylhydrazines were studied (*32*). Some of the general fragmentation modes observed in the former types of compounds are:

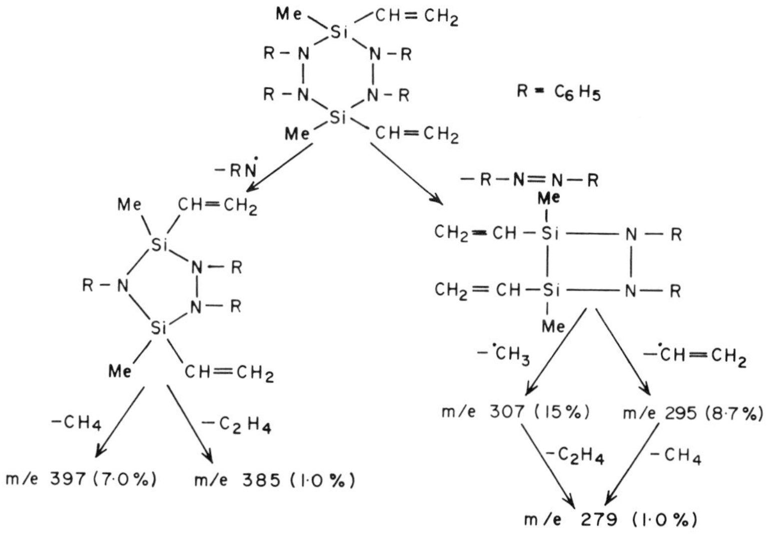

(**XV**)

Metastable-peak-supported rearrangement of the doubly charged molecular ion into two singly charged ions was observed in the spectra of tetraaza-3,6-disilacyclohexane:

$$P^{2+} \xrightarrow{\quad * \quad} [M-C_6H_5-N\!=\!N-C_6H_5]^+ + [C_6H_5-N\!=\!N-C_6H_5]^+$$
$$\text{(M.W. 504)} \qquad\qquad (m/e\ 322) \qquad\qquad (m/e\ 182)$$

The silyl hydrazines also show a number of rearrangement ions in their spectra (**XVI**).

(XVI)

G. Other Types of Compounds

Only a limited number of metal Schiff's base chelates have been analyzed spectrometrically. The mass spectra of bis[o-[N-(2-anilinoethyl)formimidoyl]-phenolato]nickel(II) and bis[o-[N-(2-methylaminoethyl)formimidoyl]pheno-lato]nickel(II) showed peaks above the molecular ion, formed by thermal reorganization processes (36). These ions are similar to the polymeric species observed with acetylacetonates (66). Recently light has been shed on the un-settled structure of the diamagnetic nickel chelate obtained by the reaction of nitrite ion with nickel acetyl acetonate (33).

Fe(sal-N-n-propyl)$_3$, Fe(sal-N-n-propyl)$_2$Cl, and [Fe(sal-N-n-propyl)$_2$]$_2$O show a base peak at m/e 380 corresponding to the Fe(sal-N-n-propyl)$_2$$^+$ ion. This ion further loses one ligand preferentially. The ligand ion and its charac-teristic fragments are significant peaks in the mass spectrum. Fe(sal-N-n-propyl)$_3$ shows peak of low intensity corresponding to the binuclear species. The chloride shows a low-intensity molecular ion that loses chlorine (92). The less volatile and air-sensitive titanium cyclopentadienyl complexes have been subjected to electron-impact studies (89). The structures were assigned to the major fragments from high-resolution data.

The fragmentation modes of a number of metal acetylacetonates have been published (4, 67, 78, 82). Even though the intense peaks correspond to the monomeric forms, peaks due to polymeric structures in the ionized vapor were reported. Evidence for the change of odd-electron ions to even-electron ions by change of valency of the metal atom in the ions was obtained for Fe and Al acetylacetonates. Successive losses of methyl radicals and methyl and phenyl migrations were also illustrated.

The mass spectra of the acetyl acetonates of Al, Be, Cd, Cr, Cu, Mn, Zn, Cr, Na, Mg, Ce, and Sr have been studied. The loss of ligand is observed. A process that is strongly dependent upon the metal present is the loss of CH_3 from $P^{+\cdot}$. Bridging structures were proposed for some of the polymeric species. Some of the characteristic fragmentation modes are loss of CH_3, ketene acetyl, and ligand radicals as shown in **XVII**.

Bruce (15) has studied the behavior of some transition-metal complexes derived from 3,3,3-trifluoroprop-1-yne. The main fragmentation reactions include the loss of carbon monoxide, loss of one fluorine atom, and expulsion of neutral metal fluorides with ligand transfer and loss of neutral fluorocarbon molecules. The scheme of **XVIII** shows the fragmentation reactions of the complex π-$C_5H_5CoC_5H_5C{\equiv}CCF_3$.

The mass spectra of many fluoroalkyl, fluoroalkenyl, and fluoroacyl complexes of manganese, rhenium, iron, and ruthenium carbonyls are reported (16). The formation of difluoro carbene-metal fluoride has been postulated in trifluoro vinyl complexes:

$$CF_3{-}CF{=}CFRe(CO)_5{}^+ \xrightarrow[\text{(stepwise)}]{-5CO} C_3F_5Re^+ \quad (m/e\ 318)$$

$$\downarrow {-}CF_2$$

$$C_2F_3Re^+ \quad (m/e\ 268)$$

Electron-impact studies on selenium compounds (63), germanocyclopentanes, germanocyclopentenes (38), indenyl metal derivatives (60), metal oxinates (68), cyclopentadienone tricarbonyl metal complexes (14), etc. have

(**XVII**)

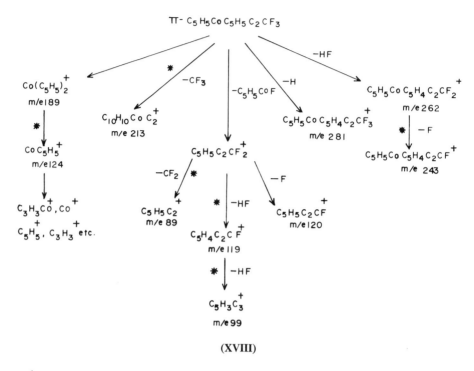

(XVIII)

been reported. It has been shown that for thermally unstable compounds the conditions of sample introduction could be of great importance both to the nature of the mass spectra and to the determination of the appearance potential.

V. An Analytical Tool

With the advent of semiconductor research, mass spectrometric analysis of inorganic materials that are solids has rapidly increased in importance and has now become one of the major applications of mass spectrometry (51). Several methods have been developed that permit bulk analysis of milligram quantities of inorganic solids for trace impurities at the parts-per-billion (ppb) level and determination of surface composition and inhomogeneity in microvolumes of solids.

Before the analysis the sample is chemically treated and "presparked" to remove any surface inpurities left as a result of the chemical treatment. For surface-impurity analysis all ions produced are recorded. The "exposure" or "total charge" recorded by the monitor is increased in steps. Spectra of very short duration, equivalent to a single spark, are usually taken to determine the exposure at which the major constituent appears.

Generally multiply charged species are observed in these spectra. The concentration of a given impurity is obtained from the exposure necessary to produce a suitable blackening on the photoplate, taking into account isotopic-abundance ratios, linewidths, and relative sensitivities. Concentrations are estimated by visual inspection to within a factor of 3 or measured on a micro-densitometer with an accuracy of $+30\%$ with the help of calibration standards. Usually mass-spectrometric analysis compares very well with neutron-activation analysis and electrical measurements. More than half of the elements of the periodic table have detection limits between 1 and 10 ppb. The other elements have poorer sensitivities (mostly between 10 and 100 ppb) because of the low abundance of the principal isotopes and interference from other spectral lines or interference from the blackened area surrounding the principal lines of the major constituent.

The conductivity of the sample under operating conditions should be sufficient to permit sparking. This is true only for metals and semiconductors, not for insulators. For the latter type, the sample is ground up to a fine powder mixed with a suitable metal powder (Ag) and compressed into rods of suitable dimensions. Sometimes known concentrations of a suitable impurity are added during mixing to serve as an "internal standard." The reproducibility is considerably poorer in the analysis of inhomogeneous materials.

Analysis of microvolumes and surface layers are made with the help of special techniques. Hickam et al. (49) have used a rotating sample and a stationary, pointed counterelectrode, which permits microvolumes to be sampled. Similarly microvolumes containing up to 10^{17} atoms are vaporized when a laser pulse is focused onto a surface as described by Honig (52). Impurities could be detected by these methods down to the parts-per-million level.

In a recent paper the possibility of estimating very small quantities of volatile metal chelates by integrating the ion current at a significant m/e value while evaporating the metal chelate directly into the ion source of a mass spectrometer was demonstrated (54, 55, 68). In the case of nickel dimethyl glyoxinate the sample size could be lowered to 10^{-12} g for measurements and 10^{-14} g for detection.

VI. High Temperature Mass Spectrometry

The mass spectrometer has been accepted as the most versatile tool for the identification of vapor species and for investigating the thermodynamics of vapor-phase and solid-phase–vapor-phase equilibria. The electric or magnetic deflection of molecular beams combined with a mass spectrometer detector allows one to decide whether a molecule has an electric or magnetic dipole moment and confirm speculations about structures of simple molecules.

The high-temperature mass-spectrometric technique employs either the Knudsen or Langmuir methods as a source of the vapor species. The work of Chupka and associates (25) on the heat of sublimation of carbon marked the birth of high-temperature mass-spectrometric technique. The experimental techniques and procedures for data treatment are primarily the results of their work over the past decade. Ionov (53) first investigated the vapor species of a number of alkali halides. Büchler and co-workers (17) have shown that the lithium halide dimers are planar. The structures of many halides were established that are in disagreement with the prediction of simple theories of molecular geometry. A number of homonuclear diatomic molecules and intermetallic molecules have also been studied. Polyatomic molecules and their thermodynamic properties are now known for the elements of Groups IV, V, and VI of the Periodic Table.

The investigation of the reaction between oxygen and molybdenum, tungsten, and tantalum as well as those of halogens with nickel and yttrium has made it possible to identify the gaseous species formed and to study the influence of time, temperature, and flux of impinging gaseous molecules on their rate of formation. Data on residence time and angular distribution of reflected molecules or reaction products were obtained by this technique.

In inorganic compounds it is seldom that one vapor species is observed in the equilibrium vapor phase, so separate quantitative fragmentation patterns cannot be obtained for the individual vapor species and the data-treatment techniques developed for routine gas analysis are not applicable. The mass spectrum produced by the equilibrium vapor phase of a given inorganic substance may contain parent and fragment ion contributions from more than one vapor species. From the spectrum one attempts to determine the partial pressures of the individual vapor species and also the various thermo-dynamic data that could be calculated from partial-pressure measurements.

The three general steps involved in getting the derived partial-pressure information from the raw data are (1) identification of the ionic species, (2) identification of the neutral vapor species, and (3) determination of the partial pressures of the individual vapor species at the various temperatures of measurement. Each of these steps is considered separately in some detail by Grimley (45).

Thus the identity of the vapor species and the partial pressures of these species as a function of temperature may be ascertained. The partial pressures so obtained have yielded a variety of thermodynamic data by the application of the second and third laws.

The equilibrium constant K_{eq} may be calculated for each temperature at which the partial pressures are measured, assuming that the pressure is equal to the fugacity. The van't Hoff equation gives the relationship between the variation of the equilibrium constant with temperature and the $\Delta H°$ for the reaction

$$\frac{dlnK_{eq}}{dT} = \frac{\Delta H}{RT^2} \quad \text{or} \quad \frac{dlnK_{eq}}{d(1/T)} = \frac{-\Delta H°}{R}$$

From the slope of the lnK_{eq} versus $1/T$ plot, we may obtain the value of $\Delta H°$.

The advantage of the second-law method is that it does not require entropy data. The disadvantages are that the value of $\Delta H°$ is for the mean temperature of the measurements and the measurement of a slope is required.

The third-law method is more accurate. Each value of the equilibrium constant permits us to calculate the free energy, $\Delta G° T = -RTlnK_{eq}$, for the reaction at temperature T. From the free-energy function $(G_T - H_0°)/T$, $\Delta H_0°$ is determined using the equation

$$\Delta H_0° = T\left[\frac{\Delta G_T°}{T} - \Delta\left(\frac{G_T - H_0°}{T}\right)\right]$$

where

$$\Delta\left(\frac{G_T - H_0°}{T}\right) = \Sigma\left(\frac{G_T - H_0°}{T}\right)_{products} - \Sigma\left(\frac{G_T - H_0°}{T}\right)_{reactants}$$

and

$$\frac{G_T - H_0°}{T} = \frac{H_T - H_0°}{T} - S_T$$

The main disadvantage is that the calculation of free-energy functions requires the structural and spectroscopic information, which is often not available.

The application of second- and third-law procedures to mass-spectrometric data has yielded heats of sublimation, heats of vaporization, dissociation energies, energies of polymerization, heats of formation, and heats of reaction.

VII. Energetics of Electron-Impact Processes

The ionization potential of a molecule (atom or radical) is the minimum energy required to ionize it, while the appearance potential is the lowest energy necessary to generate a given ion and its accompanying neutral fragment from a molecule (ion or radical).

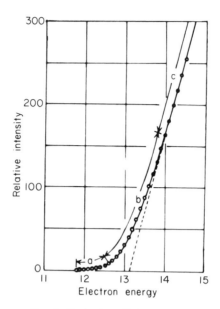

FIG. 8. Ionization-efficiency curve.

The ionization efficiency curve (Fig. 8) shows the relation between the energy of the ionizing electron beam and the relative intensity (ion current) of a particular fragment produced. By increasing the electron energy, the probability of ionization in molecules is also increased until a plateau occurs in the curve. Because of the variations in the electron energies at low energies the tail of the curve has to be extrapolated to obtain the ionization energy of molecules and appearance potential of ions. Different methods (6, 65) for extrapolation and the inherent errors have been described in detail.

The appearance potential is related to the gas-phase dissociation energy and the bond and the ionization potential:

Appearance potential = dissociation energy + ionization potential

The contributions from the kinetic energy and excitation energy are small, and hence they are dropped from the above relation. It is necessary to pick out and assign the correct process involved in the threshold measurement. Reliable values are obtained only in the case of those ions formed by simple decomposition reactions.

As the size of the ligand increases, the value of the ionization potential decreases. Winters et al. (99) have shown that for the main-group metal methyls less energy is enough for ionization as the atomic number of the metal increases. It has been concluded from the ionization potentials that in

π-cyclopentadienyl complexes and in metal carbonyls one of the electrons from a nonbonding orbital is lost in the primary ionization reaction.

It is possible to calculate the heats of formation if the appearance potential and the heat of formation of ligands are known. In a few cases (*5, 100*) evidence for the occurrence of certain fragmentation modes operating at the threshold energy has been provided from the appearance potential.

Only few absolute bond-dissociation energies have been determined from electron-impact measurements. This is because the ionization potentials of many organometallic radicals are unknown. In such cases it is reasonable to compare bond-dissociation energies and appearance-potential differences.

Stevenson's rule (*88*) states that the positive charge should reside largely on the fragment of lowest ionization potential after the bond cleavage.

VIII. Frontiers of Mass Spectrometry

It is now appropriate to apply the various recent advances together with our novel ideas and modern technology to solve new problems with this most versatile instrument. Many of the recent advances in mass spectrometry are fertile areas of further study. One of the newer areas that call for detailed investigations is the study of the neutral fragments formed by ionization and fragmentation processes. The use of a double ionization chamber and the time-of-flight study of neutrals from metastable transitions have great potential importance in this area of investigation.

Mass spectrometry is a powerful tool for the study of reaction intermediates and mechanisms. Some of the other new frontiers of mass spectrometry are study of gaseous ion chemistry in greater detail, high-energy studies, study of the structures of gaseous ions and chemiionization reactions. It is clear that mass spectrometry has a very good past record of accomplishments. This technique is an active and productive one. There is no doubt that greater advances will be made in the future. So many questions remain without answer, so many techniques are in a latent stage, that this field offers opportunities for further investigation.

References

1. Ahearn, A. J. (ed.), "Mass Spectrometric Analysis of Solids." Elsevier, New York, 1966.
2. Aston, F. W., *Phil. Mag.* **38**, 707 (1919).
3. Aulinger, F., Reerink, W., *Zeits. Für Analyt. Chemie* **197**, 24 (1963).
4. Bancroft, G. M., Reichert, C., and Westmore, J. B., *Inorg. Chem.* **7**, 870 (1968).

5. Band, S. J., Davidson, I. M. T., Lambert, C. A., and Stephenson, I. L., *Chem. Commun.* 723 (1967).
6. Barnard, G. P., "Mass Spectrometry." Institute of Physics, London, 1953.
7. Becconsall, J. K., Job. B. E., and O' Brien, S., *J. Chem. Soc.* **A**, 423 (1967).
8. Beynon, J. H., "Mass Spectrometry and Its Application to Organic Chemistry," p. 547. Elsevier, New York, 1960.
9. Bidinosti, D. R., and McIntyre, N. S., *Can. J. Chem.* **45**, 641 (1967).
10. Braterman, P. S., *J. Organometal. Chem.* **11**, 198 (1968).
11. Brion, C. E., and Paddock, N. L., *J. Chem. Soc.* **A**, 392 (1968).
12. Brown, P., and Djerassi, C., *Angew. Chem. Intern. Ed. Engl.* **6**, 477 (1967).
13. Bruce, M. I., *in* "Advances in Organometallic Chemistry" (F. G. A. Stone and R. West, eds.), Vol. 6, pp. 273–333. Academic Press, New York, 1968.
14. Bruce, M. I., *Intern. J. Mass Spectry. Ion Phys.* **1**, 335 (1968).
15. Bruce, M. I., *Org. Mass Spectry.* **1**, 687 (1968).
16. Bruce, M. I., *Org. Mass Spectry.* **2**, 63 (1969).
17. Büchler, A., Stauffer, J. L., and Klemperer, W., *J. Am. Chem. Soc.* **86**, 4544 (1964).
18. Carrick, A., and Glockling, F., *J. Chem. Soc.* **A**, 40 (1967).
19. Cavell, R. G., and Dobbie, R. C., *Inorg. Chem.* **7**, 690 (1968).
20. Chambers, D. B., Glockling, F., and Light, J. R. C., *Quart. Rev.* **22**, 317 (1968).
21. Chambers, D. B., and Glockling, F., *J. Chem. Soc.* **A**, 735 (1968).
22. Chernyak, N. Ya., Khmel'nitskii, R. A., D'yakova, T. V., Pushchevaya, K. S., and Vdovin, V. M., *J. Gen. Chem.* **37**, 867 (1966).
23. Chernyak, N. Ya., Khmel'nitskii, R. A., D'yakova, T. V., and Vdovin, V. M., *Zh. Obshch. Khim.* **36**, 89 (1966).
24. Chernyak, N. Ya., Khmel'nitskii, R. A., D'yakova, T. V., and Vdovin, V. M., *Zh. Obshch. Khim.* **36**, 96 (1966).
25. Chupka, W. A., and Inghram, M. G., *J. Chem. Phys.* **21**, 371 (1953).
26. Cooks, R. G., Howe, I., and Williams, D. H., *Org. Mass Spectry.* **2**, 137 (1969).
27. Cooks, R. G., and Gerrard, A. F., *J. Chem. Soc.* **B**, 1327 (1968).
28. Coxon, G. E., Palmer, T. F., and Sowerby, D. B., *J. Chem. Soc.* **A**, 1568 (1967).
29. Coxon, G. E., Palmer, T. F., and Sowerby, D. B., *J. Chem. Soc.* **A**, 358 (1969).
30. Damico, J. N., *J. Assoc. Offic. Agr. Chemists* **49**, 1027 (1966).
31. Das, K. G., and Saudi, S. K., unpublished observations.
32. Das, K. G., Kulkarni, P. S., Kalyanaraman, V., and George, M. V., unpublished observations.
33. Das, K. G., Sen, D. N., and Thankarajan, N., *Tetrahedron Letters* **7**, 869 (1968).
34. Dempster, A. J., *Phys. Rev.* **11**, 316 (1918).
35. Dibeler, V. H., Mohler, F. L., and Reese, R. M., *J. Chem. Phys.* **21**, 180 (1953).
36. Dudek, E. P., Chaffee, E., and Dudek, G., *Inorg. Chem.* **7**, 1257 (1968).
37. Dudek, E. P., and Barber, M., *Inorg. Chem.* **5**, 375 (1966).
38. Duffield, A. M., Djerassi, C., Mazerolles, P., Dubac, J., and Manuel, G., *J. Organometal. Chem.* **12**, 123 (1968).
39. Fehlner, T. P., *J. Am. Chem. Soc.* **90**, 6062 (1968).
40. Glockling, F., and Light, J. R. C., *J. Chem. Soc.* **A**, 717 (1968).
41. Gohlke, R. S., *J. Am. Chem. Soc.* **90**, 2713 (1968).
42. Goldsmith, D., Becker, D., Sample, S., and Djerassi, C., *Tetrahedron, Suppl.* **7**, 145 (1966).
43. Goldstein, E., *Berlin. Ber.* **39**, 691 (1886).
44. Gould, R. F. (ed.), "Mass Spectrometry in Inorganic Chemistry." American Chemical Society, Washington, D.C., 1968.

45. Grimley, R. T., *in* "The Characterization of High-Temperature Vapours" (J. L. Margrave, ed.), pp. 195–243. Wiley, New York, 1967.
46. Halmann, M., and Platzner, I., *J. Phys. Chem.* **71**, 4522 (1967).
47. Harless, H. R., *Anal. Chem.* **33**, 1387 (1961).
48. Hartwell, G. E., and Brown, T. L., *Inorg. Chem.* **5**, 1257 (1966).
49. Hickam, W. M., and Sweeney, G. S., *Ann. Conf. Mass Spectry. Allied Topics, 12th,* p. 280. Montreal (1964).
50. Hilgetag, G., and Teichmann, H., *Angew. Chem. Intern. Ed. Engl.* **4**, 914 (1965).
51. Honig, R. E., *Ann. N.Y. Acad. Sci.* **137**, 262 (1966).
52. Honig, R. E., *Appl. Phys. Letters* **3**, 8 (1963).
53. Ionov, N. I., *Dokl. Akad. Nauk. SSSR* **59**, 467 (1948).
54. Jenkins, A. E., and Majer, J. R., *Talanta* **14**, 777 (1967).
55. Jenkins, A. E., Majer, J. R., and Reade, M. J. A., *Talanta* **14**, 1213 (1967).
56. Johnson, B. F. G., Lewis, J., Williams, I. G., and Wilson, J. M., *J. Chem. Soc.* A, 341 (1967).
57. Johnson, B. F. G., Johnston, R. D., Lewis, J., and Robinson, B. H., *J. Organometal. Chem.* **10**, 105 (1967).
58. Jörg, J., Houriet, R., and Spiteller, G., *Monatsh. Chem.* **97**, 1064 (1966).
59. Khmel'nitskii, R. A., Polyakova, A. A., Petrov, A. A., *Trans. Comm. Anal. Chem. Izv. Akad. Nauk SSSR* **13**, 483 (1963).
60. King, R. B., *Can. J. Chem.* **47**, 559 (1969).
61. Kiser, R. W., "Introduction to Mass Spectrometry and Its Applications." Prentice-Hall, Englewood Cliffs, New Jersey, 1965.
62. Kostyanovsky, R. G., and Yakshin, V. V., *Izv. Akad. Nauk. SSSR Khim.*, 2363 (1967).
63. Lars-Börge Agenäs, *Acta. Chem. Scand.* **22**, 1763, 1773 (1968).
64. Lewis, J., and Johnson, B. F. G., *Accounts Chem. Res.* **1**, 245 (1968).
65. Lewis, J., Manning, A. R., Miller, J. R., and Wilson, J. M., *J. Chem. Soc.* A, 1663 (1966).
66. MacDonald, C. G., and Shannon, J. S., *Australian J. Chem.* **19**, 1545 (1966).
67. Macklin, J., and Dudek, G., *Inorg. Nucl. Chem. Letters* **2**, 403 (1966).
68. Majer, J. R., Reade, M. J. A., and Stephen, W. I., *Talanta* **15**, 373 (1968).
69. Mandelbaum, A., and Biemann, K., *J. Am. Chem. Soc.* **90**, 2975 (1968).
70. McLafferty, F. W., and Wachs, T., *J. Am. Chem. Soc.* **89**, 5044 (1967).
71. McLafferty, F. W., *Anal. Chem.* **28**, 306 (1956).
72. McLafferty, F. W., *Anal. Chem.* **28**, 314 (1956).
73. Müller, J., and D'Or, J., *J. Organometal. Chem.* **10**, 313 (1967).
74. Occolowitz, J. L., and White, G. L., *Anal. Chem.* **35**, 1179 (1963).
75. Pignataro, S., and Lassing, F. P., *J. Organometal. Chem.* **11**, 571 (1968).
76. Quayle, A., *in* "Advances in Mass Spectrometry" (J. D. Waldon, ed.), pp. 365–383, Pergamon Press, Oxford, 1959.
77. Quisenberry, K. S., Scolman, T. T., and Nier, A. E., *Phys. Rev.* **102**, 1071 (1956).
78. Reichert, C., Westmore, J. B., Gesser, H. D., *Chem. Commun.*, 782 (1967).
79. Ridder, J. J. de, and Dijkstra, G., *Rec. Trav. Chem.* **86**, 737 (1967).
80. Roberts, D. T., Little, W. F., and Bursey, M. M., *J. Am. Chem. Soc.* **89**, 4917 (1967).
81. Roboz, J., "Introduction to Mass Spectrometry Instrumentation and Techniques." Wiley (Interscience), New York, 1968.
82. Sasaki, S., Itagaki, Y., Kurokawa, T., Nakanishi, K., and Kasahara, A., *Bull. Chem. Soc. Japan* **40**, 76 (1967).
83. Schmulbach, C. D., Cook, A. G., and Miller, V. R., *Inorg. Chem.* **7**, 2463 (1968).

84. Schubert, E. H., and Sheline, R. H., *Z. Naturforsch.* **20b**, 1366 (1965).
85. Shannon, T. W., and McLafferty, F. W., *J. Am. Chem. Soc.* **88**, 5021 (1966).
86. Silbiger, J., Lifshitz, C., Fuchs, J., and Mandelbaum, A., *J. Am. Chem. Soc.* **89**, 4308 (1967).
87. Smith, J. M., Mehner, K., and Kaesz, H. D., *J. Am. Chem. Soc.* **89**, 1759 (1967).
88. Stevenson, D. P., *Discussions Faraday Soc.* **10**, 35 (1951).
89. Takegami, Y., Ueno, T., Suzuki, T., and Fuchizaki, Y., *Bull. Chem. Soc. Japan* **41**, 2637 (1968).
90. Thomas, D. W., and Biemann, K., *J. Am. Chem. Soc.* **87**, 5447 (1965).
91. Thomson, J. J., "Rays of Positive Electricity." Longmans, Green, London, 1913.
92. van den Bergen, A., Murray, K. S., O'Connor, M. J., Rehak, N., and West, B. O., *Australian J. Chem.* **21**, 1505 (1968).
93. Visser, H. D., and Olivers, J. P., *J. Am. Chem. Soc.* **90**, 3579 (1968).
94. Wada, Y., and Kiser, R. W., *Inorg. Chem.* **3**, 174 (1964).
95. Wien, W., *Ann. Physik* **65**, 440 (1898).
96. Williams, D. H., Ward, R. S., and Cooks, R. G., *J. Am. Chem. Soc.* **90**, 966 (1968).
97. Winters, R. E., and Kiser, R. W., *Inorg. Chem.* **4**, 157 (1965).
98. Winters, R. E., and Collins, J. H., *J. Phys. Chem.* **70**, 2057 (1966).
99. Winters, R. E., and Kiser, R. W., *J. Organometal. Chem.* **10**, 7 (1967).
100. Winters, R. E., and Kiser, R. W., *J. Organometal. Chem.* **4**, 190 (1965).
101. Yergey, A. L., and Lampe, F. W., *J. Am. Chem. Soc.* **87**, 4204 (1965).
102. Orlov, V. Yu., *J. Gen. Chem.* **37**, 2188 (1967).

SOFT X-RAY SPECTROSCOPY AS RELATED TO INORGANIC CHEMISTRY

William L. Baun and David W. Fischer

AIR FORCE MATERIALS LABORATORY (MAYA)
WRIGHT–PATTERSON AIR FORCE BASE
OHIO

I. Introduction

The soft x-ray region contains many lines and bands that are of interest to the inorganic chemist. For instance, the K emission spectra of the second-period elements occur in this region as well as the L spectra of the first transition-series elements. Such emission lines and bands are conventionally used for chemical analysis. In addition the fine features of these spectra may be used to further characterize a material by providing information regarding valence, coordination, and bonding. Soft x-ray spectroscopy is especially valuable for characterization of very small samples using the microprobe or of thin films and poorly crystalline materials that are difficult to characterize by conventional techniques. There seems to be general agreement that the future significance of very soft x-rays lies in the interpretation of spectra rather than the determination of concentration (*16*, *53*). This review will cover, within the limited space available, discussion of soft x-ray methods and instrumentation, spectral characteristics, and practical examples of the use of low-energy x-ray spectroscopy.

II. Principles, Methods, and Instrumentation

Significant progress and improvements have been made in soft x-ray spectroscopy in the last decade. Naturally the basic principles of x-ray excitation are the same for soft x-rays as for the harder or more energetic x-rays commonly used for chemical analysis. The reader is referred to a basic physics text for details of the excitation process or to the book by Birks (*13*), where a simplified explanation is given of the fundamentals of excitation, dispersion, detection, and applications of conventional x-rays. Techniques and instrumentation for soft x-ray spectroscopy are significantly different from conventional x-ray spectroscopy because the low-energy x-rays are strongly or completely absorbed in very thin layers of matter.

Because of this high absorption of soft x-rays, work in the soft x-ray region must be done in vacuum. In addition, when primary excitation with electrons or protons is employed, a very good hydrocarbon-free vacuum is required to minimize contamination of the target surface. The condition of the sample surface is extremely important in soft x-ray spectroscopy because of the low penetration power of both the exciting particles and the x-rays that are produced within the sample. Carbon contamination of the sample surface is very difficult to avoid, since it originates from decomposition of hydrocarbons from oil diffusion pumps, rough pumps, rubber "O" rings, and cleaning solvents. Possible solutions to these vacuum problems are to use well-trapped mercury diffusion pumps, molecular sieves, and metal gaskets throughout the system. A better solution may be to use ion and titanium sublimation pumps, which may be brought to rough vacuum by sorption pumps. Contamination may be further minimized by placing a cold finger very close to the sample. Holliday (*45*) has found ion bombardment of the sample surface to be very effective in cleaning a sample. Holliday uses high-purity argon passed over hot titanium chips at a pressure of about 10^{-3} Torr with a target potential of -400 to -600 V.

A. Excitation of Soft X-Rays

1. SECONDARY EXCITATION

A decade ago the usual secondary-excitation source was a conventional side- or end-window x-ray tube having a tungsten target. Windows on these tubes were usually 0.030- or 0.040-in. beryllium. These tubes are quite efficient in the hard x-ray region, but are very poor in the soft and ultrasoft regions. One change in these tungsten target tubes, the substitution of a 0.010-in. beryllium window for the 0.030-in. window, improves excitation efficiency by a factor of 2 in the Cl K and S K region. The substitution of chromium or silver for tungsten further improves efficiency by another factor of 2. A titanium-target tube with a 0.010-in. beryllium window is some five times as effective for excitation in the 5–10-Å region as a tungsten tube, but efficiency drops dramatically on the short-wavelength side of TiK. Dual-anode tubes of, for instance, chromium and platinum prove valuable for efficient excitation over the range of 0.5–10 Å. Recently Kirkendall and Varadi (51) described a platinum-target x-ray tube with a 0.005-in. beryllium window that uses the Pt L lines for hard x-ray excitation and the Pt M lines for excitation of soft x-rays. Even more recently Kirkendall and co-workers (52) described a chromium-target, 0.002-in. beryllium-window tube, which was used for excitation of K spectra of carbon through sodium. This tube, however was a low-power tube used in energy-dispersion spectroscopy. In conventional scanning spectrometry where high power is needed, beyond 10 Å or so, absorption in the beryllium window necessitates going to a windowless or ultrathin-window tube of the Henke type. In the Henke (40) design the tube is continuously ion pumped and is isolated from the spectrometer by a sliding-gate valve that blanks off the tube when the spectrometer is exposed to atmospheric pressure, as shown in Fig. 1 (43). When the spectrometer is pumped to rough vacuum ($\sim 50~\mu$), an opening that is covered by a thin window material is slid into place over the port. Since the window only separates the high-vacuum side (x-ray tube) from the low-vacuum side (spectrometer), it experiences very little mechanical stress and can be extremely thin. Windows prepared by Henke (40) consisting of stretched polypropylene and a thin film of Formvar transmit about 60% of the radiation on the long-wavelength side of the carbon absorption edge. Tubes of the Henke type are designed to operate using very high filament current. Typical settings shown by Henke are 6 kV at 330 mA. This high power gives a very strong beam of x-rays,

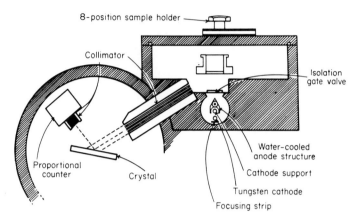

FIG. 1. Henke high-power demountable tube and spectrograph (*43*).

which efficiently excite the characteristic x-rays in the sample without the heating effects that are obtained with direct electron excitation. This is a definite advantage in that unstable materials may be investigated using secondary excitation. The target of the Henke tube is chosen to most efficiently excite the target material. Table I shows tube targets, the wavelength of the characteristic exciting line from the target, and the K- and L-series elements that are most efficiently excited by each target material. Here target materials are chosen having characteristic lines just shorter than the characteristic lines of the sample material.

A convenient x-ray source is one described by Mattson and Ehlert where one is able to switch back and forth between electron and x-ray excitation (*57*). Their source, which may be pumped separately with an ion pump, has been used for x-ray excitation, electron excitation, electron excitation of a heated wire, and electron excitation of gases. In the x-ray excitation mode the sample and cathode are maintained at ground potential while the anode

TABLE I

DEMOUNTABLE X-RAY TUBE TARGETS

Target and series	Wavelength (Å)	K-series excitation of	L-series excitation of
Al (K)	8.34	Mg, Na, F	Se, As, Ge
Cu (L)	13.33	F, O, N, C	Ni, Co, Fe, etc.
Cr (L)	21.71	O, N, C	V, Ti, Sc, etc.
C (K)	44.8	B, Be	Cl, S, P, etc.

assumes a positive potential up to 16 kV. For electron excitation the sample is maintained positive with respect to the filament, while the anode acts as a bias electrode to deflect the electron beam onto the sample.

2. DIRECT EXCITATION

The increasing use of the electron microbeam probe and the introduction of three commercial instruments using electron excitation has rapidly made direct excitation methods important. In the early history of x-ray spectroscopy nearly all measurements were made using direct excitation. The resurgence of the technique has come about because of increased interest in analysis of low-atomic-number elements by x-ray techniques. The major advantage of electron excitation is that it is far more efficient than photon (x-ray) excitation. A disadvantage of the technique is that electron excitation requires the use of high vacuum. In addition, decomposition of unstable compounds is a possibility under electron bombardment. Cold-cathode tubes do not require high vacuum and may be operated windowless, but this technique has not been used in modern instruments except by Solomon and Baun (71) and Davidson and Wyckoff (18).

Excitation by protons and alpha particles has not yet been applied extensively, but these techniques offer certain advantages over electron excitation. Proton excitation of soft x-rays is more efficient than electron excitation, and the intensity of the bremsstrahlung radiation produced by protons is much smaller than that produced by electrons. Therefore, with protons, virtually monochromatic radiation is produced. The use of radioactive sources emitting alpha and beta particles and x- and gamma rays are useful for specific purposes, especially when using high-efficiency energy-dispersion systems. The major advantage of radioisotope excitation is that no high-voltage power supply is required.

Other special sources such as the spark source or plasma source prove useful for x-ray absorption where a large flux of white radiation is needed or for reference purposes where a sharp line is desired for alignment.

B. Dispersion of Soft X-Rays

1. CRYSTAL SPECTROMETERS

The crystal spectrometer is a versatile instrument that can best be used in the conventional and soft x-ray region but is also usable in a portion of the ultrasoft region. Crystal spectrometers may take many forms, depending on source geometry or the degree of sensitivity or resolution desired. Basically, crystal spectrometers are classified according to the configuration of the crystal

—i.e., "flat" for use with Soller slits, "cylindrical" for use with line-focus sources, and "toroidal" for use with point-focus sources, as shown in Fig. 2, taken from the summary by Henke (43). The two-crystal spectrometer generally uses two plane crystals to achieve very high resolution.

There are many inorganic and organic single crystals that may be used as dispersing elements in crystal spectrometers. Table II contains the name, 2d spacing, and notes on advantages and disadvantages of crystals useful in the 1–100-Å region. There are many other crystals that may be used. For instance, four or five cuts of quartz have been used. As can be seen from this table, there are several stable crystals that may be used in place of gypsum ($CaSO_4 \cdot 2H_2O$), which dehydrates very rapidly to plaster of paris in high vacuum.

Potassium acid phthalate (KAP) represents a family of compounds where potassium may be replaced by sodium, rubidium, or cesium if potassium fluorescence in the crystal should be a problem in a particular application. KAP and others in the family cleave in thin sections along the (001), which makes crystal preparation fairly easy.

A natural crystal that gives higher intensity in some applications than KAP is clinochlore (6). Clinochlore has a (001) spacing of 14.196 Å or a 2d of 28.392 Å, which allows it to be used at somewhat longer wavelengths than KAP. Clinochlore gives particularly good results for Ti, L and O K spectra.

The very-long-spacing crystals (OHM and OHS) shown at the bottom of Table II are extremely difficult to grow, and they have very long growth cycles in order to attempt to reduce spontaneous nucleation, which is a problem with many organic crystals (64). With these crystals it is not uncommon to take six months or more to grow a 1 × 1-in. crystal, and the temperature must be controlled to $\pm 0.001°C$ or better while dropping the temperature in the bath 0.01°C or less per day.

Since it is so difficult to obtain single crystals of x-ray analyzers, it has been necessary to go to the use of soap films. Metallic salts of aliphatic acids

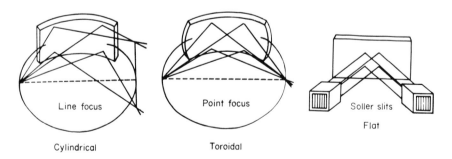

FIG. 2. Types of crystal spectrometers (43).

TABLE II

ANALYZING CRYSTALS (1–100 Å)

Name	$2d$ (Å)	Notes
Topaz	2.71	Natural mineral
Lithium fluoride	4.02	Durable, strong reflection
Sodium chloride	5.64	Cleavage, easily bent
Calcite	6.06	Durable natural mineral, cleavage
Silicon	6.28	Semiconductor slices may be used, suppressed second order
Fluorite	6.30	Cleavage
Germanium	6.53	Semiconductor slices may be used, suppressed second order
Potassium bromide	6.58	No cleavage, easily bent
Ammonium phosphate	7.50	Large natural growth face
Pentaerythritol	8.74	Soft, difficult to grow, intense reflections
Ammonium tartrate	8.80	Cleavage, easy to grow from H_2O solution
Ethylene diamine d-tartrate	8.80	No cleavage
Ammonium phosphate	10.64	Easy to grow from H_2O solution
Ammonium tartrate	14.15	Large natural growth face, deammoniates
Ammonium citrate	14.92	Natural growth face
Sucrose	15.12	Difficult to grow, spontaneous nucleation
Gypsum	15.18	Dehydrates in high vacuum, easily bent
Beryl	15.96	Natural mineral
Itaconic acid	18.50	Cleavage, soft
Mica	19.91	Natural mineral, cleavage, easily bent
Potassium acid phthalate	26.63	Growth from H_2O solution, cleavage
Clinochlore	28.39	Natural mineral, cleavage, easily bent
Bismuth titanate	32.81	Platelets, may be bent, first order weak, multiple orders strong
Octadecyl hydrogen maleate	63.5	Both difficult to grow, long growth cycles, spontaneous nucleation
Octadecyl hydrogen succinate	96.9	

deposited one layer at a time on a flat or curved substrate have become very popular for dispersion of soft x-rays. In this technique a trough is filled with water containing (usually) a small concentration of lead ions and a buffer. A dilute solution of stearic acid in hexane is spread as a monomolecular layer on the water, the hexane evaporates, and the lead ions react with the stearic acid to form a soap. This monomolecular soap film is then placed under tension from a floating barrier. The film is then picked up one layer at a time on a plane or curved surface such as glass or freshly cleaved mica at a rate of about one double layer per minute. When about twenty layers have been deposited, the trough is cleared off and the process begun again. This is continued until a sufficient number of layers are deposited, usually 50–100 for lead stearate. For shorter or longer soaps it will be necessary to have more or fewer layers, respectively, to obtain comparable performace of the multilayer films. Soap films are very versatile in that nearly any spacing between about 60 Å (2d) and 160 Å may be obtained with the correct choice of metal ion and aliphatic acid.

A list of soap films is given in Table III and shows the number of carbon

TABLE III

SOA– FILM MULTILAYER ANALYZERS

Name	Number of carbon atoms	$2d$ spacing (Å)	Notes
Lead laurate	12	70	More difficult to prepare and possible degradation in vacuum
Lead myristate	14	80.5	More difficult to prepare
Lead palmitate	16	90	Easy to prepare
Lead stearate	18	100.79	Easy to prepare
Lead arachidate	20	110	Easy to prepare
Lead lignocerate	24	131.45	More diffiuclt to prepare
Lead mellissate	30	165	More difficult to prepare and usually poor quality

atoms in a single chain in the original acid, along with the $2d$ spacing of the lead salt. The soaps near the stearate are easiest to prepare, while the very longest and shortest are the most difficult to make. It has generally been considered an art to make good soap-film dispersing elements. Very recently, however, Whatley (76) reported on a systematic study of the effect of pH, aqueous cation concentration, and the mechanics of the dipping process.

2. GRATING SPECTROMETER

The grating spectrometer is basically a high-resolution instrument. Generally the grating spectrometer is much more difficult to align and use than crystal spectrometers. Many of the data in the literature taken by means of gratings are no better (with respect to resolution) than those which soap-film crystal spectrometers provide, but recent improvments in gratings and the grating spectrometer, made by Holliday (45), obviously advance the state of the art.

The grazing-incidence spectrometer used by Holliday (45) is seen in Fig. 3. Slits S_1 and S_2 and the concave grating are on the Rowland circle, which has one-half the radius of curvature of the grating. The x-rays from the target are directed by slit S, at a glancing angle to the grating. The x-rays are then focused on the Rowland circle. The analyzer is a blazed replica grating. Holliday has demonstrated that there are definite advantages to the blazed replica grating

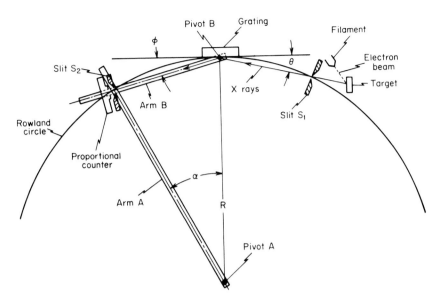

FIG. 3. Grazing-incidence grating spectrometer (*45*).

compared to conventional gratings that have been used in the past for x-ray spectroscopy. Holliday has extensively investigated such factors as the effect of the number of lines scribed on a grating and the blaze angle on resolution and intensity of x-ray spectra. Generally, in the very soft x-ray region, the grating spectrometer gives higher resolution than a crystal spectrometer. However, count rates may be somewhat lower with the grating instrument, and for analytical purposes the crystal spectrometer may be preferable. In the shorter-wavelength region, such as for first transition-series L spectra, the crystal spectrometer gives much better results than the grating spectrometer. For instance, Ti L_{II} and L_{III} from titanium metal recorded in the author's laboratory using a clinochlore crystal showed complete separation, while the same spectrum shown by Holliday using a grating (*46*) showed L_{II} as a shoulder on the L_{III} band, but self-absorption effects make direct comparison of spectra difficult. Intensities were significantly better with the crystal spectrometer.

C. *Detection of Soft X-Rays*

1. FLOW PROPORTIONAL DETECTOR

The most widely used detector for soft x-ray spectroscopy at this time is the flow proportional counter. This versatile detector is supplied with commercial x-ray spectrometers and electron microbeam probes as well as being used in most custom equipment installations in the United States. The major difficulty with the flow proportional detector is found in installing and maintaining a leak-free window between the counter and the spectrometer vacuum, especially when the counter is operated at or near a pressure difference of one atmosphere. The best solution appears to be the use of thin double Formvar films supported on copper or nickel grids having 60–80% transmission, as shown in the exploded view in Fig. 4 of a flow proportional detector for soft x-rays (7). It is useful to evacuate the proportional counter as the spectrometer chamber is evacuated and then operate the counter at a pressure of only 100–200 mm Hg above the chamber pressure. A thin conductive coating may be necessary to minimize charging effects on the window, but usually such a a coating is not necessary when using a supporting grid. Absorption in the window must be taken into consideration. For instance, if cellulose nitrate is used as window, absorption edges will appear at about 23, 31, and 44 Å, corresponding to oxygen, nitrogen, and carbon, respectively. Holliday (47) has shown that in his cellulose nitrate window there is 85% absorption at the short-wavelength limit of the oxygen K absorption edge and 8% at the long-wavelength limit, which shows that any measurements taken in this region would be in serious error if not corrected for absorption.

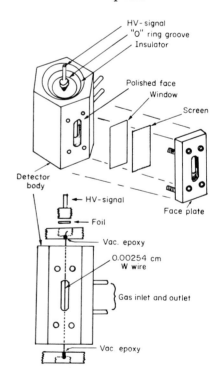

FIG. 4. Exploded view of flow proportional detector for soft x-rays (7).

2. Magnetic Electron Multiplier (MEM)

The magnetic electron multiplier is a device using a continuous surface of high-resistance semiconductor material under the action of crossed electric and magnetic fields to obtain high-gain electron multiplication. The device was originally developed for ion detection in the time-of-flight mass spectrometer. Although the MEM has been applied successfully to detection of low-energy photons, no real effort seems to have been made to optimize this type of detector for use with soft x-rays. Just the choice of the photocathode materials can have a marked effect on the intensities realized from this detector. There is a very large change in efficiency with a change in the angle of the photocathode to the photobeam. In general, efficiencies are much better for near-grazing incidence. Unfortunately, commercial MEMs built for mass spectrometry have the photocathode 90° to the beam. A detector built especially for x-ray spectroscopy with a replaceable and rotatable cathode has proven very useful for soft x-ray detection. A major limitation of this detector, even when it is optimized for x-rays, would appear to be that the MEM (and the CEM) signal is not pulse-height dependent and cannot be used with pulse-height analysis.

3. Channel Electron Multiplier (CEM)

The continuous channel photomultiplier, developed by Bendix Corporation, is in the form of a hollow glass tube with a highly resistive inner surface. It has an internal bore diameter of typically a few tenths of a millimeter, and has a length-to-bore diameter ratio of about 50. A potential difference of about 2000 V is maintained between the ends, causing a current on the inner surface. The initial electrons are generated either from the coating at the input end or from an external photocathode material. The emitted electrons cascade down the tube, producing gains of 10^5 or more. The multiplier is curved to allow higher gain and minimize feedback. These detectors can be made very small, and multiple arrays of straight channels may be fabricated for special purposes (72). The channels, along with photocathode, anode, and resistance–capacitance networks, may be combined into a package much smaller than conventional detectors.

4. The Photographic Plate

The photochemical effect on silver halide salts occurs throughout the soft x-ray region, which accounts for the widespread use of the photographic plate in early work. Because of the high absorption of soft and ultrasoft x-rays the emulsion must necessarily be very thin, with virtually no gelatin in the emulsion. The photographic plate suffers from the disadvantage that the information must be extracted from the plate by a microdensitometer. Since the densitometer measures only photographic blackening, which is not always linearly related to intensity, the operator is forced into making complicated corrections and calibration curves for quantitative work.

D. The Electron Microbeam Probe

A special application of x-ray spectroscopy is found in the use of the electron microbeam probe. In this technique a finely focused beam of electrons is directed at a specimen surface, causing characteristic x-rays to be produced in a volume of only a few cubic microns. These x-rays are analyzed by high-efficiency curved-crystal spectrometers. The device allows nondestructive qualitative and quantitative analysis of elements of atomic number 5 (boron) and greater. In addition, most microbeam probes provide sufficient resolution so that the fine features of x-ray spectra described in Section III may be used to determine chemical combination effects.

E. Electron Spectroscopy for Chemical Analysis (ESCA)

The technique ESCA pioneered by Siegbahn and co-workers (70) is based on magnetic or electrical high-resolution analysis of the electrons that are expelled from a material when it is irradiated with photons (x-rays) of sufficient energy. ESCA allows the energy levels to be measured exactly, as compared to x-ray spectroscopy where we are measuring energy differences between two energy levels. It has the further advantages that all elements may be studied and sensitivities for low-atomic-number elements are much higher than for x-ray spectroscopy. Although ESCA will not be considered in this review, it is mentioned here to indicate the future importance of this technique. It is felt that ESCA by itself and in conjuction with x-ray spectroscopy will be a most important and indispensable tool for the inorganic chemist of the future. A very recent publication by Hamrin et al. (38) on the correlation of the binding energy of the inner electrons of sulfur and the chemical bond serves to show the power of this kind of analysis.

III. Spectra

A. Background

Any serious attempt to describe all the ways in which soft x-ray spectra can be applied to inorganic chemistry could not possibly be confined to one short chapter. The spectra discussed here are therefore necessarily limited in scope. For the most part we will concentrate on very recent work in the 5–120 Å wavelength region. References to and discussions of much of the earlier work can be found in review articles by Siegbahn (*69*), Herglotz (*44*), Faessler (*22*), Sandstrom (*65*), Shaw (*68*), Tomboulian (*74*), and Blokhin (*14*). Appleton (*3*) and Thompson and Kellen (*73*) have summarized most of the soft x-ray results from alloys. We will confine ourselves mainly to the changes that occur in spectra when going from pure element to a simple compound and discuss their interpretations and applications. Most of the elements referred to are low-atomic-number elements, since these are the ones whose spectra show the most dramatic changes as a function of chemical state.

In soft x-ray spectroscopy it is common to distinguish between emission lines and emission bands. Emission lines result from transitions between two sharp inner levels. Emission bands result from transitions of a valence or conduction-band electron to a vacancy created in an inner shell. Much of the soft x-ray emission from the light elements consists of bands rather than lines. The wavelengths of the lines and bands and the energy values of the various electronic shells for each of the elements can be found in the compilations of Bearden (*11*) and Bearden and Burr (*12*).

B. General Chemical Effects

From a practical standpoint, soft x-ray spectroscopy can be used not only to identify the presence of a particular element but also very often to give an indication of how it is combined chemically. This usefulness stems from the large influence that chemical bonding and valence state can have on various parts of the spectra. In general, emission spectra can show one or more of the following effects as a result of chemical combination: (1) energy shift of intensity maximum, (2) change in shape, (3) change in relative intensities, and (4) appearance or disappearance of certain spectral components.

The energy shift of an emission line when going from pure element to compound can be to either higher or lower energy. The direction of shift is dependent on the direction of charge flow in the chemical bond. If an atom is behaving as an electron donor (charge flow away from the atom), an emission line will generally shift to higher energy. For an electron acceptor the shift will be to lower energy. Shifts of emission bands, on the other hand, cannot be so readily classified. Often the emission band from a compound does not represent the same electronic transition as it does in the pure element. This will become apparent shortly, in the discussion of crossover transitions. A shift of an x-ray emission line measures only the change in the energy separation of two inner levels. Both of these levels could shift significantly, but if the shift of each is in the same direction and by the same amount the resultant x-ray line will not shift at all. X-ray absorption spectroscopy and electron spectroscopy (23, 70) can be used to measure shift of individual inner levels.

The shape of an emission band is determined in large part by the energy distribution of the electrons in the valence or conduction band. It is these electrons that are involved in the chemical bond, and the emission-band shape can therefore be a very sensitive indicator of a change in bond character. From a more basic standpoint, much valuable information about the electronic band structure of materials can be deduced from properly interpreted emission-band spectra.

Large changes in the relative intensity of certain emission lines and bands also can signify variations in bonding character. The reasons for some intensity changes, such as the K_α satellites of the third-period elements are not well understood. Nevertheless these changes can be used as practical aids in determining bond character. Often the change in the intensity of emission bands is directly related to the change of electron population in the valence band as a result of the chemical bond. In this case the intensity change is intimately tied in with the change in shape and energy shift of the band.

In the soft x-ray spectra of many compounds we find the presence of emission bands that are not seen at all in the spectrum of the pure element. Over the years most of these extra emission components (e.g., $K_{\beta'}$ in oxides of Mg, Al, and Si) have been rather vaguely labeled "satellites" with no clear explanation as to their origin. Recent work indicates, however, that many of these extra bands in the compounds result from crossover transitions between the anion and cation. Such transitions can occur because of strong anion–cation orbital overlap and can therefore be used to provide significant information about the chemical interaction.

One must also be aware of the problem of chemical stability of the specimen under investigation. Some materials will decompose under electron-beam bombardment so that the resulting emission spectrum is no longer characteristic of the original material.

Despite these problems, however, soft x-ray spectroscopy can be an invaluable tool if treated properly.

C. K-Series Spectra

1. NORMAL LINE EMISSION

The only normal single-vacancy soft x-ray K-emission-line changes that have been applied with much success are the $K_{\alpha_{1,2}}$ lines of the third-period elements, particularly Mg, Al, and Si. Even here the applications are limited because the main K_α line shift is not very great. Usually the satellite lines and emission band will undergo much greater change, and therefore much more has been done with them.

The $K_{\alpha_{1,2}}$ line arises from a transition from the $L_{2,3}$ shell to a vacancy created in the K shell. For the third-period elements its wavelength varies from 4.7 Å at chlorine to 11.9 Å at sodium. The Al K_α line at 8.3 Å has been studied more thoroughly than any of the others. Its usefulness has been demonstrated by White *et al.* (*77*), Day (*19*), and Arrhenius (*4*). Day, for instance, showed that the amount of shift in the Al K_α line could be correlated with the coordination site of aluminum in minerals (*19*).

Schnell has observed the K_α emission lines from various compounds of S, P, and Si (*67*), showing only a very small effect of bonding. The K_α line of Cl, however, does show a measurable shift with oxidation state (*66*).

2. Satellite Emission

Immediately to the high-energy side of the $K_{\alpha_{1,2}}$ line one finds several weaker lines called satellites. In the first-row transition metals these lines are quite weak and difficult to separate from the parent line. In the third-period elements, however, these satellites are relatively intense and can be well resolved from the parent K_α line. The three main satellite lines are $K_{\alpha'}$, K_{α_3}, and K_{α_4}, which are shown in Fig. 5 for Mg, Al, Si and their oxides. These satellites are of particular interest for several reasons. For one thing they can be easily generated, dispersed, and detected using rather simple and readily available instrumentation such as the commercial vacuum-path or helium-path spectrometers. Also, all three elements form a wide variety of interesting compounds, and the satellite lines are quite sensitive to changes in chemical bonding. This sensitivity is manifested by energy shifts and large changes in the $K_{\alpha_4}/K_{\alpha_3}$ intensity ratio (*8, 9, 25, 26, 63*).

From the normal energy-level diagrams it is not readily apparent what transitions could give rise to the $K_{\alpha'}$, K_{α_3}, and K_{α_4} lines. Actually, they originate in doubly-ionized atoms, also called KL double-hole states. $K_{\alpha'}$, K_{α_3}, and K_{α_4} all arise from $L_{2,3} \rightarrow K$ transitions in the presence of two simultaneous electron vacancies, one vacancy being in the K shell, the other in one of the L shells. Recent theory based on sudden-approximation calculations has succeeded quite well in explaining both the origin and relative intensities of these satellites (*1, 75*). It has been widely believed that such satellites are found in electron-excited spectra but not in x-ray-excited spectra. This is now known to be untrue (*75*). K_α satellites of the third-period elements are found with equal intensity using either method of excitation. It is not well understood, however, why a change in chemical bonding has such a large effect on these lines.

The effect of oxidation on the aluminum lines, for instance, is apparent from Fig. 5 and the data in Table IV. Notice that the energy shift of the K_{α_3}

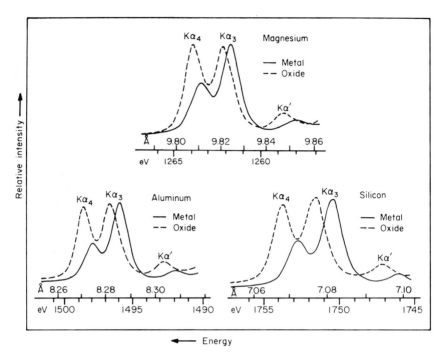

FIG. 5. $K_{\alpha'}$, K_{α_3}, and K_{α_4} satellite lines obtained from pure element and oxide for magnesium, aluminum, and silicon.

and K_{α_4} lines is twice as great as the $K_{\alpha_{1,2}}$ parent line. Not only is there an energy shift, but the $K_{\alpha_4}/K_{\alpha_3}$ intensity ratio changes significantly also. Such readily observable effects make these satellite lines valuable tools for determining changes in bonding and coordination number in compounds of magnesium, aluminum, silicon, and other third-period elements. Some practical applications of these spectra have been shown by Baun and Fischer (*10, 29, 31, 32*). In general, the more ionic the bond, the greater the $K_{\alpha_4}/K_{\alpha_3}$ intensity ratio and the higher the energy position of the K_{α_4} and K_{α_3} lines. A typical example of this is shown in Fig. 6 for the Al–Ni binary system. As the nickel concentration is increased the AlK_{α_4} and K_{α_3} lines shift to higher energy and the $K_{\alpha_4}/K_{\alpha_3}$ intensity ratio becomes larger. From this it was concluded that the bonding on the aluminum atoms becomes more ionic in character as the nickel concentration is increased (*29*).

3. K-Band Emission

Some of the largest chemical effects on x-ray spectra can be observed in the K emission bands of the light elements. Many different investigators have studied K bands, and we will here show some typical examples of the more

TABLE IV

ALUMINUM K LINES AND BANDS FROM ALUMINUM AND Al_2O_3

Line	λ (Å)	E (eV)	Intensity[a]	ΔE (eV) [b]
Aluminum metal				
$K_{\alpha_1\alpha_2}$	8.3393	1486.3	1000	—
$K_{\alpha'}$	8.3080	1492.0	13	—
K_{α_3}	8.2854	1496.0	78	—
K_{α_4}	8.2744	1498.0	39	—
K_{α_5}	8.2284	1506.3	5.0	—
K_{α_6}	8.2098	1509.8	3.9	—
K_β	7.9590	1557.3	6.5	—
Aluminum oxide, Al_2O_3 (anodized film)				
$K_{\alpha_1\alpha_2}$	8.3380	1486.6	1000	+0.3
$K_{\alpha'}$	8.3037	1492.7	18	+0.7
K_{α_3}	8.2820	1496.6	64	+0.6
K_{α_4}	8.2718	1498.6	60	+0.6
K_{α_5}	8.2240	1507.2	4.6	+0.9
K_{α_6}	8.2050	1510.7	3.5	+0.9
$K_{\beta'}$	8.0618	1537.5	1.3	—
K_β	7.9819	1552.9	7.0	−4.4

[a] Peak intensity.
[b] Shift between metal and oxide.

recent work. The bands shown in the figures are the as-recorded spectra with no corrections applied to them. In some cases satellite emission and self-absorption effects may have a slight bearing on the band shape at the absorption-edge position, but this will not significantly affect the discussion on chemical effects. These effects are important, however, in certain L and M emission bands, as will be shown later.

Examples of K-band emission from beryllium, boron, carbon, and nitrogen are shown in Fig. 7. These are the entire x-ray emission spectra of these elements and represent a valence-band → K-shell transition. Since the valence electrons are the ones involved in the chemical bond, it is hardly surprising that these emission bands show large changes with a change in the chemical environment. All the spectra of Fig. 7 were obtained using some of the soap-film "crystals" listed in Table III as dispersing devices.

The beryllium K band was obtained using a lead lignocerate "crystal" (21, 33). Better resolution at this wavelength (120 Å) can be obtained with a grating spectrometer (48), and such an instrument would be preferable for accurate band-shape studies. For analytical work, however, the crystal

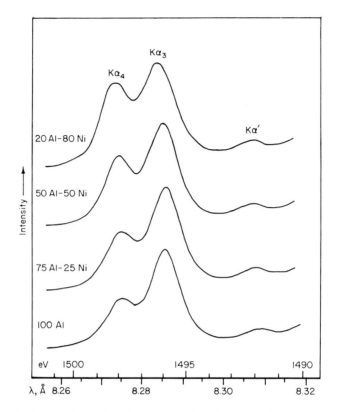

FIG. 6. Aluminum $K_{\alpha'}$, $K_{\alpha3}$, and $K_{\alpha4}$ lines from some Al–Ni alloys.

spectrometer is much more convenient, and the resolution is adequate to observe any significant chemical effects on the spectrum. Notice in Fig. 7 that there is quite a large wavelength shift and band-shape change when going from pure Be to BeO. This reflects the change in the bond character from metallic to ionic.

A similar effect is observed in the boron K band, also shown in Fig. 7 for pure boron, BN, and B_2O_3. These bands were obtained with a lead stearate "crystal" (21, 30, 32). Others have shown these spectra from a grating spectrometer (48) with perhaps better resolution, but the band shapes are in very good argeement. The boron K bands in Fig. 7 illustrate three pronounced but quite different effects caused by chemical combination: (1) A dramatic change in the shape of the band, (2) a large wavelength shift, and (3) the appearance of two new bands in BN and B_2O_3 that are not present in the pure element. The changes in shape and the wavelength shift are rather characteristic of the light-metal K bands when going from metallic to ionic

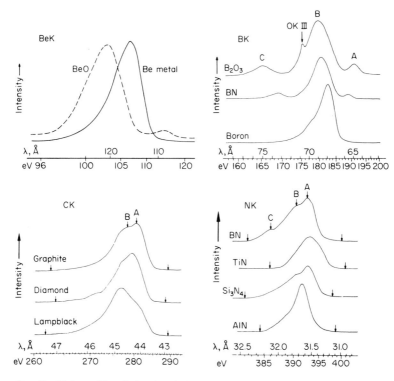

FIG. 7. Valence K emission bands from some low-atomic-number elements.

bond type. The extra bands that appear are also characteristic of several of the light-element K spectra from oxides and nitrides. It now appears that the low-energy band, labeled C in Fig. 7, is due to a crossover transition from the anion $2s$ band to a vacancy in the boron K shell. Similar transitions are found in the K-band spectra from certain compounds of Mg, Al, Si, P, Cl, and S and will be discussed later. Such an $s \to s$ transition occurs because of the large anion–cation orbital overlap, which introduces other symmetries into the anion $2s$ level. The band labeled A had been explained as being due to reemission of electrons that have been excited into higher, normally unoccupied, levels (*37*). Considerable information about the chemical environment is therefore present in these spectra if properly interpreted.

Of all the light-element K bands, the carbon K band has been the most widely studied. Most of the data in the literature have been recorded using grating spectrometers and photographic registration. The carbon K bands in Fig. 7 are second-order reflections from lead stearate with a flow proportional counter (*27*). These band shapes do not agree too well with some of Holliday's recent work on a grating spectrometer (*49*). Some of his results are shown in Fig. 8. Perhaps the differences are due to the dispersion method, although excitation conditions may also have something to do with it.

Holliday (*49*) has studied the carbon K band from some metal carbides and has related the intensity distribution of the band to the bonding character (*49*). Several of these bands are shown in Fig. 8. Mattson and Ehlert (*58*) have shown some unusual carbon spectra from various gases. Most of these spectra are considerably different from that obtained from solids. Similar molecular species appear to yield similar spectra. Mattson and Ehlert have also found a correlation between the peak position and the first ionization potential.

The nitrogen K band has been studied recently by Fischer and Baun (*27*) and Holliday (*48*). As can be seen in Fig. 7, there is a wide variation in nigrogen K-band shapes from different nitrides. Unfortunately, many nitrogen containing compounds that would be expected to show even greater changes, such as nitrates and nitrites, are unstable under electron-beam bombardment.

Mattson and Ehlert have shown the fluorine K band and K_{α_3} and K_{α_4} satellite lines for a large number of fluorine-containing compounds (*57*). A change in chemical bonding appears to have two major effects on the spectrum: a shift in energy of the main band and a considerable variation in

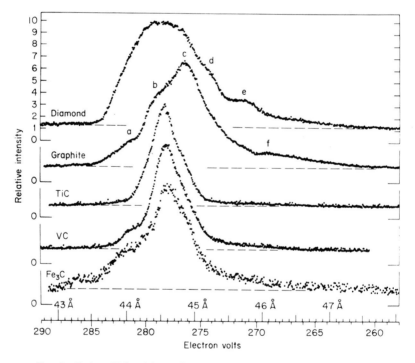

FIG. 8. Carbon K band from diamond, graphite, TiC, VC, and Fe_3C (*49*).

intensity of the satellite lines. The spectra from the fluorides are grouped by crystal-structure types with a fair degree of correspondence. It was also shown that the spectra excited by x-rays did not differ significantly from those excited by electrons.

Little has been done with sodium in the way of determining chemical effects on the K emission band. From some preliminary studies in our own laboratory it appears that the band changes are rather similar to those observed in Mg, Al, and Si.

Magnesium, aluminum, and silicon are grouped together because the chemical effects on their spectra are very much alike. Some of these effects were shown earlier for the satellite lines (Fig. 5). The K emission bands from these elements and their oxides are shown in Fig. 9. For the pure elements this band is very asymmetric in shape, having a rather sharp high-energy edge. In each of the oxides the band becomes much more symmetrical and shifts considerably to lower energy. The intensity maximum of the aluminum band, for instance, shifts 4.4 eV when going from metal to oxide, as indicated in Table IV (*9, 24, 26, 66*). The K band of Mg, Al, and Si is therefore extremely sensitive to any changes in the chemical bond. These large alterations in

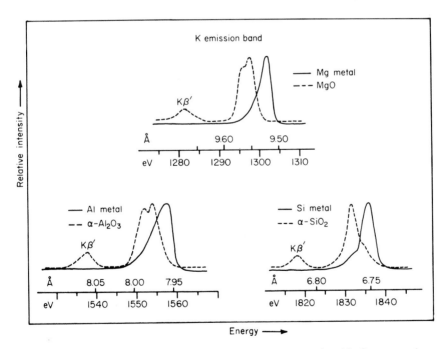

FIG. 9. K emission band obtained from pure element and oxide for magnesium, aluminum, and silicon.

shape and energy position have resulted in the use of these bands in analyzing a wide range of alloys, minerals, and many other inorganic compounds.

Fischer and Baun have shown the K-band changes in simple compounds of aluminum (*10, 26*) and in many aluminum binary alloys (*29, 31, 32, 33*). It was shown that the chemical effect on the K band is quite periodic in nature. Elements that belong to the same subgroup each have virtually the same effect on the Al K band. Elements from different subgroups have different effects on the Al K band, and these effects can be correlated with the relative position of the subgroups in the periodic table. A few examples of this are shown in Fig. 10. As the ionicity of the bond increases, the bands become more symmetrical in shape and shift to a lower-energy position. Furthermore, within any one binary alloy system the energy position of the Al K-band intensity maximum is linearly dependent on the alloy composition (*32*). A rather curious relationship is also found between the Al K-band position and the $K_{\alpha_4}/K_{\alpha_3}$ satellite intensity ratio in each of the aluminum compounds studied (*35*). Over 200 different compounds were studied, and in each case it was found that if the Al K band shifts in energy, there is always a corresponding change in the Al $K_{\alpha_4}/K_{\alpha_3}$ intensity ratio. It makes no difference whether the compound is a conductor, semiconductor, or insulator (*35*).

There is not universal agreement on the interpretation of the energy shifts and shape changes in the K bands of Mg, Al, and Si. It is argued that any model that successfully explains the observed changes is a good model. Most attempts at interpretation have invoked the band model. Dodd and Glenn (*20*), on the other hand, have used molecular orbital theory to explain these band changes. Most of these theoretical interpretations, however, have overlooked a portion of the spectrum that now appears to be an important key to the evaluation of the main K band. The portion of the spectrum referred to is the weak emission band labeled $K_{\beta'}$ in Fig. 9. Notice that this band is present in the spectra from the oxides but not in the spectra from the pure elements. $K_{\beta'}$ has usually been identified as a satellite, but with no adequate explanation of its origin. It has also been identified as one of several "forbidden" transitions and as a transition from an excited state. Recent work, however, indicates that $K_{\beta'}$ is probably a crossover transition from the anion 2s level to a vacancy in the metal K shell. This resulted from work in our own laboratory on the 3d metal L bands from oxides (to be discussed later) and, independently, by Mendel (*59*). The main K emission band in the oxides of Mg, Al, and Si would therefore be the crossover transition from the anion 2p band to a vacancy in the metal K shell. This is not too surprising, because if one assumes purely ionic bonding in the oxides, the metal-ion 3s3p band would be empty and the normal band emission such as found in the pure elements could not occur. The normally forbidden $s \rightarrow s$ transition used to explain the presence of $K_{\beta'}$ can occur because of the strong anion–cation

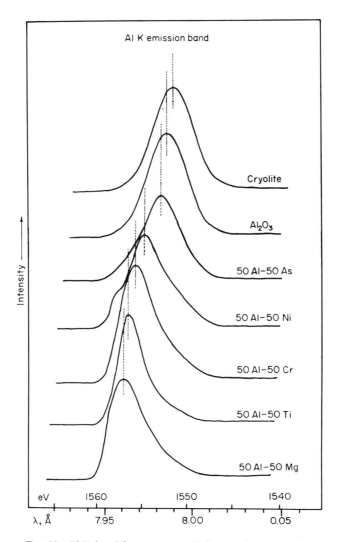

FIG. 10. Al K band from a group of alloys and compounds.

orbital overlap, which introduced other symmetries into the anion $2s$ band. The relative intensity of $K_{\beta'}$ would then be a direct indication of the degree of orbital overlap between anion and cation. $K_{\beta'}$ is also found in the spectra from nitrides, fluorides, phosphides, and possibly other compounds of Mg, Al, and Si. Dodd and Glenn (20) had to assume a considerable covalent character to the bond in the oxides for their molecular-orbital interpretation of the main K band.

$K_{\beta'}$ is also found in the K spectra of certain compounds of phosphorous, sulfur, and chlorine. Since these elements are much more electronegative than magnesium, aluminum, and silicon, their K emission bands do not show as large a change as those in Fig. 9 (*66, 67*). Schnell's results show that for the elements aluminum through chlorine, $K_{\beta'}$ occurs in the spectrum when the element is bonded with oxygen. In general, the higher the oxidation state the higher the intensity of $K_{\beta'}$ with respect to the main K band. Wilbur (*80*) has done a rather thorough investigation of the sulfur K band and its dependence on chemical bonding. His spectra show large changes in the band shape and energy position, but there is no apparent correlation with bonding. He also observed $K_{\beta'}$ but found that it occurred only when sulfur was bonded to at least two oxygen atoms. From this work Wilbur suggests that $K_{\beta'}$ is due to a $3s \rightarrow 1s$ transition, which can occur because of the admixture of p character into the $3s$ level.

4. K ABSORPTION SPECTRA

Very little has been done in the way of measuring chemical effects on soft x-ray absorption spectra. Much has been done in the hard x-ray region, but there are serious problems when going to the long-wavelength region. The main problem is sample preparation. The sample must be thin enough to transmit the soft x-ray continuum and homogenous enough to give an absorption spectrum truly characteristic of the material. Films can be prepared for metals by vacuum deposition or by using very thin foils, but it becomes very difficult for most compounds. As a result soft x-ray absorption is not generally considered applicable to routine analysis problems. An example of one of the few experimental investigations made in this area is the work of Nagakura (*62*). He measured the aluminum K absorption spectrum from metal and two oxides, showing a large shift in the edge position. Considerably more work has been done in soft x-ray absorption for certain L spectra, and this will be discussed in the next section.

D. L-Series Spectra

Many soft x-ray L emission spectra show very large changes with chemical bonding and are easily observed with available commercial instrumentation. One of the most striking spectral changes is observed for sulfur. It is found that the sulfur L spectrum is exceptionally sensitive to bonding effects (*28*). This is especially evident when going from the -2 to $+6$ valence state as shown in Fig. 11. These spectra were obtained using a lead stearate crystal and cover the 75–85 Å wavelength region (*28*). Such gross changes obviously make the sulfur L emission spectra a most valuable means of determining the valence

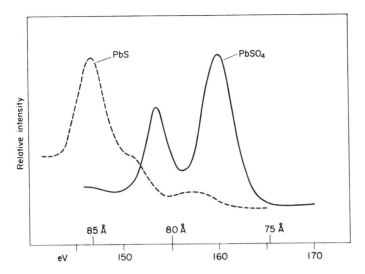

FIG. 11. Sulfur L emission spectra obtained from PbS and PbSO₄.

state of sulfur in a compound. One must be very careful, however, when using electron-beam bombardment. Many sulfur-containing compounds are quite unstable under these conditions, and the resulting spectra are no longer characteristic of the starting material. This has been demonstrated by Fischer and Baun (*33*) and Merritt and Agazzi (*60*).

The chlorine L spectrum also shows significant changes with chemical bonding, as demonstrated by Fischer and Baun (*28*) and Henke and Smith (*41*). Fischer and Baun have shown that the energy shift of the main peak shows a strong correlation with the character of the bond in the chlorides. There are many other chlorine-containing compounds that would be interesting to study, such as chlorites, chlorates, perchlorates, etc., but unfortunately most of these materials are not very stable under electron-beam bombardment. Henke and Smith (*41*), however, have obtained the chlorine L spectra from some of these compounds by using x-ray excitation. Some of their results are shown in Fig. 12 (*43*). These spectra certainly demonstrate a significant dependence on the chemical bonding. As the compounds become more complex, so does the spectrum. To date, no theories have appeared that adequately explain the spectral changes.

The L_{II} and L_{III} emission bands have been investigated for most of the $3d$ transition metals and their compounds. The most notable recent work has been that of Chopra for nickel (*17*); Liefeld for copper, nickel, and zinc (*54, 55*); Bonnelle for iron, nickel, cobalt, and copper (*15*); Hanzely for iron, cobalt, nickel, copper, and zinc (*39*); Fischer for titanium and vanadium (*24, 36*). All of these investigators have made a point of stressing how badly

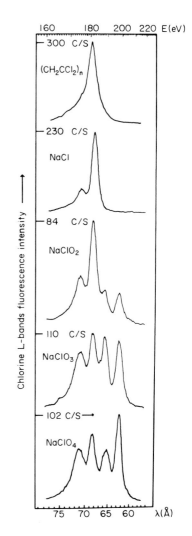

FIG. 12.   Chlorine L emission spectra (43).

satellite emission and self-absorption can distort the L_{III} band shape. It is apparent that to obtain meaningful and reproducible results from these spectra one must account for these gross distorting effects. How they can actually affect the appearance of a spectrum is shown in Fig. 13 for the Ti L_{II} and L_{III} bands (36). The numbers in the lower right-hand corner indicate the electron-beam voltage and takeoff angle under which the individual spectra were obtained. As the penetration depth of the bombarding electron beam is increased, there is a severe intensity loss beginning at the high-energy

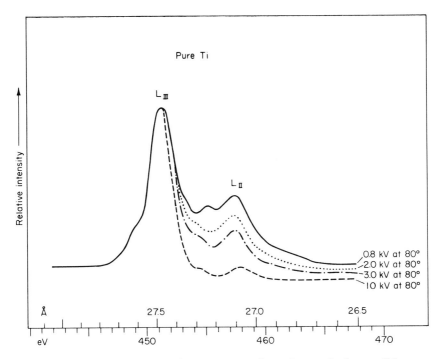

FIG. 13. Titanium $L_{II,III}$ emission spectrum under various excitation conditions.

side of the L_{III} band and extending beyond the L_{II} intensity maximum. It can readily be appreciated that if two different investigators obtained these spectra under widely different excitation conditions, they would not agree at all on the shape or intensities of the bands. This effect is caused in large part by overlap of emission structure by the L_{III} absorption edge. This emission-absorption overlap occurs because much of structure at the high-energy side of the L_{III} emission band is satellite emission from multiply ionized atoms. We therefore have a rather curious phenomenon occuring as the beam voltage is increased. Satellite emission is growing, and at the same time it is being suppressed by the self-absorption effect (24, 36). Much of the loss in intensity of the L_{II} band is due to the $L_{III} \rightarrow L_{II}$ Auger transition (36). Similar distorting effects have been shown to occur for other $3d$ metal spectra as as well (15, 17, 39, 54, 55).

Although self-absorption causes some problems, it has a good side too. In the long-wavelength region it is usually very difficult to obtain accurate absorption spectra, especially from compounds. The self-absorption effect provides us with a practical method of obtaining absorption-edge replicas in many cases. The procedure is to use two emission-band spectra afflicted with widely different amounts of self-absorption. The spectrum with the lesser amount of self-absorption is used as I_0 and the spectrum with the larger amount of self-absorption is used as I. A point-by-point $\ln I_0/I$ curve is then constructed, which gives an absorption replica that is virtually identical in appearance to the normal photon absorption curve (36). It should be pointed out, however, that these two types of curves are not equivalent. Nevertheless, in the absence of normal photon-absorption spectra, the self-absorption curves can provide very useful replicas. If one is mainly interested in determining the energy position of an absorption edge, then the self-absorption curve serves the purpose quite well. This has been done with considerable success for many transition-metal compounds (24, 36). Liefeld (54) first demonstrated this technique for the L_{III} spectra of nickel, copper, and zinc.

Liefeld (55) has also pointed out that it is impossible to correct certain emission-band spectra for both satellite and self-absorption distortions except at threshold voltage. It is very difficult to obtain spectra near threshold, but Liefeld had done so for several $3d$ metal L_{III} bands. Such distortion-free spectra are of special interest for attempting to relate band shapes to the theoretically determined energy-band structure. So far, good agreement between theory and experiment has been achieved for a few of the $3d$ metals using the soft x-ray L_{III} emission band.

The appearance of the $L_{II, III}$ emission spectrum from several of the $3d$ elements is affected significantly by changes in chemical composition. An example of this is shown in Fig. 14, for some oxides of titanium (36). These spectra were all obtained under conditions of negligible self-absorption, although L_{II} and L_{III} satellite emission is at saturation. The most noticeable change in these spectra is the appearance of a new emission band in the oxides, labeled A in Fig. 14, which is not present in the spectrum from the pure metal. This band grows in intensity and shifts to a higher energy position as the oxidation state increases. Bands B and C also shift to higher energy with an increase in oxidation state. In the spectrum from the pure metal, band B is the normal L_{III} band and band C is the normal L_{II} band. This is not necessarily the case in the spectra from the oxides, however. Fischer and Baun (36) have shown that band A in the oxides is due to the $02p \rightarrow Ti\ L_{III}$ crossover transition. This crossover transition evidently occurs only in compounds where the $3d$ shell is half filled or less than half filled (24). Such a $p \rightarrow p$ transition is normally forbidden by the dipole selection rules, but it probably occurs here because of strong cation–anion orbital overlap, which admixes other symmetries into the anion $2p$ states. The presence of this transition in the spectrum may therefore provide a direct measure of the anion–cation (d_ϵ, p_π) overlap. It is believed that this type of anion–cation interaction has an important bearing on whether a transition-metal compound acts as insulator or metal (61). There is probably much significance in the fact that the anion $2p \rightarrow$ metal

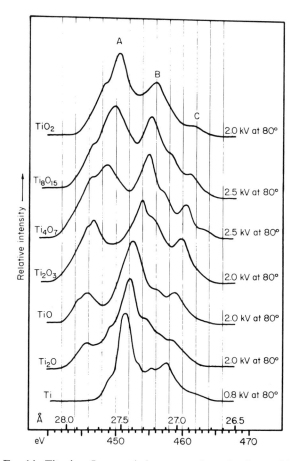

FIG. 14. Titanium $L_{II,III}$ emission spectra from titanium oxides.

L_{III} crossover transition occurs for the compounds of titanium and vanadium, which are conductors or show semiconductor-to-metal transitions, but does not occur for the compounds of heavier $3d$ elements such as Fe, Co, and Ni.

Each oxidation state of titanium yields an $L_{II, III}$ emission spectrum that is immediately distinguished from the spectra of other oxidation states, as can be seen in Fig. 14. This holds even under conditions of considerable self-absorption and instrumental broadening. From the practical application standpoint, then, these spectra could prove quite useful to the inorganic chemist.

It is found that the Ti L_{III} absorption-edge spectra of these same oxides are also significantly affected by changes of oxidation state. Other compounds of titanium and of other $3d$ elements also show considerable chemical effects in their L_{III} absorption spectra. The amount of energy shift in the edge position is an indication of the degree of charge transfer from cation to anion. From a routine analysis standpoint, however, low-energy absorption spectra are of questionable value mainly because they are so difficult to obtain. In fact the best way to obtain most of them is by the differential self-absorption method described earlier.

The spectra discussed above are just a few typical examples of what has been done recently. Many other soft x-ray L spectra have been studied by various investigators and have yielded significant results. A discussion of all these results, however, is far beyond the space limitation of this chapter.

E. M-Series Spectra

Very few investigators have studied soft x-ray M spectra in recent years. Most M emission bands are quite weak and difficult to observe with the necessary resolution. Holliday (47–49), however, has studied some M lines and bands from the elements Y, Zr, Nb, and Mo. An example of his results are shown in Fig. 15 for the Zr M_{IV} and M_V bands. Obviously there is a

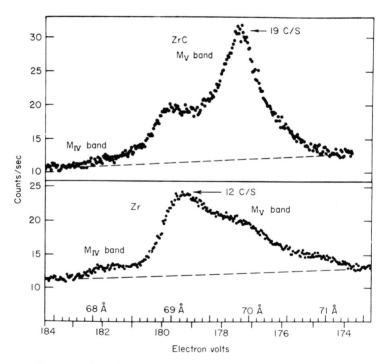

Fig. 15.   Zirconium $M_{IV,V}$ emission bands from Zr and ZrC (49).

significant change in band shape when going from Zr to ZrC. The intensity of these bands is so weak, however, that they would be difficult to use analytically. Also, at the wavelength region involved (60–100 Å) a crystal spectrometer cannot prvide the resolution required to separate the individual intensity maxima. A grating spectrometer, such as used by Holliday, must be employed.

Holliday has also attempted to relate the M_V band shapes from the pure metals Y, Zr, Nb, Mo to density-of-states curves obtained from specific-heat measurements and energy-band calculations. The comparisons show some agreement, but there is still considerable room for improvement. It is not clear whether the difficulty lies with the theoretical or experimental results or both.

Fischer and Baun (34) have demonstrated some rather unusual effects in the M_α and M_β emission lines of the rare-earth elements. These lines fall in the 7–15-Å wavelength region and are easily observed with a crystal spectrometer. The rare earths are a unique series of elements insofar as they have an inner shell of electrons (the 4f shell) that is unfilled and that apparently plays little part in chemical bonding. This fact accounts for the almost identical chemical properties of the rare earths and also forms the basis for the unusual phenomenon observed in their M_α and M_β x-ray emission spectra. The M_α $(4f \rightarrow 3d_{5/2})$ and M_β $(4f \rightarrow 3d_{3/2})$ emission lines arise from transitions involving the unfilled shell. Theoretically they would be expected to show a rather complicated emission structure because of coupling between the incomplete 4f shell and the singly ionized $3d_{5/2}$ (M_V) or $3d_{3/2}$ (M_{IV}) shell. These spectra were obtained between 35 and 40 years ago and were found to have a very complicated structure (56). Experiment and theory agreed, but we now know that this agreement was right for the wrong reasons. This is demonstrated by the curves in Fig. 16. The spectra on the left (a) are the M_α and M_β emission lines of each of the rare-earth elements, obtained using a bombarding-beam voltage of 10 kV at a 30° takeoff angle. They look complicated just as the theory says they should. This apparently convincing agreement becomes quickly shattered, however, by observing the spectra shown on the right side (b) of Fig. 16. These are the same spectra as those on the left side (a) except that they were obtained at beam voltages only 200–300 V in excess of M_{IV} threshold with a high takeoff angle. The striking change in spectral shape, therefore, is a function of the excitation conditions. Fischer and Baun have shown that the complicated multiplet structure observed in the spectra (Fig. 16a) is due to target self-absorption (34). Therefore, it is not true emission structure, but rather, it is absorption structure. This was further proved by obtaining the $M_{IV, V}$ absorption spectra of each of the rare earths and comparing them to the emission spectra. It was found that the emission and absorption spectra fall at exactly the same energy positions.

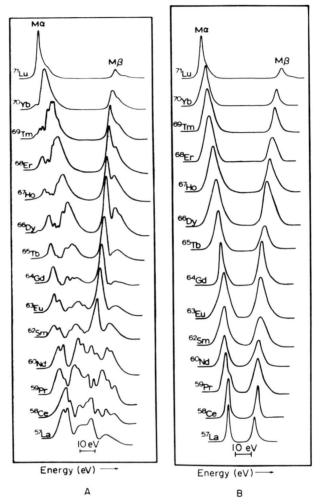

FIG. 16. $M_{\alpha, \beta}$ Lines for rare-earth elements under conditions of maximum (a) and minimum (b) self-absorption.

For each of the rare earths the M_{IV} absorption completely overlaps the M_{β} emission and the M_V absorption completely overlaps the M_{α} emission. In all cases, the absorption fine structure is precisely mirrored in the emission spectra. The true M_{α} and M_{β} emission spectrum consists of single lines for both M_{α} and M_{β} with some overlapping high-energy satellite structure, as shown in Fig. 16b. Emission and absorption peaks fall at the same energy because transitions can occur in both directions between the unfilled $4f$ shell and the $3d$ subshells.

Although the rare-earth M spectra present a rather extreme example of self-absorption effects, they do point up the dangers of putting too much faith in emission spectra without also investigating the conditions under which the spectra were obtained. Unfortunately, many of the spectra shown in the literature are unreliable for this reason.

F. Other Series

A number of N-series spectra have been reported. However, because of the weak intensities and long wavelengths involved, N lines have not been very well characterized, and sometimes lines that have been given N designations do not involve the N shell at all. Frequently, higher orders of M spectra are identified as N spectra when voltages are high enough to excite M lines. Other series such as the O series are even more in doubt.

IV. Applications

Applications of x-ray spectroscopy are found primarily in the following areas: Qualitative and quantitative analysis of the light elements; determination of bonding and coordination by use of the fine features of x-ray spectra; distribution and intensity of solar and stellar spectra; and analysis of man-made high-temperature plasmas. Since stellar and solar spectra and distribution in plasmas are not of direct interest to conventional inorganic chemistry, these areas will not be considered.

A short time ago it was considered to be a major accomplishment to be able *just to detect* radiation from elements such as oxygen, nitrogen, carbon, boron, and beryllium. Now it is possible to accomplish quantitative analysis on these elements. Henke (42) has shown the experimental conditions necessary to analyze very small quantities of low-atomic-number elements. Henke defined the minimum detectable limit as that concentration of an element that gives rise to a signal equal to or greater than three times the standard deviation of the background. In the referenced work Henke showed conditions for obtaining minimum detectable limits (in wt. %) for oxygen (0.03 %), nitrogen (0.11 %), carbon (0.04 %), boron (0.01 %), and beryllium (0.2 %).

A further illustration of the use of soft x-rays in quantitative analysis is the analysis of FeO in Fe_2O_3 (81). In this work, the intensities of the O K and Fe L lines were measured from standard physical mixtures and a calibration curve was prepared relating the Fe L/O K ratio to the iron-to-oxygen atomic ratio. Using this calibration curve the authors were able to routinely determine relative amounts of FeO and Fe_2O_3 in mixtures.

Rocks, which contain premarily low-atomic-number elements, are being analyzed by x-ray emission spectroscopy. Baird and Henke (5) have shown that the precision and speed of x-ray analysis is better than with wet chemical techniques.

In bulk analysis of common silicates the x-ray spectrometer can now be used for all major and minor element determinations, often with just one sample preparation. The determination of oxygen allows elemental summations approaching 100% by weight of all elements without using the usual difference calculations for oxygen.

The electron microprobe is fast becoming an indispensable tool in many areas, and mineralogical analysis is no exception. In geologic samples the elements of most interest are of atomic number 25 and lower. Light-element (long wavelength) capability is therefore becoming very important. Qualitative analyses can be carried out on the electron microprobe using either a scanning or a stationary electron beam. The scanning-beam method allows recording x-rays, backscattered electrons, and sample current. Electron-beam scanning provides quick qualitative and semiquantitative information on the distribution of elements, homogeneity, texture, and grain-growth phenomena (50). The resolution of x-ray images, which may be as good as 0.3 μ when using low accelerating voltages and L and M lines of high-Z elements instead of K lines, is superior to the resolution of the optical microscope (50).

The electron microprobe is being used for low-atomic-number analysis in many other areas. For instance in the analysis of hypervelocity impact phenomena it is desired to know the distribution and the degree of interaction between the projectile and the target.

Figure 17a shows the crater in a copper target produced by a $\frac{1}{16}$-in. aluminum pellet traveling in the direction of the arrow. The aluminum K x-ray picture obtained in the electron microprobe at point X is shown in picture b. Further analysis in this area shows at least three binary Al–Cu alloys formed in the crater. Outside the crater considerable material has deposited. Picture (c) is a specimen-current picture of the area marked Y. Picture (d), another aluminum K x-ray picture, shows that much of the aluminum pellet has splashed over the edge of the crater and has been collected in the cusp of the once-molten copper. Pictures (b), (c), and (d) appear to correspond to the side opposite that marked in (a) because they were electronically reversed on the oscilloscope when they were photographed. From a complete analysis of such results many conclusions may be reached concerning the kinetics and local temperatures involved in the impact area.

Quantitative analyses using the electron microprobe are carried out using a static electron beam. An example of such analysis may be found in the work of Andersen et al. (2). In this work the mineral sinoite (silicon oxynitride), from a meteorite, was discovered and quantitatively analyzed for Si, O, and

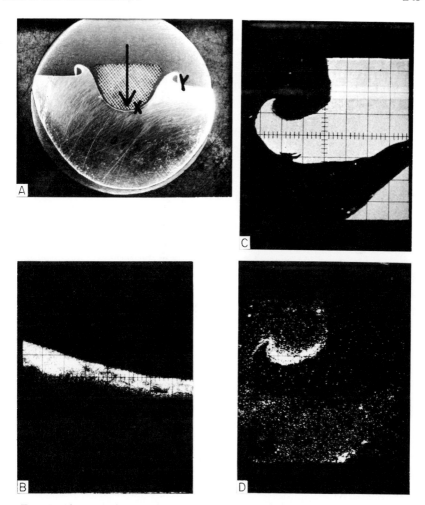

FIG. 17. Hypervelocity-impact sample: aluminum pellet fired into copper target. (a) Light micrograph of crater; (b) Al K x-ray distribution in crater; (c) specimen current presentation in area of Y; (d) Al K x-ray distribution in cusp at Y.

N, using the electron microprobe. The mineral was qualitatively scanned for all elements between $Z = 5$ and $Z = 92$ and only Si, O, and N were detected. Quantitative analysis was carried out by moving the mineral grains in 3-μ steps under a stationary electron beam, thus covering the sample with rows of point integrations. Such a quantitative analysis and identification of single grains in a mineral specimen would be virtually impossible using conventional techniques. This particular analysis is important because it implies availability of nitrogen in the environment in which the material formed.

Fine features in x-ray spectra and changes in emission spectra with chemical combination are beginning to be used for practical purposes. The potential uses for such a use of x-ray spectra are nearly unlimited, and certainly increasing emphasis will be placed in this area in the future. Perhaps the greatest potential use of the technique lies in the characterization of poorly crystalline and glassy materials that are difficult to work with using conventional techniques such as x-ray diffraction. Other possible uses include characterization of thin films, oxidation and reduction phenomena, and valence and coordination determination.

An example of this technique is found in the work of White and Gibbs (*79*). They found that there was a strong correlation between K_β wavelength and the mean Si—O bond distance in a large group of well-characterized silicates. In framework silicates, where the Si—O bond distance is essentially constant, the wavelength of Si K_β could be correlated with the Al/Si ratio. The authors conclude that this technique will be valuable for studies on glasses, gels, and very-fine-grained powders, which are not amenable to precise structural refinements. In other work using Si K_β, White and Roy (*78*) attempted to deduce the structure of very thin "SiO" films on glass. They interpreted their results to show that "SiO" consists of an initimate mixture of Si and SiO_2.

Many other examples of the use of soft x-ray spectroscopy could be cited in the areas of analysis and characterization of inorganic materials. Those briefly mentioned here are meant to show typical results and the types of samples that may be investigated by soft x-rays.

References

1. Aberg, T., *Phys. Rev.* **156**, 35 (1967).
2. Andersen, C. A., Keil, K., and Mason, B., *Science* **146**, 256 (1964).
3. Appleton, A., *Contemp. Phys.* **6**, 50 (1964).
4. Arrhenius, G., *in* "Optique des Rayons X et Microanalyse." Hermann Press, Paris, 1966.
5. Baird, A. K., and Henke, B. L., *Anal. Chem.* **37**, 727 (1965).
6. Baun, W. L., and White, E. W., *Anal. Chem.* **41**, 831 (1969).
7. Baun, W. L., *Rev. Sci. Instr.* **40**, 1101 (1969).
8. Baun, W. L., and Fischer, D. W., *Phys. Letters* **13**, 36 (1964).
9. Baun, W. L., and Fischer, D. W., *Nature* **204**, 642 (1964).
10. Baun, W. L., and Fischer, D. W., AFML-TR-64-350 (December, 1964).
11. Bearden, J. A., *Rev. Mod. Phys.* **39**, 78 (1967).
12. Bearden, J. A., and Burr, A. F., *Rev. Mod. Phys.* **39**, 125 (1967).
13. Birks, L. S., "X-Ray Spectrochemical Analysis," pp. 1–96. Wiley (Interscience), New York, 1959,

14. Blokhin, M. A., "The Physics of X-Rays," 2nd ed. State Publ. House of Techn.-Theoret. Literature, Moscow, 1957.
15. Bonnelle, C., *Ann. Phys. (Paris)* **1**, 439 (1966).
16. Campbell, W. J., and Brown, J. D., *Anal. Chem.* **40**, 346R (1968).
17. Chopra, D. R., PhD Dissertation, New Mexico State Univ. (1964).
18. Davidson, F. D., and Wyckoff, R. W. G., "*Advanc. X-Ray Anal.* **9**, 344 (1966).
19. Day, D. E., *Nature* **200**, 649 (1963).
20. Dodd, C. G., and Glenn, G. L., *J. Appl. Phys.* **39**, 5377 (1968).
21. Ehlert, R. C., and Mattson, R. A., *Advan. X-Ray Anal.* **9**, 456 (1966).
22. Faessler, A., 1508-Rontgenspektrum und Bindungszustand, "Landoldt–Bornstein Tables, Zahlenwerte und Funktionen," 6th ed., Vol. I, part 4, pp. 769–808. 1955.
23. Fahlman, A., Hamrin, K., Hedman, J., Nordberg, R., Nordling, C., and Siegbahn, K., *Nature* **210**, 4 (1966).
24. Fischer, D. W., *J. Appl. Phys.* **40**, 4151 (1969).
25. Fischer, D. W., and Baun, W. L., *Spectrochim. Acta* **21**, 443 (1965).
26. Fischer, D. W., and Baun, W. L., *J. Appl. Phys.* **36**, 534 (1965).
27. Fischer, D. W., and Baun, W. L., *J. Chem. Phys.* **43**, 2075 (1965).
28. Fischer, D. W., and Baun, W. L., *Anal. Chem.* **37**, 902 (1965).
29. Fischer, D. W., and Baun, W. L., *Phys. Rev.* **145**, 555 (1966).
30. Fischer, D. W., and Baun, W. L., *J. Appl. Phys.* **37**, 768 (1966).
31. Fischer, D. W., and Baun, W. L., *J. Appl. Phys.* **38**, 229 (1967).
32. Fischer, D. W., and Baun, W. L., *Advan. X-Ray Anal.* **10**, 374 (1967).
33. Fischer, D. W., and Baun, W. L., *Norelco Reporter* **XIV** (3–4), 92 (1967).
34. Fischer, D. W., and Baun, W. L., *J. Appl. Phys.* **38**, 4830 (1967).
35. Fischer, D. W., and Baun, W. L., *J. Appl. Phys.* **38**, 2404 (1967).
36. Fischer, D. W., and Baun, W. L., *J. Appl. Phys.* **39**, 4757 (1968).
37. Fomichev, V. A., and Rumsh, M. A., *J. Phys. Chem. Solids* **29**, 1015 (1968).
38. Hamrin, K., Johansson, G., Fahlman, A., Nordling, C., and Siegbahn, K., *Chem. Phys. Letters* **1**, 557 (1968).
39. Hanzely, S., PhD Dissertation, New Mexico State Univ. (1968).
40. Henke, B. L., *Advan. X-Ray Anal.* **8**, 269 (1965).
41. Henke, B. L., and Smith, E. N., *J, Appl. Phys.* **37**, 922 (1966).
42. Henke, B. L., *Advan. X-Ray Anal.* **9**, 430 (1966).
43. Henke, B. L., *Proc. Symp. Low Energy X- and Gamma Sources and Appl. 2nd, Austin, Texas* March 27–29, 1967. (ORNL-IIC-10, p. 523, Sept., 1967).
44. Herglotz, H., Einflusse der Bindung auf das Rontgenspektrum, *Trans. Doc. Center Techn. Economy (Vienna)* **13**, (1955).
45. Holliday, J. E., *J. Appl. Phys.* **33**, 3259 (1962).
46. Holliday, J. E., *Advan. X-Ray Anal.* **9**, 365 (1966).
47. Holliday, J. E., *Develop. Appl. Spectr.* **5**, 77 (1966).
48. Holliday, J. E., *in* "Handbook of X-Rays," Chapter 38. McGraw-Hill, New York, 1967.
49. Holliday, J. E., *J. Appl. Phys.* **38**, 4720 (1967).
50. Keil, K., *Fortchr. Mineral.* **44**, 4 (1967).
51. Kirkendall, T. D., and Varadi, P. F., *Anal. Chem.* **39**, 1342 (1967).
52. Kirkendall, T. D., Varadi, P. F., and Naill, R. F., Pittsburgh Conf. on Anal. Chem. and Appl. Spectry., paper no. 5 (March 2, 1969). Cleveland, Ohio (abstracts of papers).
53. Liebhafsky, H. A., Wilkins, D. H., Bernstein, F., *Proc. Colloq. Spectry. Intern., 13th,* p. 61. Adam Hilger, Ltd., London (1968).
54. Liefeld, R. J., *Bull. Am. Phys. Soc.* **10**, 549 (1965).

55. Liefeld, R. J., *in* "Soft X-Ray Band Spectra," pp. 133–151, Academic Press, London 1968.
56. Lindberg, E., *Nova Acta Reg. Soc. Sci. Upsaliensis* **7**, 7 (1931).
57. Mattson, R. A., and Ehlert, R. C., *Advan. X-Ray Anal.* **9**, 471 (1966).
58. Mattson, R. A., and Elhert, R. C., *J. Chem. Phys.* **48**, 5465 (1968).
59. Mendel, H., *Koninkl. Ned. Akad. Wetenschap. (Amsterdam)* **70B**, 276 (1967).
60. Merritt, J., and Agazzi, E. J., *Anal. Chem.* **38**, 1954 (1966).
61. Morin, F. J., *J. Appl. Phys.* **32**, 2195 (1961).
62. Nagakura, I., *Sci. Rept. Tohoku Univ.* **48**, 166 (1965).
63. Nordfors, B., *Arkiv Fysik* **10**, 279 (1956).
64. Ruderman, I. W., Ness, K. J., and Lindsay, J. C., *Appl. Phys. Letters* **7**, 17 (1965).
65. Sandstrom, A., Experimental methods of X-ray spectroscopy: ordinary wavelengths, "Encyclopedia of X-Rays," Vol. XXX, pp. 78–245. Springer-Verlag, Berlin, 1957.
66. Schnell, E., *Monatsh. Chem.* **93**, 1383 (1962).
67. Schnell, E., *Monatsh. Chem.* **94**, 703 (1963).
68. Shaw, C. H., "The X-Ray Spectroscopy of Solids," pp. 13–62, Theory of Alloy Phases. Am. Soc. for Metals, Cleveland, 1956.
69. Siegbahn, M., "Spektroskopie der Röntgenstrahlen." Springer-Verlag, Berlin, 1931.
70. Siegbahn, K., Nordling, C., Fahlman, A., Nordberg, R., Hamrin, K., Hedman, J., Johansson, G., Bergmark, T., Karlsson, S. E., Lindgren, I., and Lindberg, B., *Nova Acta Reg. Soc. Sci. Upsaliensis, Ser. IV*, **20** (1967).
71. Solomon J. S. and Baun, W. L., *Pittsburgh Conf. Anal. Chem. Appl. Spectry.*, paper no. 7 (March 2, 1969). Cleveland, Ohio (abstracts of papers)
72. Somer, T. A., and Graves, P. W., *IEEE Nucl. Sci. Proc.* (Feb. 1969) (paper presented at 15th Nuclear Science Symposium).
73. Thompson, B. J., and Kellen, P. F., *Develop. Appl. Spectry.* **4**, (1965).
74. Tomboulian, D. H., The experimental methods of soft X-ray spectroscopy and the valance band spectra of the light elements,' "Encyclopedia of X-Rays," Vol. XXX, pp. 246–304. Springer-Verlag, Berlin, 1957.
75. Utriainen, J., Linkoaho, M., Rantavuori, E., Aberg, T., and Graeffe, G., *Z. Naturforsch.* **23**, 1178 (1968).
76. Whatley, T. A., *Pittsburgh Conf. Anal. Chem. Appl. Spectry.*, paper no. 6 (March 2, 1969). Cleveland, Ohio (abstracts of papers)
77. White, E., McKinstry, H., and Bates, T., *Advan. X-Ray Anal.* **2**, (1959).
78. White, E. W., and Roy, R., *Solid State Commun.* **2**, 151 (1964).
79. White, E. W., and Gibbs, G. V., *Am. Mineralogist* **52**, 958 (1967).
80. Wilbur, D. W., Relationship Between Chemical Bonding and the X-Ray Spectrum: Studies with the Sulfur Atom, UCRL-14379, T1D-4500, AEC Contract No. W-7405-eng.-48.
81. Zingaro, P. W., and Croke, J. F., paper presented at Mid-America Symp. Spectry., Chicago (June, 1967).

HIGH-RESOLUTION NUCLEAR MAGNETIC RESONANCE

A. Chakravorty

DEPARTMENT OF CHEMISTRY
INDIAN INSTITUTE OF TECHNOLOGY
KANPUR, INDIA

I. Introduction

It has been known for more than forty years that certain nuclei possess spin angular momenta and magnetic moments in their ground states (Table I). In the presence of an external magnetic field the angular-momentum vector can take several discrete orientations differing in energies. In a nuclear magnetic resonance (NMR) experiment, one studies transitions among these energy levels. The first NMR signal in matter of normal density was observed in late 1945. In the two decades since then this branch of spectroscopy has grown explosively. In line with this a number of books have appeared on the subject (*1, 34, 45, 63, 115, 118, 126*). One of these (*45*) deals specifically with applications to inorganic chemistry, as does the review article of Muetterties and Phillips (*102*).

An astonishingly wide variety of problems can be solved by using NMR techniques. It is not feasible to illustrate all types of applications in a chapter of this size. Instead, we shall choose a few topics and examine them in some detail. Today the majority of NMR experiments are carried out with commercial instruments using straightforward techniques. These techniques will not be described.

II. Basic Concepts and Parameters

A. The NMR Phenomenon

The angular momentum p and magnetic moment μ of a nucleus can be described by the equations

$$p = [I(I + 1)]^{1/2}\hbar \tag{1}$$

$$\mu = \gamma p \tag{2}$$

In Eq. (1) I is the nuclear spin quantum number, which can be zero, a half-integer, or an integer. In Eq. (2) γ is the characteristic magnetogyric ratio of the nucleus. γ is usually expressed in radians per gauss per second. Nuclei with $I = 0$ do not give rise to NMR signals.

In the presence of an external magnetic field H_0 (along z direction),

$$\mu_z = \gamma p_z = \gamma M \hbar \tag{3}$$

and

$$E = -\mu_z H_0 \tag{4}$$

or

$$E = -\gamma M \hbar H_0 \tag{5}$$

M, the magnetic quantum number, can have the values $I, I-1, I-2, \ldots,$ $-I+2, -I+1, -I$; E is the interaction energy between μ and H_0. μ_z and p_z are the z components of μ and p respectively. Equation (5) shows that each value of M is characterized by a definite energy E, and there are in effect $2I+1$ energy levels. In the absence of H_0 these levels are degenerate.

In order to induce magnetic-dipole transitions among these levels, it is necessary to supply energy in the form of an oscillating magnetic field polarized at right angles to H_0. In practice this can be achieved by passing an alternating current through a coil mounted perpendicular to H_0. The selection rule $\Delta M = \pm 1$ leads to the transition energy

$$\Delta E = h\nu = \gamma\hbar H_0 \tag{6}$$

or

$$\nu = \gamma H_0/2\pi \tag{7}$$

where ν is the NMR frequency, usually expressed in Hz (1 Hz = 1 cycle per second). In the classical model ν corresponds to the frequency of Larmor precession of μ about H_0. It is not surprising that the NMR phenomenon can be described classically since Planck's constant h does not occur in Eq. (7).

B. Relaxation Times and Linewidths

In magnetic fields of the order of 10^4 G that are normally used in NMR experiments the resonance frequency ν falls in the radio-frequency (rf) region (Table I). During NMR absorption, nuclei are raised from a lower to an upper spin state. As a result the very small (since $\Delta E \ll kT$) Boltzmann population difference between the states tends to decrease further. The intensity of NMR absorption depends directly on this population difference. Unless the upper state is continuously vacated by some mechanism, the NMR signal will progressively decrease in intensity and finally disappear ("saturation").

In most cases, however, this problem does not arise unless high rf power is used. An important reason, particularly in fluid media, is the thermal motion of molecules containing the magnetic nuclei. Such motions create a random magnetic field whose fluctuations will normally include the Larmor frequency. This stimulates emission from the excited spin states. Energy flows from the spin system into the molecular degrees of freedom. The process is exponential in time, and the corresponding time constant is called the spin–lattice or longitudinal relaxation time T_1.

The second relaxation time is the spin–spin or transverse relaxation time T_2. Its origin may be traced to the direct dipole–dipole interaction among the nuclear spins. This interaction leads to flip-flop nuclear transitions and

TABLE I

SOME PROPERTIES OF A FEW SELECTED^a NUCLEI

Isotope	NMR frequency (MHz) for a 10-kG field	Natural abundance (%)	Relative sensitivity for equal number of nuclei		Observable magnetic moment^c (nuclear magnetons)^d	Spin^e I (h/2π)	Nuclear quadrupole moment ($e \times 10^{-24}$ cm^2)
			At constant field	At constant frequency^b			
^{1}H	42.577	99.9844	1.000	1.000	2.79270	$\frac{1}{2}$	—
^{2}H	6.536	1.56×10^{-2}	9.64×10^{-3}	0.409	0.85738	1	2.77×10^{-3}
^{9}Be	5.983	100	1.39×10^{-2}	0.703	−1.1774	$\frac{3}{2}$	2×10^{-2}
^{10}B	4.575	18.83	1.99×10^{-2}	1.72	1.8006	3	0.111
^{11}B	13.660	81.17	0.165	1.60	2.6880	$\frac{3}{2}$	3.55×10^{-2}
^{13}C	10.705	1.108	1.59×10^{-2}	0.251	0.70216	$\frac{1}{2}$	—
^{14}N	3.076	99.635	1.01×10^{-3}	0.193	0.40357	1	2×10^{-2}
^{15}N	4.315	0.365	1.04×10^{-3}	0.101	−0.28304	$\frac{1}{2}$	—
^{17}O	5.772	3.7×10^{-2}	2.91×10^{-2}	1.58	−1.8930	$\frac{5}{2}$	-4×10^{-3}
^{19}F	40.055	100	0.834	0.941	2.6273	$\frac{1}{2}$	—
^{27}Al	11.094	100	0.207	3.04	3.6385	$\frac{5}{2}$	0.149
^{29}Si	8.460	4.70	7.85×10^{-2}	0.199	−0.55477	$\frac{1}{2}$	—
^{31}P	17.235	100	6.64×10^{-2}	0.405	1.1305	$\frac{1}{2}$	—
^{33}S	3.266	0.74	2.26×10^{-3}	0.384	0.64274	$\frac{3}{2}$	-6.4×10^{-2}

TABLE I (continued)

^{55}Mn	10.553	100	0.178	2.89	3.4610	$\frac{5}{2}$	0.5
^{59}Co	10.103	100	0.281	4.83	4.6388	$\frac{7}{2}$	0.5
^{77}Se	8.131	7.50	6.97×10^{-3}	0.191	0.5333	$\frac{1}{2}$	—
^{117}Sn	15.77	7.67	4.53×10^{-2}	0.356	-0.9949	$\frac{1}{2}$	—
^{119}Sn	15.87	8.68	5.18×10^{-2}	0.373	-1.0409	$\frac{1}{2}$	—
^{125}Te	13.45	7.03	3.16×10^{-2}	0.316	-0.8824	$\frac{1}{2}$	—
^{127}I	8.519	100	9.35×10^{-2}	2.33	2.7939	$\frac{5}{2}$	-0.75
^{129}Xe	11.78	26.24	2.12×10^{-2}	0.277	-0.7726	$\frac{1}{2}$	—
^{131}Xe	3.490	21.24	2.77×10^{-3}	0.410	0.6868	$\frac{3}{2}$	-0.12
^{133}Cs	5.585	100	4.74×10^{-2}	2.75	2.5642	$\frac{7}{2}$	≤ 0.3
^{195}Pt	9.153	33.7	9.94×10^{-3}	0.215	0.6004	$\frac{1}{2}$	—
Free neutron	29.165	—	0.322	0.685	-1.9130	$\frac{1}{2}$	—
Free electron	27.994	—	2.85×10^{8}	658	-1836	$\frac{1}{2}$	—

ᵃ Only those nuclei that appear in this chapter are included. A more complete list may be found elsewhere (115).

ᵇ Nuclei with high values of magnetic moments and spins are more sensitive.

ᶜ This quantity is equal to $\gamma \hbar I$.

ᵈ The nuclear magneton is equal to $eh/4\pi M_p c$, where e and M_p are proton charge and mass respectively and c is the velocity of light. The numerical value of the nuclear magneton is 5.049×10^{-24} erg/G.

ᵉ If the nuclear mass number is odd, I is a half-integer; if the mass number is even but the atomic number is odd, the spin is integral. (If the nuclear mass number and atomic number are both even, I is zero).

also creates a spread of precessional frequencies. As a result, if all nuclei precess initially in phase, they will gradually lose phase. T_2 is a measure of the time taken for this to occur. Dipolar interactions are particularly important in cases where the relative orientation of molecules changes only slowly with time. Examples are solids (117), highly viscous liquids, and the nematic phase of liquid crystals (13). In gases and in mobile liquids dipolar interactions average more or less to zero. In fact when molecular tumbling is so rapid that

$$2\pi v\tau \ll 1$$

it is no longer necessary to consider T_2 explicitly; T_1 and T_2 become equal in this limit. τ, the correlation time, is the time taken for the molecule to re-orient itself through an angle of one radian. The above limit is applicable to a majority of solution NMR work.

The width of an NMR line is intimately related (by the uncertainty principle) with relaxation times. In general a single well-resolved NMR line is Lorentzian in shape and the full width at half-height is equal to $(\pi T_2)^{-1}$. When $T_1 = T_2$, the width can be written as $(\pi T_1)^{-1}$. Additional broadening may result from instrumental reasons—e.g., field inhomogeneity.

There are two important factors that can decrease T_1 considerably with a consequent increase in linewidth. First, the presence of paramagnetic species in the system may be a factor. For example, T_1 of water protons is a few seconds in pure water, whereas in a molar solution of CrF_6^{3-} it is $\sim 10^{-3}$ sec. This effect can be traced to the magnetic moment of the unpaired electron, which is very much larger than nuclear moments (Table I). The intensity of the random magnetic field produced by molecular tumbling is consequently much higher. When $2\pi v\tau \ll 1$, T_1 is inversely proportional to the mean-square value of this intensity. A second factor that decreases T_1 is the quadrupole moment possessed by all nuclei having $I \geqslant 1$. In such nuclei the charge dis-tribution about the spin axis is spheroidal rather than spherical ($I = \frac{1}{2}$). Quadrupole moments interact electrically with the electric-field gradient of an insufficiently symmetrical electronic environment. Fluctuations (e.g., due to vibration) of the field gradient provide a dominant path for nuclear relaxation of the quadrupolar nucleus itself and also for other nuclei coupled to it. T_1 of heavy water is much shorter than T_1 of ordinary water. This is primarily due to the quadrupole moment of ^2H. The ^{14}N resonance signal of NH_3 is much broader than that of NH_4^+. The field gradient in the latter species alone is vanishingly small.

C. The Chemical Shift

In atoms and molecules the electrons shield the nuclei from the external magnetic field. The magnitude of this shielding effect is sensitive to the nature of the electronic environment of the nucleus in question. Each chemically

distinct environment for the nucleus is therefore associated with a characteristic resonance frequency. In fluid media linewidths are generally quite small,
and these frequencies can be picked up separately using magnets of good field
stability. In essence this is high-resolution NMR. In solids (*117*) resonances
are generally too broad (because of the small value of T_2) for observation
of such subtle effects. The high-resolution NMR spectrum of methyl azide
(*134*) is shown in Fig. 1. Three distinct ^{14}N signals can be clearly seen. The
signals are of unequal widths, but the area under each signal is the same.

For a shielded nucleus the resonance Eq. (7) is modified to

$$v = \frac{\gamma H_0(1 - \sigma)}{2\pi} \tag{8}$$

where σ is the shielding constant. $H_0(1 - \sigma)$ is the effective field at the nucleus.
In a fixed-frequency instrument, v is kept constant and H_0 is varied to obtain
resonance. The chemical-shift parameter δ_i for the ith environment is defined
as

$$\delta_i = \frac{H_i - H_r}{H_r} \times 10^6 \tag{9}$$

here H_i and H_r are the characteristic resonance fields for the ith and a
reference environment respectively. Experimental determination of chemical
shifts, e.g., by the sideband technique (*115*), often involves measurement of
the frequency equivalent of $H_i - H_r$. The chemical shift then becomes

$$\delta_i = \frac{v_i - v_r}{v_0} \times 10^6 \tag{10}$$

where v_0 is the fixed instrument frequency. δ_i is a dimensionless quantity and
is expressed in parts per million. δ_i is independent of the instrument frequency,
since any change in v_0 results in a corresponding change in $v_i - v_r$. In spite
of this very desirable property of δ_i, many workers report chemical shifts
simply as $v_i - v_r$ expressed in units of Hz. Referring to Fig. 1, the ^{14}N
chemical shifts are referred to CH_3NO_2, arbitrarily taken as the zero of the
ppm scale.

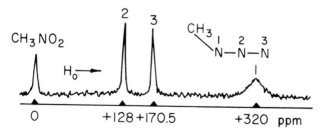

FIG. 1. ^{14}N NMR spectrum of methylazide, referred to nitromethane (*134*).

The frequency shift $v_i - v_r$ increases with increasing v_0. At higher v_0, NMR signals are better separated, the spin–spin fine structure tends toward simple first-order behavior (Section II.D), and the sensitivity of the NMR experiment becomes higher. Fixed-frequency spectrometers with v_0 equal to 60 or 100 MHz are now routinely used for proton-resonance measurements. Recently the 220-MHz spectrometer has been introduced. It uses a super-conducting solenoid (at liquid-helium temperature) as the magnet. A brief description of this system is available (8).

Sometimes the determination of chemical shifts may become difficult be-cause the signals are too weak to be detected. The substance under examina-tion may not be available in large amounts or its solubility may be low or the inherent sensitivity (Table I) of the NMR experiment for the nucleus in question may be low. Under such circumstances an invaluable tool (commer-cially available) is the CAT (computer of average transients), which stores spectral intensity data for repeated scans. When the average spectrum is displayed, only the random noise cancels out in part. The signal-to-noise ratio is thus enhanced.

D. Spin–Spin Splitting: An Example

The electronic environment in molecules does one more subtle thing. It couples magnetic nuclei present in the same molecule. Such a coupling is experimentally observed as a splitting of NMR lines. A typical example (74) is the BeF_4^{2-} ion. The ^9Be and ^{19}F resonances of commercial ammonium tetrafluoroberyllate are shown in Fig. 2. The ^9Be spectrum consists of five equally spaced lines. This is due to the coupling of ^9Be with the ^{19}F nuclei. The net spin $(\sum M_F)$ of the four ^{19}F nuclei can be 2, 1, 0, -1, -2. Each of these couples with the spin levels of ^9Be in a characteristic way. Each level of ^9Be is thus split into a quintuplet. NMR transitions can occur from the compo-nents of one quintuplet to those of another ($\Delta M_{Be} = \pm 1$) keeping the net ^{19}F spin unaffected. It can be easily seen that the observed spectrum should contain five lines, as is actually observed. The ion BeF_4^{2-} belongs to the general spin system AX_n in which the A resonance consists of $2nI_X + 1$ equally spaced lines of relative intensities given by the nth binomial coefficients. The BeF_4^{2-} lines should have the intensities 1 : 4 : 6 : 4 : 1. Although intensities were not measured (74) quantitatively, it can be seen from Fig. 2a that the observed order is qualitatively correct. The spacing between any two adjacent lines in such a multiplet is called the spin–spin splitting constant J_{AX} and is expressed in Hz. J_{AX} is a measure of the interaction of the nuclear moments of A and X through the intervening electron clouds. Its magnitude depends on the nature of the connecting bond(s) and not on the strength of H_0. From the ^9Be spectrum J_{BeF} is found to be 33 ± 2 Hz.

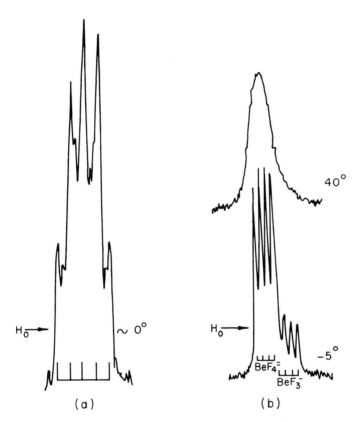

FIG. 2. (a) ^{9}Be NMR spectrum of commercial ammonium tetrafluoroberyllate; (b) ^{19}F spectrum of the same sample. (74).

The ^{19}F spectrum of BeF_4^{2-} consists of four lines (Fig. 2b). This arises from equal coupling of each of the four symmetrically equivalent ^{19}F nuclei with ^{9}Be of the tetrahedral anion. The predicted number of lines is $2I_{Be} + 1$ —i.e., four, as observed. J_{BeF} is found to be 33.2 ± 1.2 Hz. The observed spectrum carries no information about the coupling of the ^{19}F nuclei among themselves. Theory also requires that coupling of equivalent nuclei should have no observable effect on NMR spectra.

Commercial ammonium tetrafluoroberyllate contains a small amount of the $BeF_3 \cdot H_2O^-$ ion, as seen in the ^{19}F spectrum (Fig. 2b). However, no evidence of $BeF_3 \cdot H_2O^-$ is found in the ^{9}Be spectrum. This is probably due to the virtual absence of chemical shifts for beryllium-containing compounds (74). On warming the solution the spin–spin splitting disappears (Fig. 2b). This is indicative of a rapid fluorine exchange, probably between BeF_4^{2-} and

$BeF_3 \cdot H_2O^-$. If the Be—F bond is broken and formed very frequently, each fluorine will be exposed randomly and rapidly to the various spin states of ^9Be. Because of this averaging effect the spin–spin structure collapses. In general, collapse will occur when the frequency of exchange exceeds the coupling constant.

The simple rules relating to intensities and multiplicities of the AX_n spectra apply only when $|v_A - v_X| \gg |J_{AX}|$. Such spectra are called first-order spectra. When A and X are nuclei of different species (as in BeF_4^{2-}), the spectra are generally of the first-order type. When A and X are nuclei of the same kind, differing only in chemical shifts, the spectra may or may not be first-order in character. In the latter case the nuclei are designated as A and B rather than A and X. Methods are available (18, 115) for analysis of spectra that are not first-order.

III. The Nature of Chemical Shifts and Spin–Spin Splittings

A. Chemical Shifts

Chemical shifts arise from induced circulation of electrons under the influence of the external magnetic field and the field of the nucleus. In the case of atoms or ions having noble-gas configuration (^1S), this circulation can be simply described as a rotation of the spherical charge cloud ("free diamagnetic circulation") about the direction of the external field. The corresponding induced magnetic field opposes the external field at the nucleus. This shielding is called diamagnetic shielding (σ_d). Theoretical calculation of σ_d of atoms is a relatively simple matter (122).

Molecules do not have spherical symmetry, and the concept of free diamagnetic circulation is no longer appropriate. The shielding (σ) of a nucleus in a molecule is usually written in the form

$$\sigma = \sigma_d + \sigma_p \tag{11}$$

where σ_p —the paramagnetic deshielding term—essentially corrects for the excess circulation implicit in σ_d. The word *paramagnetic* is used in the sense that the circulations corresponding to σ_d and σ_p are in opposite directions. The presence of unpaired electrons is not implied. For the purpose of theoretical calculation of σ it is certainly desirable to consider the molecule as a whole. However, except in the simplest molecules (79), this cannot be achieved in practice because of lack of our knowledge of molecular eigenfunctions and eigenvalues. The next best thing to do is to investigate the effects of various electrons separately (120). In many cases it may suffice to consider electrons "local" to the nucleus of interest. Elsewhere the electrons of neighboring atoms or bonds and delocalized electrons may make significant contributions to σ. With this semiempirical approach one can hope to explain only trends rather than absolute magnitudes of chemical shifts.

In the large majority of cases variation of σ_p rather than σ_d from system to system is at the root of observed trends of chemical shifts. The filled inner electronic shells around a nucleus are more or less spherical, and accordingly their contribution to σ_p is relatively small. In the special case of hydrogen the same applies to the nearly spherical valence shell ($1s$). For most other nuclei the valence shell (p, d, f) will normally be associated with considerable paramagnetic circulation.

Nonspherical charge clouds become distorted under the influence of the magnetic field. The new ground state can in general be constructed by mixing field-free ground and excited states. This mixing manifests itself in the form of the paramagnetic term σ_p. This interpretation of σ_p is readily amenable to semiquantitative development. A particularly simple case arises when the mixing of only one low-lying excited state with the ground state is important. In this case both σ_p and the extent of mixing are inversely proportional to the energy gap (ΔE) between the two states. In a series of octahedral cobalt(III) complexes the ^{59}Co chemical shifts are found to be (54) approximately linear with the inverse of the energy gap between the $^1A_{1g}$ and $^1T_{1g}$ states. These are indeed the states that are expected to be mixed by the magnetic field. This treatment was recently extended to include some organo–cobalt complexes (81). A second example (43) is set out in Table II. The ^{17}O chemical

TABLE II

^{17}O CHEMICAL SHIFTS IN SOME OXYANIONS AND OXIDES[a] (43)

Compound	δ (ppm)	ΔE (cm^{-1})
Na_3VO_4	-571 ± 4	36,900
Na_2CrO_4	-835 ± 5	26,800
K_2MoO_4	-540 ± 2	44,000
Na_2WO_4	-420 ± 2	50,300
$NaMnO_4$	-1219 ± 8	18,300
$NaTcO_4$	-749 ± 7	34,000
$NaReO_4$	-569 ± 4	43,500
RuO_4	-1119 ± 10	26,400
OsO_4	-796 ± 3	33,500
$Na_2Cr_2O_7$	-1125 ± 6	22,700
CrO_2Cl_2	-1460 ± 8	18,000

[a] H_2O is used as the standard.

shifts are found to be linear with the inverse of ΔE, the energy of the lowest-energy electronic transition.

With these examples featuring the importance of σ_p may be contrasted the ^{14}N chemical shifts (10) shown in Table III. In the complexes the resonance positions are more or less independent of the nature of the central metal atom. Variations of σ_p from system to system are therefore unimportant. A second point to note is that in the complexes the ^{14}N resonance positions are at considerably higher fields than those of the free ligands. The "lone pairs" in the free ligands are likely to undergo considerable paramagnetic circulations. On tight engagement of the pairs in the M—N bonds, these circulations apparently become very much less important.

We have briefly mentioned above the success achieved in correlation of chemical shifts with σ_p in the case of [59]Co. This has recently led to similar investigations on [195]Pt and [55]Mn chemical shifts. In planar Pt(II) complexes of idealized D_{4h} symmetry, the $^1A_{2g}$ and ^1E_g states mix with the ground state ($^1A_{1g}$) under the influence of the magnetic field. The paramagnetic contribution (using an ionic crystal field model) is of the form (109),

$$\sigma_p = -\frac{4e^2\hbar^2\langle r^{-3}\rangle}{3m^2c^2}[2\Delta E_1^{-1} + \Delta E_2^{-1}] \qquad (12)$$

where $\langle r^{-3}\rangle$ is the average value of the inverse cube radius of Pt(II) $5d$ orbitals; ΔE_1 and ΔE_2 are the energies of the $^1A_{1g}\to\,^1A_{2g}$ and $^1A_{1g}\to\,^1E_g$ transitions respectively. Some [195]Pt chemical shifts (25) of compounds of the type trans-PtHL(PEt$_3$)$_2$ are shown in Table IV. When L is varied, the shielding of the [195]Pt nucleus increases in the order $NO_3 < NO_2 < Cl < Br < CN < I$.

TABLE III

[14]N CHEMICAL SHIFTS OF SOME
COMPLEXES[a] (10)

Compound	δ (ppm)
NH$_3$	+159
Co(NH$_3$)$_6$Cl$_3$	+286 ± 6
Rh(NH$_3$)$_6$Cl$_3$	+295 ± 4
Ru(NH$_3$)$_6$Cl$_2$	+316 ± 4
NaNO$_2$	−385
Na$_2$Pd(NO$_2$)$_4$	−134 ± 4
Na$_2$Pt(NO$_2$)$_4$	−136 ± 4
Na$_3$Co(NO$_2$)$_6$	−135 ± 2

[a] Shifts are measured from acetonitrile.

TABLE IV

^{195}Pt (25) AND ^{1}H (HYDRIDIC) (4)
CHEMICAL SHIFTS^a
OF *Trans*-PtHL(PEt$_3$)$_2$

L	δ_{Pt} (ppm)	δ_H (ppm)
NO$_3$	00.0	+23.6
NO$_2$	+60.3	+19.4
Cl	+137.3	+16.8
Br	+249.3	+15.6
CN	+408.3	+17.6
I	+442.9	+12.7

^{a} The ^{1}H shifts are referred to tetramethylsilane.

This order differs grossly from the spectrochemical series. Clearly changes in transition energies ΔE_1 and ΔE_2 do not dictate the trend of chemical shifts in ^{195}Pt complexes. Interestingly, the observed order has a strong parallelism to the nephelauxetic series (69), except that cyanide is misplaced. Roughly the nephelauxetic series represents the order of increasing covalent character. *A posteriori* this has been taken (25) as an indication that covalency changes determine the trend of ^{195}Pt chemical shifts. Equation (12) can be modified (25) to include covalency. When this is done the predicted order of chemical shifts agrees reasonably well with the observed order.

The ^{55}Mn shifts (15) shown in Table V can be qualitatively interpreted along similar lines. It has been speculated (25) that the low oxidation states of platinum and manganese are much more polarizable than the trivalent state of cobalt. As a result covalency effects become very significant in the former cases alone. Hopefully, as more and more nuclei are investigated, a general pattern of the origin of chemical shifts of transition-metal ions in their compounds will emerge.

TABLE V

SOME ^{55}Mn CHEMICAL SHIFTS^a (15)

Compound	$\delta \pm 20$ (ppm)	Compound	$\delta \pm 20$ (ppm)
ClMn(CO)$_5$	+1005	CH$_3$Mn(CO)$_5$	+2265
BrMn(CO)$_5$	+1160	CF$_3$Mn(CO)$_5$	+1850
IMn(CO)$_5$	+1485	NaMn(CO)$_5$	+2780
HMn(CO)$_5$	+2630	Mn$_2$(CO)$_{10}$	+2325

^{a} The reference is aqueous KMnO$_4$.

An important side-effect of the large paramagnetic circulations in transition-metal ions is the characteristic high-field shift (52) of protons bound directly to the metal. Some results (4) are included in Table IV. The relevant theory is given by Buckingham and Stephens (14). Unlike the proton, the ^{19}F isotope, when attached to a transition metal, is chemically shifted very considerably either to lower or to higher fields than when it is attached to a nontransitional element. The data (24, 88a, 103) in Table VI highlight this point. Metal d–d transitions and/or mixing of fluorine p and metal d orbitals are probably important, but detailed theoretical treatment is lacking.

The special case of chemical shifts in systems containing unpaired electrons will be discussed in Section VII.

TABLE VI

^{19}F CHEMICAL SHIFTS OF SOME MF_6^{2-} AND MF_6
SPECIES^a (24, 88a, 103)

MF_6^{2-}	δ (ppm)	MF_6	δ (ppm)
SiF_6^{2-}	+51.0	SF_6	−127
GeF_6^{2-}	+46.4	SeF_6	−128
SnF_6^{2-}	+79.3	TeF_6	−20.6
TiF_6^{2-}	−152.2	MoF_6	−355
ZrF_6^{2-}	−73.9	WF_6	−242
HfF_6^{2-}	−33.0	—	—
NiF_6^{2-}	+243	—	—
PdF_6^{2-}	+274	—	—
PtF_6^{2-}	+287	—	—

^{a} The reference is trifluoroacetic acid.

B. Spin–Spin Splittings

There are several ways by which intramolecular spin–spin coupling among nuclei can arise. The most important of these is the Fermi contact interaction. An s electron has a finite probability density at the nucleus. As a result the spin of such an electron is polarized directly by "contact" with the nuclear spin. The polarization by one nucleus is felt by another nucleus when the valence s electrons are delocalized over both nuclei. In other words, the nuclei can become coupled via s electrons.

In the case of two general nuclei N and N′ connected by a covalent chemical bond, an approximate LCAO–MO approach leads to the following expression (114) for the coupling constant:

$$J_{NN'} = \tfrac{16}{9}h\beta^2\gamma_N\gamma_{N'}(\Delta E)^{-1}|s_N(0)|^2|s_{N'}(0)|^2C_N^2C_{N'}^2 \tag{13}$$

In Eq. (13) β is the Bohr magneton; ΔE is an average singlet–triplet excitation energy; $|s_N(0)|^2$ and $|s_{N'}(0)|^2$ are the valence s-electron densities at the nuclei N and N' respectively; C_N and $C_{N'}$ are respectively the coefficients of s-orbitals in the σ-bonding wave functions localized on atoms N and N'.

The coupling constant depends sensitively on the effective nuclear charge (Z_{eff}). This is because the magnitude of s density at the nucleus is proportional to Z_{eff}^3. In the similarly constituted molecules CH_4, SiH_4, and SnH_4 the coupling constant of the proton with the central atom increases in the order $CH_4 < SiH_4 < SnH_4$, paralleling the increase in Z_{eff} in the order C < Si < Sn. When comparing various coupling constants involving different nuclei, as in the above example, it is more appropriate to use the *reduced* coupling constant $K_{NN'}$ (expressed in units of cm^{-3}) defined by

$$K_{NN'} = \frac{2\pi}{\hbar\gamma_N\gamma_{N'}} J_{NN'} \tag{14}$$

$K_{NN'}$ is independent [compare Eqs. (13) and (14)] of the nuclear moments of N and N'. It therefore reflects the characteristics of the intervening charge cloud alone. In Table VII are collected $J_{NN'}$ and $K_{NN'}$ values for CH_4, SiH_4, and SnH_4 and for SF_6, SeF_6, and TeF_6 (*116*).

The second important factor on which the magnitudes of coupling constants depend is the s character of the bonding orbitals. This aspect has often been discussed in the last ten years. It can be seen from Eq. (13) that $J_{NN'}$ is proportional to $C_N^2 C_{N'}^2$ provided that the other quantities remain more or less constant. This holds remarkably well (*84*) for the $^{13}C-^1H$ coupling constants in C_2H_6 (carbon hybridization, sp^3, $J = 124.9$ Hz), C_2H_4 (sp^2, 156.4 Hz), and C_2H_2 (sp, 248.7 Hz). Another interesting series is provided by the $^{31}P-^1H$ coupling constants in PH_2^- (139 Hz), PH_3 (185 Hz), and PH_4^+ (548 Hz), where the phosphorus hybridizations are $\sim p^2$, $\sim p^3$, and sp^3 respectively (*90*). Unfortunately, in most cases changes in hybridization and Z_{eff} occur together, and the correlation of the variations of coupling constants

TABLE VII

COUPLING CONSTANTS IN SOME MOLECULES

Molecule	NN'	J (Hz)	$K \times 10^{-20}$ (cm^{-3})
CH_4	$^{13}C\,^1H$	125	41.4
SiH_4	$^{29}Si\,^1H$	202.5	84.9
SnH_4	$^{119}Sn\,^1H$	1931	431.5
SF_6	$^{33}S\,^{19}F$	254	146.7
SeF_6	$^{77}Se\,^{19}F$	1400	649.1
TeF_6	$^{125}Te\,^{19}F$	3688	1034.0

with one of these factors alone is not warranted. A case in hand is provided by the halomethanes. The interested reader is referred to the original literature (26, 51).

Every coupling constant has a characteristic sign. $J_{NN'}$ is defined to be positive when the antiparallel as opposed to the parallel arrangement of the coupling nuclear magnetic moments is energetically more favorable. If one of γ_N and $\gamma_{N'}$ is negative, $J_{NN'}$ and $K_{NN'}$ will have opposite signs. For example, the N–H coupling constants in $^{14}NH_4^+$ and $^{15}NH_4^+$ have opposite signs simply because the magnetic moments of ^{14}N and ^{15}N are of opposite signs. However, both species have identical reduced coupling constants.

In an important paper, Pople and Santry (114) made an LCAO–MO study of the sign of coupling constants for directly bonded light atoms (H, B, C, N, O, F). Their conclusion is that all the reduced coupling constants should be positive except when bonding to F and possibly O is involved. The relative signs of a large number of coupling constants are now known, and many more are being determined currently. Observed results are generally in agreement with the predictions of Pople and Santry. In the future it will certainly be possible to extend this approach to include heavier elements.

The magnitude of coupling constants generally diminishes as the number of bonds separating the two nuclei increases. Thus in the pyrophosphite ion, $HP(O_2)$—O—$(O_2)PH$, $^1J_{PH}$ and $^3J_{PH}$ are respectively 668 and 2 Hz. The superscript on left refers to the number of bonds separating the two coupling nuclei (^{31}P and ^1H). The magnitude of nonbonded coupling is also dependent on the s character and Z_{eff} of the atoms concerned. The various ^{205}Tl–^1H coupling constants in the series Ph_3Tl, Ph_2Tl^+ and $PhTl^{2+}$ correlate well (85) when changes in s character and Z_{eff} are both taken into account.

The reader is referred to the article of Barfield and Grant (7) for further discussion on spin–spin coupling.

IV. NMR and Chemical Equilibria

In later sections, systems involving chemical equilibria will be frequently encountered. NMR spectra of such systems often contain features that are characteristic of the particular equilibrium process(es) involved.

In the simplest case a given nuclear species may be considered to be exchanging positions between two distinct environments A and B of mean lifetimes τ_A and τ_B respectively. When the rate of exchange is neglibible (i.e., τ_A and $\tau_B \approx \infty$), two separate signals having frequency positions v_A^0 and v_B^0 will be observed. The width of these signals will be determined by the relaxation times (Section II.B) T_{2A} and T_{2B} respectively. When the exchange rate becomes significant, linewidths are no longer determined by T_{2A} and T_{2B} alone.

In the case of *slow exchange*—i.e., when

$$\tau_A \text{ (and } \tau_B) \gg \frac{1}{2\pi(v_A{}^0 - v_B{}^0)}$$

two broadened signals are observed in the vicinity of $v_A{}^0$ and $v_B{}^0$. The width of the A line can be described in terms of a new "relaxation time" T'_{2A}:

$$\frac{1}{T'_{2A}} = \frac{1}{T_{2A}} + \frac{1}{\tau_A} \tag{15}$$

A similar expression can be written for the B signal. As the rate of exchange increases, the signals broaden further and they begin to merge. At the completion of merger a single broad resonance is obtained. As the rate becomes still faster, the signal sharpens, and in the limit of extremely fast exchange the frequency and width of the signal become equal to quantities that are weighted averages of the two environments. The validity of Eq. (15) ceases when the linewidths become comparable to $(v_A{}^0 - v_B{}^0)$ in magnitude. Approximate equations can be written for various other limiting conditions (*115*).

The quantity that chemists are often interested in is τ_A (and τ_B). This can be obtained either by using approximate limiting equations such as Eq. (15) or by a line-shape analysis. The latter process is tedious and usually requires a computer. Nevertheless, it is of general applicability and is being increasingly used. The review article of Johnson is recommended for further details (*68*).

Once τ_A can be determined over a range of temperature and the kinetic order of the exchange process A → B is known, the activation enthalpy ($\Delta H^{\ddagger}$) and entropy ($\Delta S^{\ddagger}$) parameters can be obtained from appropriate plots. In the specific case of a first-order process the rate constant is equal to τ_A^{-1}. The Eyring equation can then be written as

$$\tau_A = \left(\frac{kT}{h}\right)^{-1} \exp\left[\frac{\Delta H^{\ddagger}}{RT} - \frac{\Delta S^{\ddagger}}{R}\right] \tag{16}$$

where k and R are respectively the Boltzmann and gas constants.

V. Solvation

The nature of solutions of salts in aqueous and nonaqueous solvents has long fascinated chemists (*58, 62*). In recent years the NMR technique has made many inroads into this area. In the solution phase, salts usually exist in the form of solvated ions. Solvation may extend to several shells, the primary shell being most tightly held to the ion. It would be normal to expect a difference in the chemical shift between solvent molecules that are free ("bulk" solvent) and molecules that are bound, particularly in the primary solvation shell ("bound" solvent). Two separate signals for the two types of solvent molecules may or may not be observed, depending upon the rate at which they undergo exchange with each other. Often this exchange process can be kinetically characterized, using variable temperature NMR data.

A. Aqueous Solutions

Historically, the first direct identification of cationic solvation shells was made on aqueous solutions using ^{17}O resonance of partially enriched water. In acidic aqueous solutions containing Be^{2+}, Al^{3+}, and Ga^{3+} ions the chemical shifts of the bound and bulk signals do not differ very significantly. However, addition of paramagnetic ions (Fe^{2+}, Co^{2+}, Dy^{3+}, etc.) to such solutions selectively shifts the bulk signal to lower fields (64). This arises as follows. The solvation shells of the paramagnetic ions in question are labile (lifetime, 10^{-6} sec); fast exchange with bulk water produces a relatively large (due to contact interaction, Section VII) average shift of the whole bulk signal; water molecules bound (lifetime, 10^{-4} sec) to the diamagnetic ions are left practically unshifted. Area measurements of bound and bulk signals using 11.5% enriched $H_2^{17}O$ samples have led to the identification (41) of the species $Be(H_2O)_4^{2+}$, $Al(H_2O)_6^{3+}$, and $Ga(H_2O)_6^{3+}$ in solution phases. Identical results are obtained by using the molal-shift method (3), which requires measurements of chemical shifts rather than signal areas, and accordingly a less enriched $H_2^{17}O$ sample suffices. The basis of this method is that the ^{17}O chemical shift of the *bulk* signal is proportional to the molar ratio of the added paramagnetic ions and the *exchangeable* water. Chemical shifts are measured in the absence and in the presence of a known amount of the diamagnetic ion (e.g., Be^{2+}) under investigation. The change in the shift gives a measure of the amount of water held tightly by the diamagnetic ion. The hydration number of the cation Me_3Pt^+ is found to be 2.9 ± 0.1 from ^{17}O molal-shift data (50).

Fiat and Connick (42) have recently studied the exchange of bound and bulk water in acidic solutions of Al^{3+} and Ga^{3+}. Widths and shapes of ^{17}O signals were analyzed as functions of temperature. τ_c, the lifetime of the ^{17}O nucleus of bound or coordinated water, was obtained from these data. A plot of τ_c versus the reciprocal of the absolute temperature yielded a straight line in each case. The $\Delta H^{\ddagger}$ and $\Delta S^{\ddagger}$ for the first-order reaction

$$M(H_2O)_6^{3+} + H_2O^* \rightarrow M(H_2O)_5(H_2O^*)^{3+} + H_2O$$

were then determined, using Eq. (16). Results are shown in Table VIII. The relatively high positive values of $\Delta H^{\ddagger}$ and $\Delta S^{\ddagger}$ in the case of $Al(H_2O)_6^{3+}$ imply

TABLE VIII

PARAMETERS FOR THE EXCHANGE OF BULK AND BOUND WATER
IN SOLUTIONS OF Al^{3+} AND Ga^{3+}

Ion	Radius (Å)	τ_c, 25°C (sec)	$\Delta H^{\ddagger}$ (kcal/mole)	$\Delta S^{\ddagger}$ (eu)
Al^{3+}	0.51	7.5	27	28
Ga^{3+}	0.62	5.5×10^{-4}	6.3	—22

an S_N1 mechanism in which the coordination number of the activated complex is less than six. On the other hand, the much lower enthalpy of activation for the $Ga(H_2O)_6^{3+}$ ion and the negative value of the entropy of activation may indicate that in the transition state the coordination number is higher than that in the octahedral ground state (S_N2 mechanism). The differences in the behavior of the two ions can be largely attributed (42) to size effects. The original paper (42) is recommended for further study.

Experimentally it is much simpler to study ^1H than ^{17}O NMR. Even then the latter technique was first applied because rapid proton exchange between bound and bulk water leads to a single ^1H signal at room temperature. Very recently it has been possible to resolve this difficulty by cooling solutions to sufficiently low temperatures (47a, 89). For example, a concentrated aqueous solution of $Mg(ClO_4)_2$ does show two broad but distinct ^1H signals at $-70°C$. The area of the signal at lower field corresponds to the composition $Mg(H_2O)_6^{2+}$. Above $-65°C$ only a single averaged resonance is observed, and below $-85°C$ the bulk and bound signals broaden and overlap considerably because of the high viscosity of the solutions.

Use of aqueous acetone instead of pure water as solvent is another important development in this area (47a, 89). Acetone breaks water structure by disrupting hydrogen bonds. The bulk water signals are consequently displaced to higher magnetic fields, and this magnifies the chemical shift between bound and bulk signals. In addition, lower viscosity and slower proton exchange in the mixed solvent lead to sharper resonance lines. Lastly, a lower temperature can be attained (due to lower freezing point) in aqueous acetone than in pure water. Acetone is a very poor donor, and normally only aquated species are present in aqueous acetone. Some representative ^1H spectra of a solution of $Mg(ClO_4)_2$ in aqueous acetone is shown in Fig. 3 (89). A thorough line-shape analysis (132) as a function of temperature yielded the following parameters for the proton-exchange process in a solution of $Mg(ClO_4)_2$ in 1:2 aqueous acetone: τ_c at 25°C, 2.6×10^{-7} sec; $\Delta H^{\ddagger}$, 14.5 kcal/mole; and ΔS, 19.8 eu. For $Mg(NO_3)_2$, the corresponding figures are 2×10^{-7}, 12.7, and 14.3. In these systems τ_c is found to be significantly

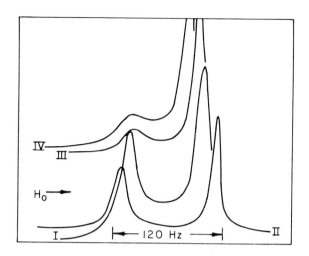

FIG. 3. ^{1}H NMR spectra of water protons in solutions of $Mg(ClO_4)_2$ in aqueous acetone at $-80°C$. Solution compositions in moles of $Mg^{2+} : H_2O :$ acetone are (I) 0.499 : 6.48 : 11.83; (II) 0.257 : 3.18 : 12.10; (III) 0.293 : 6.83 : 12.35; (IV) 0.452 : 8.31 : 11.28. The spectra were recorded at different spectrum amplitudes. The signal at lower field represents bound water in each case (*89*).

decreased by addition of acids. An acid-catalyzed proton transfer from the cationic primary hydration sphere to the bulk is indicated. Acids are known to play a similar role in the case of other ions (*132*). The coordination numbers of Al^{3+}, Ga^{3+}, and In^{3+} in cold aqueous or aqueous acetone solutions are found (^{1}H spectra) to be uniformly six (*47a*). Various other cations will undoubtedly be subjected to similar studies in the near future.

Until now we have not considered the fate of the anion in solutions. In no case has a separate signal been obtained from solvent molecules bound to an anion. Such binding is therefore very weak. From measurements of chemical shifts as functions of temperature and concentration it is feasible to determine (*86*) a "total effective hydration number" of a salt (cation plus anion). For $Al(NO_3)_3$ a value of 13.4 is obtained (*86*). However, unambiguous interpretation of such results remains to be accomplished. Future NMR studies on aquation will hopefully unfold the role of anions, hydration beyond the primary shell, and various mechanistic aspects of solvent exchange.

In common with several other ions, the solvation shells of alkali-metal and alkaline-earth-metal (except Mg^{2+}) ions are found to be too labile to be directly observable (^{17}O and ^{1}H NMR) even at very low temperatures. However, the cation (e.g., $^{133}Cs^+$ in CsCl) and anion (e.g., $^{127}I^-$ in NaI) NMR signals shift to higher fields in going from H_2O to D_2O solutions (*80*). The origin of this isotope shift may lie in the interaction of the ions with the hydration shells. This interaction (and hence the chemical shift of the ion) will alter by changes in molecular vibration amplitudes caused by isotopic substitution.

B. Nonaqueous Solutions

One of the solvents that has received considerable attention is methanol. In cold ($<0°C$) and feebly acidic solutions of $Mg(ClO_4)_2$ the bound OH signal (^1H) is a neat $1:3:3:1$ quartet (due to coupling with Me) appearing downfield from the bulk OH resonance (105). The intensity of the bound signal corresponds to $Mg(MeOH)_6^{2+}$. The bound Me signal can also be observed when the bulk signal is selectively broadened by addition of Cu^{2+}. On warming the solution bound OH and Me signals broaden parallelly. This means that the exchange process ($\Delta H^\ddagger$, 16.7 kcal/mole; $\Delta S^\ddagger$, 14 eu) involves the whole of the methanol molecule and not merely the OH proton. A similar situation obtains in methanolic solution of $Co(ClO_4)_2$ (82). Addition of water to such solutions produces species such as $Mg(MeOH)_5(H_2O)^{2+}$, $Co(MeOH)_5(H_2O)^{2+}$, and $Co(MeOH)_4(H_2O)_2^{2+}$ (two isomers). In these species, the rates of exchange of bound and bulk methanol are faster than those in the corresponding anhydrous solvates.

^1H resonance has established the existence of, for example, $Al(DMF)_6^{3+}$ and $Co(DMF)_6^{2+}$ in N,N-dimethylformamide (DMF) solution (88, 99). In general, DMF coordinates through its oxygen atom rather than through the nitrogen atom. The ion $Al(DMSO)_6^{3+}$ is known to exist in dimethylsulphoxide (DMSO) solution (131). Acetonitrile is an extremely weak donor. In a solution of $Al(ClO_4)_3$ in acetonitrile, the cation coordination shell contains (130) both MeCN and ClO_4^-. From a study of competitive solvations in various aqueous organic solvents the following order of decreasing solvating ability can be deduced (46):

$$DMSO > alcohols > amides > acetone > acetonitrile$$

In conclusion we shall briefly refer to solvent coordination to neutral molecules. For example, BX_3 (X = halogen) combines with a variety of Lewis bases to form $(base)\cdot BX_3$. In presence of excess base, separate bulk and bound signals can be distinguished (47). Solutions of $GaCl_3$ in diethyl ether or sulphide undoubtedly contain the adducts $Et_2O\cdot GaCl_3$ and $Et_2S\cdot GaCl_3$ respectively. However, because of rapid exchange, separate bulk and bound methyl or methylene signals (^1H) are not observable at room temperature. Instead the observed signal is shifted downfield compared to that of the pure solvent. When $GaCl_3$ is added to a mixture of the ether and the sulphide, the respective signals undergo shifts in proportion to the adduct-to-donor equilibrium molar ratio. By comparing these shifts with those of the separately studied $GaCl_3$–Et_2O and $GaCl_3$–Et_2S systems a straightforward determination of the equilibrium constant K for the reaction

$$Et_2O\cdot GaCl_3 + Et_2S \rightleftharpoons Et_2O + Et_2S\cdot GaCl_3$$

can be achieved (53). At 25°C K is 2.07. The sulphide is thus a better donor than the ether, a result in accord with thermochemical studies.

VI. Stereochemistry

Inorganic molecules present a wide spectrum of stereochemical patterns. Nuclear magnetic resonance is undoubtedly the most popular technique for exploration of stereochemical problems in the solution phase. Geometries of rigid molecules can often be identified from examination of characteristic resonance frequencies and/or spin–spin fine structure. However, straightforward conclusions about stereochemistry may not be warranted when the molecule concerned is nonrigid, undergoing some exchange process(es). In a case such as this, a variable-temperature experiment is imperative, and this will often diagnose the nature of the dynamic process(es).

A. *Rigid Structures*

By NMR criterion a rigid structure is one in which a given nuclear configuration is maintained for a time that is large compared to the NMR time scale under the conditions of measurement (Section IV). The molecules discussed below have rigid structures at room temperature.

The cyclotriborazane $(MeNHBH_2)_3$ can be obtained (*11*) in two isomeric forms by pyrolysis of the amine borane, $MeNH_2 \cdot BH_3$. The methyl proton resonance of one of the isomers consists of a single doublet, while that of the other isomer has two such doublets of relative intensity $1:2$. The doublet structures $(J \sim 5 \text{ Hz})$ arise from coupling with N—H protons. On the basis of analogy with the cyclohexanes, the most stable conformer may be expected to be a chair with all methyl groups in equatorial positions, while the next stable isomer should have two equatorial and one axial methyl groups. The NMR results suggest the skeleton **I** for the former isomer and skeleton **II**

(I)   (II)

for the latter isomer. The energy difference between **I** and **II** is probably small. However, the barrier to interconversion is quite large, and both the structures are rigid at room temperature, in contrast to the cyclohexanes. A similar Ga–N system is also known (*129*).

The magnitude of the spin–spin coupling constant and/or the pattern of spin–spin splitting are useful parameters for geometrical identification. The structures of the two products of the photochemical reaction

$$CFBr_3 + N_2F_4 \longrightarrow$$

$$J_{FF} , 17.6 \text{ Hz} \qquad J_{FF} , 218 \text{ Hz}$$

were established (27) simply from the magnitudes of J_{FF}. The geometrical assignment is based on the previous experience that carbon–carbon double-bonded structures with fluorine atoms in *trans* positions display coupling constants of consistently greater magnitudes than those of the corresponding *cis* case.

The molecular nitrogen complex $CoHN_2(PPh_3)_3$ is a model for nitrogen binding sites in nitrogenase enzymes. Its hydride resonance (δ, 19 ppm from tetramethylsilane; J_{PH}, 50 Hz) consists (96) of a $1 : 3 : 3 : 1$ quartet. This suggests coupling of ^1H with three equivalent ^{31}P nuclei. The skeleton structure **III** is in accord with this and with x-ray crystallographic data. Shaw and

(III)

co-workers have shown the utility of spin–spin patterns in geometrical identification of transition-metal complexes containing at least two phosphine ligands (65, 125). The methyl proton-resonance spectrum of a chloroform solution of the planar complex $PdCl_2(PPhMe_2)_2$ consists of a $1 : 1$ doublet and a $1 : 2 : 1$ triplet. These are assigned respectively to cis and trans isomers existing in equilibrium. The splittings are due to the coupling with ^{31}P nuclei. The nature of the observed spectrum is governed by the following relations:

$$(J_{PP})_{cis} \ll (J_{PP})_{trans} \gg (^2J_{PH} - {}^4J_{PH})$$

In the trans isomer the methyl resonance is a triplet because of the inequality shown on the right-hand side ("virtual coupling"). The observed doublet in the cis isomer results from the small value of $(J_{PP})_{cis}$. These two systems are actually "deceptively simple" limiting cases of the general spin system (56), $X_n AA'X_n'$ (X = H, A = P, n = 3).

A few examples from the chemistry of metal chelates will now be quoted. Tris chelates of the type $M(A—B)_3$ (where A—B is an unsymmetrical bidentate ligand) can exist, in principle, in cis (**IV**) and trans (**V**) forms. The cis

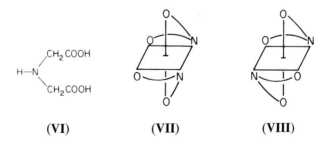

cis

(**IV**)

trans

(**V**)

isomer has a threefold axis of symmetry, the trans isomer has no symmetry at all. The three chelate rings are therefore magnetically equivalent only in the former isomer. A given magnetic nucleus (or a group of equivalent nuclei) should, in principle, have the same chemical shift for the three rings in the cis isomer but different chemical shifts in the trans isomer. Following the elegant work of Fay and Piper (40) this idea has been used in several cases (70) for stereochemical identifications. The case of an arylazooximato chelate (71) is shown in Fig. 4. Interestingly the corresponding cobalt(III) chelate exists exclusively in the trans form, probably because of steric factors (70).

A typical tridentate ligand is iminodiacetic acid, H_2IDA (**VI**). The anionic chelates $M(IDA)_2^-$ (M = Rh, Co) can be isolated in two isomeric forms, which can be assigned cis (**VII**) and trans (**VIII**) structures (17, 127). The plane of symmetry through N—M—N atoms places each acetate —CH_2— group in equivalent environment in **VIII**. In **VII** the —CH_2— groups are in two different environments. In both **VII** and **VIII** the two protons in a given —CH_2—

CH₂COOH / H—N \ CH₂COOH

(**VI**)

(**VII**)

(**VIII**)

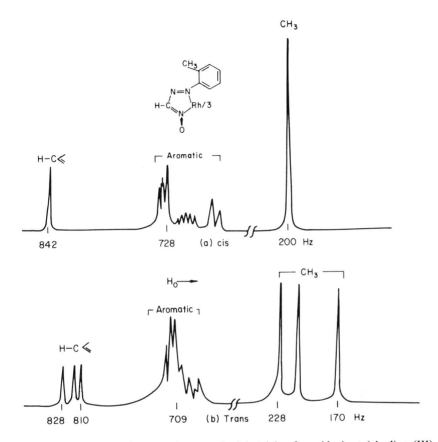

FIG. 4. ^{1}H spectra of the two isomers of tris(o-tolylazoformaldoximato)rhodium (III) in CDCl$_3$ at 100 MHz. Numerical figures refer to chemical shifts in Hz, downfield from tetramethylsilane (71).

group are nonequivalent and constitute an AB spin system. Thus in effect **VII** and **VIII** should produce two and one AB patterns respectively. This was observed to be the case for the two isomers of M(IDA)$_2$$^-$, when the additional complexity due to spin–spin coupling with the N—H proton was removed by deuteration. The bis chelates derived from N—Me analogue of **VI** exist only in the form **VIII**. Apparently steric repulsion of two adjacent methyl groups is prohibitively high in **VII**. NMR has been put to use in a variety of other studies of the chelates of various multidentate ligands, including ethylene-diaminetetraacetic acid (128). Further examples of the application of ^{1}H NMR to stereochemical problems of metal chelates may be found elsewhere (91, 121).

B. *Nonrigid Systems*

In 1965 Muetterties (*100*) called attention to the importance and possible widespread occurrence of nonrigid, highly dynamic inorganic molecules. Such molecules will often (but not always) have two or more nuclear configurations, each corresponding to a minimum on the molecular potential-energy surface. The minima are separated by barriers whose dimensions are permissive of rapid passage (tunneling, thermal excitation over barrier, or intermediate case) of the molecule from one potential well into another under easily accessible conditions. Familiar examples of this general intramolecular phenomenon are inversions of ammonia and cyclohexane. In the last few years there has been an upsurge in the NMR study of nonrigid inorganic and organometallic molecules.

The case of PF_5 will be considered first. Infrared and electron diffraction data show that the molecule has a trigonal bipyramidal geometry in the ground state. However, the ^{19}F NMR spectrum consists of a single doublet (J_{PF}, 916 Hz) down to $-160°C$. Apparently a fast intramolecular (note that J_{PF} is maintained) process that exchanges axial and equatorial fluorines is operative. What is fast by the NMR time scale can of course be slow by the time scales of infrared or electron-diffraction experiments. The last two experiments directly "see" the trigonal bipyramidal geometry; NMR does not.

A particularly attractive and more or less accepted mechanism for the exchange process is Berry pseudorotation. The transition state is probably square pyramidal (*61*). The rearrangement process leaves the molecule in a

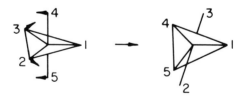

rotated, permuted form of its original state. The vibrational mode dominating the exchange process is primarily an axial bending mode, a relatively low-frequency vibration with considerable amplitude in the ground vibration state. The weakness of the axial bonds compared to equatorial bonds is an important contributing factor. The rate constant and activation energy of pseudorotation in PF_5 are estimated to be $1.7 \times 10^7 \ sec^{-1}$ and 7.9 kcal/mole respectively at $27°C$. Fast axial–equatorial exchange is also observed in AsF_5 and in derivatives of PF_5—e.g., $MePF_4$, PCl_2F_3, H_2PF_3, etc. In the last two molecules separate axial and equatorial ^{19}F resonances can be observed at low temperatures. The ion SiF_5^- (isoelectronic with PF_5) and several of its derivatives are similarly nonrigid, as indicated by ^{19}F NMR data (73). $Fe(CO)_5$ (single ^{13}C resonance down to $-63°C$) also appears to be nonrigid, although in this case the possibility of an insignificant difference in axial and equatorial bonding cannot be ignored (123). Nonrigidity is a fairly common feature of pentacoordination (104). Nonrigidity is also likely to be widespread in hepta- (e.g., IF_7), octa- (e.g., $Mo(CN)_8^{4-}$), and nona- (e.g., ReH_9^{2-}) coordination (100).

The octahedron is a sterically excellent coordination polyhedron. Even then nonrigidity is not uncommon in this geometry. $Ti(acac)_2X_2$ (X = F, Cl, Br; acac = anion of acetylacetone) is a typical example (38). At low temperatures (Fig. 5) two distinct methyl signals are observed, corresponding to a *cis* geometry (**IX**). As the temperature is raised sufficiently, fast *intramolecular* (38) rearrangement makes all methyl groups NMR-equivalent. The activation energy for this process is calculated to be around 11 kcal/mole. The Zr(IV) and Sn(IV) analogs behave in a qualitatively similar fashion. Tris(β-diketonato)M(III), (M = In, Ga, Al) are also known to be nonrigid (39). Normally twist (Bailer or Rây–Dutt type) and/or bond-rupture mechanisms are considered as alternatives for the rearrangement processes in octahedral complexes. However, it is rarely possible to arrive at a definite conclusion from NMR data alone. Several papers (39, 101, 124) that highlight this point are recommended for further study.

As an example of a nonrigid binuclear structure, the case of $Ta_2(OMe)_{10}$

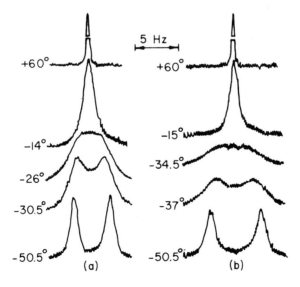

FIG. 5. Methyl region of ^{1}H NMR spectra of (a) $Ti(acac)_2Cl_2$ and (b) $Ti(acac)_2Br_2$ in dichloromethane solution (38).

(IX)   (X)

(**X**) may be presented (9). At 100°C (in octane) this molecule shows a single averaged methyl ^{1}H signal. Below 40°C two signals (area 1:4) are seen. Presumably only the α and β sites are being averaged at this stage. Finally, at $-58°C$ three distinct signals (area 1:2:2) can be recognized. When methanol is added to the system, a separate methyl resonance is observed. The exchange of α, β, and γ sites is thus an intramolecular process. The exchange of a γ site with an α or β site must of course involve the scission of a bridge. No other definitive statements can be made about exchange mechanisms at present. $Nb_2(OMe)_{10}$ behaves similarly.

C. *Fluxional Organometallic Molecules*

This name is given to a class of organometallic molecules that were recently reviewed by Cotton (19). Detailed references to most of the results described in this section may be found in Cotton's article.

The molecules $(\pi$-$C_5H_5)Fe(CO)_2C_5H_5$ (**XI**) $(C_2H_5)_3PCuC_5H_5$ (**XII**), and $(C_5H_5)_2Hg$ (**XIII**) were so formulated by Piper and Wilkinson in 1956 on the basis of infrared and chemical reactivity data. However, in each case the room-temperature ^{1}H spectra contained only one signal for the σ-bonded C_5H_5 moeity. The authors proposed that a rapid shifting of the metal–carbon σ bond was responsible for the observed behavior. This turned out to be correct when Cotton, Davison, and co-workers examined the variable-temperature ^{1}H spectrum of **XI** a decade later. Their results are shown in Fig. 6. The π-C_5H_5 protons produce the expected sharp signal throughout the

(**XI**)   (**XII**)   (**XIII**)

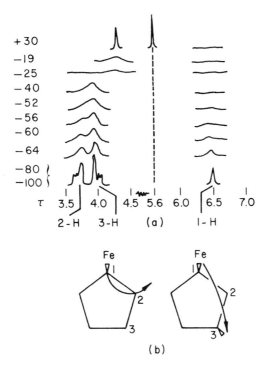

FIG. 6. (a) The ^{1}H NMR spectra of $(\pi\text{-}C_5H_5)Fe(Co)_2C_5H_5$ in CS_2 ; the position and width of the $(\pi\text{-}C_5H_5)$ resonance at $\tau \sim 5.6$ are essentially invariant. The τ scale is defined as $10 - \delta$, where δ is measured from tetramethylsilane taken as standard. (b) Schematic representation of 1,2 and 1,3 shifts (*19*).

temperature range. However, at low temperatures ($-80°$ and below) three types of protons for the σ-bonded C_5H_5 ring can be distinguished. In this limit, the bond-shifting process has stopped or is taking place at a slow rate. The weakest signal at the highest field is immediately assigned to the unique 1-proton. On the basis of certain features in the spin–spin structure, the signal at lowest field is assigned to 2-protons and the remaining signal to 3-protons.

As the temperature is raised, the signals broaden and finally collapse. The collapse of the 2- and 3-protons is unsymmetrical. This eliminates any mechanism ("rearrangement pathway") in which exchange occurs either randomly or via a configuration in which all protons become equivalent. Two reasonable possibilities that can explain the unsymmetrical collapse are a succession of 1,2 shifts or a succession of 1,3 shifts (Fig. 6b). In either case the interaction between the metal and the ring is continuously maintained.

It can be shown that a sequence of 1,2 shifts should cause a faster collapse of the 2- over the 3-protons, whereas the 1,3 sequence should do just the opposite. The experimental facts clearly support the 1,2 rearrangement pathway. A claim that low-temperature ($-70°$) ^{1}H NMR of **XIII** in liquid sulphur dioxide is also compatible with the 1,2-shift mechanism remains to be confirmed (*87*). (π-C_5H_5)$Cr(NO)_2C_5H_5$ is isoelectronic with **XI**, and it behaves similarly. A 1,3 pathway has been suggested in the case of **XII**, but this is open to question.*

Structures **XI**, **XII**, and **XIII** (and, e.g., PF_5) are examples of *fluxional* molecules. The necessary condition is that the several minima on the molecular potential-energy surface be equivalent. The stable structure (in its various equivalent nuclear permutations) is called the instantaneous structure.

Many other examples of fluxional organometallic molecules are now known. Many more are likely to be discovered. To mention a few more cyclopentadienyl complexes: $Me_nM(C_5H_5)_{4-n}$ and $Me_nM(C_5H_4Me)_{4-n}$(M = Si, Ge, Sn; $n = 0, 1, 2, 3$) are known to be fluxional (*23*). The important point here is that methyl substitution on the cyclopentadiene ring does not arrest nonrigidity. The molecule $C_{15}H_{15}MoNO$ contains three cyclopentadienyl rings in two different states of bonding (*15a*). It shows an intricate fluxional pattern (*20*).

Often a cyclopolyene will form a π complex with a metal, such that some but not all of the π electrons are involved in bonding. Such a molecule is likely to be fluxional. Thus $C_8H_8Fe(CO)_3$ shows a single-proton signal at room temperature, even though in the crystal lattice the structure is known to be of the (1,3-diene)–metal type (**XIV**). On cooling the solution to $-144°C$ several features become observable, but the rearrangement process is still too fast for unambiguous conclusions about the instantaneous structure to be drawn. The corresponding ruthenium complex $C_8H_8Ru(CO)_3$, however, becomes rigid at $-145°C$. Four equally intense signals are discernible (instantaneous structure **XIV**). The pattern of collapse at higher temperatures is in agreement only with a 1,2 pathway (as opposed to 1,3, 1,4, and random shifts). Low-temperature ($-40°C$) spectra of $C_8H_8M(CO)_3$ (M = Cr, Mo, W) are in accord with the instantaneous structure **XV**. However, because of excessive spin–spin structure, the spectra are quite complex and the rearrangement pathway cannot be traced. This difficulty is absent in $Me_4C_8H_4M(CO)_3$.

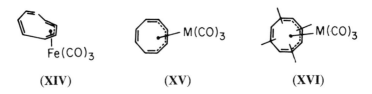

        (**XIV**)   (**XV**)   (**XVI**)

---

* The structure **XII** has recently been shown to be incorrect by F. A. Cotton and T. J. Marks, *J. Am. Chem. Soc.* **91**, 7281 (1969).

The limiting low-temperature spectrum consists of four sharp ring-proton resonances and four sharp methyl resonances (**XVI**). The spectrum collapses in such a way that there are clearly at least two successive stages of rearrangement. The most facile one is a 1,2 shift past an unsubstituted carbon atom. The inference is that in $C_8H_8M(CO)_3$ also the rearrangement pathway is most probably a succession of 1,2 shifts.

For a discussion of more complex fluxional molecules containing cyclopolyenes the review (*19*) of Cotton should be consulted.

D. Redistribution Reactions

In Section VI.B were given examples [PF_5, $Ta_2(OMe)_{10}$, etc.] in which substituents on a central moeity exchange places with each other intramolecularly. Intermolecular redistribution of substituents is more commonly encountered, and a few examples of this phenomenon will be presented in this section. The exhaustive review of Moedritzer (*97*) is recommended for further study.

In a mixture of $MeGeCl_3$ with $MeSiBr_3$ facile scrambling of chlorine and bromine atoms occurs. In the final mixture a maximum of eight species may be expected: $MeGeBr_3$, $MeGeBr_2Cl$, $MeGeBrCl_2$, $MeGeCl_3$, $MeSiCl_3$, $MeSiBr_2Cl$, $MeSiBrCl_2$, and $MeSiBr_3$. Methyl proton NMR spectra of the equilibriated mixtures indeed exhibit up to eight different resonances, depending upon the overall composition of the mixture (*97*).

The above system belongs to the class QZ_3 versus MT_3, where Q and M represent the trifunctional moeties MeSi and MeGe respectively; Z and T are halogen atoms. A minimum of two equilibrium constants is needed (*98*) to represent the scrambling of the substituents Z and T on each of the single trifunctional central moeities. In addition one other equilibrium constant is needed to define the overall mixture. This constant—the intersystem constant K_I—relates the molecules based on central moeity Q with those of central moeity M. A convenient format of K_I is

$$K_I = \frac{[QT_3][MZ_3]}{[QZ_3][MT_3]}$$

K_I can be determined from experimental ^1H spectra, usually with the help of computer programs (*97*). Results for a few pairs are as follows:

Z	T	K_I (at 120°C)
Cl	Br	$(9.7 \pm 4.0) \times 10^{-6}$
Br	I	$(4.2 \pm 0.9) \times 10^{-5}$
Cl	I	$(1.5 \pm 0.7) \times 10^{-12}$

For a random redistribution, K_1 would be equal to unity. The observed figures are much smaller, indicating that there is a pronounced preference of a given central moeity for one of the two substituents involved in the scrambling process. It can be seen that the methylsilicon moeity is preferentially associated with the heavier halogen, leaving the lighter halogen to the methylgermanium moeity.

A redistribution reaction occurs when Al_2I_6 and $(n\text{-}Pr)_4N^+I^-$ are dissolved in anhydrous dichloromethane, which functions both as a solvent and as a source of Cl^-. Such solutions contain all the five possible $AlCl_nI_{4-n}^-$ ($0 \leqslant n \leqslant 4$) ions. These species were identified (Fig. 7) using ^{27}Al NMR resonances (72).

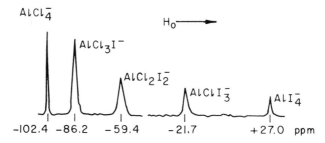

FIG. 7. The ^{27}Al NMR spectrum of various $AlCl_nI_{4-n}^-$ ions in dichloromethane solution (72).

^{27}Al ($I = \frac{5}{2}$) has a relatively high electric quadrupole moment. However, in the tetrahedral ions $AlCl_4^-$ and AlI_4^-, the electric-field gradient is more or less zero, and accordingly they give rise to the narrowest lines (Section II.2). The AlI_3Cl^- and $AlICl_3^-$ ions with C_{3v} symmetry have lines twice as wide, while the $AlI_2Cl_2^-$ ion of lowest symmetry (C_{2v}) gives the widest signal of the five. The differences in the linewidths of $AlCl_4^-$ and AlI_4^- and of $AlCl_3I^-$ and $AlClI_3^-$ can be understood on the basis of size differences (72).

In metal chelate chemistry mention may be made of the ^1H and ^{19}F NMR studies (2, 113) on the exchange of eight-coordinated metal acetylacetonates, $M(acac)_4$, and trifluoroacetylacetonates, $M(tfac)_4$, where $M = Zr$, Hf, Ce, and Th. All possible mixed complexes can be detected in solution.

VII. Contact Shifts in Paramagnetic Molecules

Most molecules discussed heretofore are diamagnetic. Presence of unpaired electrons in a compound can profoundly influence NMR parameters—e.g., linewidths and chemical shifts. It has already been noted that paramagnetic molecules are likely to give broad NMR lines due to fast nuclear relaxation (Section II.B). However, many systems are now known in which electronic relaxation is so fast that nuclei see a constant averaged electron moment rather than a fluctuating moment. NMR lines in such systems—and these concern us in this section—are reasonably sharp and show large frequency shifts.

These shifts can be of two types, corresponding to the two types of magnetic nucleus–electron interactions. First is the through-space dipolar interaction between nuclear spin and electronic spin plus orbit. This interaction has a nonzero average value only in cases where the electron magnetic moment is anisotropic. The resultant shift (in the solution phase) is called the isotropic pseudocontact shift. Second is the isotropic contact shift. This arises when the unpaired electron has a finite probability of being on the nucleus. Note that a basically similar (no unpaired electrons are involved, however) interaction is often the major contributor to nuclear spin–spin splitting (Section III.B). Contact-shift data contain the key to many subtle questions about electronic structures of paramagnetic species. For example, the cyanide ^{14}N resonance of the paramagnetic ion $Cr(CN)_5NO^{2-}$ is practically at the same frequency as that of the diamagnetic species $Fe(CN)_4^{4-}$ (57). The conclusion is that the unpaired electron is more or less localized on the NO group. On the other hand, both ^{13}C and ^{14}N resonances show (22, 57) large shifts in $Fe(CN)_6^{3-}$. The unpaired electron is therefore delocalized from the metal into the cyanide orbitals in the system. Similarly, in tetrakis[1-phenyl-1,3-hexanedionato] U(IV) isotropic shifts are observed for the phenyl and alkyl protons, indicating that the 5f shell is involved in covalent bonding (133a). An excellent review of this general area is available (31).

A. Some Basic Equations

The contact interaction will be considered first. This is of the form $\mathbf{I} \cdot \mathbf{S}$, where $\mathbf{I}$ and $\mathbf{S}$ refer to the nuclear and electronic spin vectors respectively. The corresponding coupling constant for the ith nucleus will be designated as a_i. This interaction produces the familiar hyperfine splitting in electron-spin resonance (ESR) spectra of paramagnetic species. However, if the electron-spin relaxation (T_{1e}) or exchange (T_{ex}) time is so short that T_{1e}^{-1} or $T_{ex}^{-1} \gg a$, the ESR signals will generally be quite broad and difficult to observe. On the other hand, sharp and easily observable NMR signals are obtained under these conditions. The contact shift Δf_i of the NMR signal is given by

$$\frac{\Delta f_i}{f} = -a_i \frac{\gamma_e}{\gamma_N} \frac{g\beta S(S+1)}{3kT} \tag{17}$$

In Eq. (17) f is the fixed instrument frequency; γ_e and γ_N are the magnetogyric ratios of the electron and nucleus respectively; g is the electronic g-factor characteristic of the system under study; β is the Bohr magneton; S is the net electron spin of the system; k is the Boltzmann constant; and T is the absolute temperature. Δf_i is generally measured with respect to a suitable diamagnetic material. It is counted as negative if the signal is at a lower field in the paramagnetic species. It can be seen from Eq.(17) that Δf_i is negative if a_i is positive and vice versa. Thus an extremely powerful technique for determination of the sign of a_i is at hand.

In the next section we shall consider several cases in which unpaired electron spin is located in the π orbital of an sp^2-hybridized carbon atom to which a proton (whose contact shift is being measured) is attached by a sigma bond. Spin finds its way into the hydrogen $1s$ orbital via indirect π–σ polarization (excited-state mixing). In such a situation the McConnell equation

$$a_H = Q_{C-H}\rho_C/2S \tag{18}$$

holds. ρ_C measures the density of the unpaired spin at the carbon p_π orbital. When ρ_C is positive (α spin—i.e., spin aligned with the external field at the carbon p_π orbital), the corresponding spin density at the hydrogen nucleus is negative (β spin). As a result, the constant Q_{C-H} of Eq. (18) is negative. Its value is usually taken as -22.5 G. A combination of Eqs. (17) and (18) readily converts experimental contact shift data into p_π spin densities. Once the experimental spin density is mapped, the next step is to compare it with theoretical maps obtained by calculations on the basis of suitable π-delocalization models. The McConnell–Dearman (92) valence bond approach and the McLachlan MO method (94) have often been employed with considerable success for paramagnetic metal chelates.

In molecules containing a group X attached to an sp^2 carbon it is convenient to write

$$a_X = Q_{C-X}\rho_C/2S \tag{19}$$

Q_{C-X} is no longer a simple quantity, and its value depends on the nature of the molecule concerned and on the position of substitution. When $X = Me$, Q_{C-Me} is known to vary from $+5$ to $+31$ G. The important point is that its sign is opposite to that of Q_{C-H}. This happens because the π-spin density can travel directly with unchanged sign into the methyl hydrogen $1s$ orbitals by, for example, a hyperconjugative mechanism. The experimental observation of opposite signs of a_H and a_{Me} in the unsubstituted and the methyl substituted compounds respectively is very often taken as a criterion for the existence of spin density in a π orbital. The question of spin density in σ orbitals will be discussed later.

The pseudocontact shift of the ith nucleus in rapidly tumbling (e.g., in solution) molecules with an axially symmetric g tensor is of the form

$$\frac{\Delta f_i}{f} = -\frac{(3\cos^2\theta_i - 1)}{r_i^3}F(g_\parallel - g_\perp)\frac{\beta^2 S(S+1)}{3kT} \tag{20}$$

In this equation r_i is the distance of the nucleus from the paramagnetic center; θ_i is the angle between r_i and the symmetry axis; $g_{\parallel}$ and $g_{\perp}$ are the electronic g tensors parallel and perpendicular to the axis; F is a linear function of $g_{\parallel}$ and $g_{\perp}$. For a molecule whose geometry plus $g_{\parallel}$ and $g_{\perp}$ are known it is a relatively simple matter to calculate the pseudocontact shift from Eq. (20). More generally, however, information on $g_{\parallel}$ and $g_{\perp}$ is not available. Even then Eq. (20) is useful because in a given molecule the relation

$$\Delta f_i : \Delta f_j : \cdots = \frac{3\cos^2\theta_i - 1}{r_i^3} : \frac{3\cos^2\theta_j - 1}{r_j^3} : \cdots \qquad (21)$$

holds. Thus if the geometric factors [right-hand side of Eq. (21)] are known or can be estimated, the *ratio* of the shifts of various nuclei can be computed. Conversely, from the observed ratio of shifts, the geometry of the molecule can be inferred.

B. Isotropic Contact Shifts

Widespread interest of inorganic chemists in contact-shift studies was aroused by the very elegant works of Phillips, Eaton, and co-workers (*33*) on nickel(II) aminotroponeimineates [Ni(ATI)$_2$] of structure **XVII**. This system

(XVII)   (XVIII)   (XIX)

has great synthetic flexibility allowing wide variations of the substituents R and X. In the solution phase, magnetic moments and electronic spectral data establish the existence of the equilibrium

planar $(S = 0) \rightleftharpoons$ tetrahedral $(S = 1)$.

^{1}H NMR spectra (down to $-100°$C), however, show an averaged signal for each type of proton. The interconversion is therefore fast by the NMR timescale. The observed signals are quite sharp and often show the expected first-order spin–spin splitting. This greatly facilitates the assignment of spectra.

The ^{1}H NMR signals of Ni(ATI)$_2$ are all contact-shifted [shifts are measured with respect to the free ligand or the diamagnetic tetrahedral complex, Zn(ATI)$_2$]. For example, the approximate ring-proton shifts of the complex with X $=$ H and R $=$ Et ($\sim 99\%$ paramagnetic) in CDCl$_3$ at $23°$C (40 MHz) are α–H, $+4200$; β–H, -2000; γ–H, $+5600$ Hz. To relate these shifts to the corresponding coupling constants a_i it is necessary to modify Eq. (17) to

$$\frac{\Delta f_i}{f} = -a_i \frac{\gamma_e}{\gamma_N} \frac{g\beta S(S+1)}{3kT[\exp(\Delta F/RT)+1]} \tag{22}$$

where ΔF is the free-energy difference between the two conformations. The constants a_i determined from Eq. (22) are the true coupling constants of the paramagnetic tetrahedral configuration. From a study of Δf_i as a function of temperature together with a single bulk susceptibility measurement it is possible to determine the temperature dependence of the free-energy change ΔF characterizing the process planar $\rightarrow$ tetrahedral. In $Ni(ATI)_2$ complexes, ΔH and ΔS are found to lie in the region 1–5 kcal/mole and 10–15 eu respectively.

The fact that ring protons undergo contact shifts is a clear indication that unpaired electrons have "leaked" from the metal ion into the ligand system. The mechanism of this "leak" will be discussed a little later. For the present it is assumed that α spin is delocalized into the highest filled π orbital of the ligand. This delocalization process can be described by writing suitable valence bond structures (**XVIII, XIX,** etc.). α spin can be placed only at α and γ positions of the ring. The π system is odd-alternant, and alternation of the sign of spin density is a common feature of such systems. Positions that do not show any spin in the valence-bond structures attain negative (β) spin density by spin correlation effects. The observed alternation in the sign of contact shifts is thus rationalized.

Several other tetrahedral nickel(II) complexes that behave similarly are known (specific examples are **XX** (*108*), **XXI** (*37*), **XXII** (*60*). In all these cases the spin densities alternate in sign over the ligand π frame.

By assuming that one complete α spin is delocalized into each ligand, it is possible to calculate [by using, for example, the McConnell–Dearman VB method (*92*)] the theoretical spin densities at various positions. Comparison with the observed (from contact-shift data) spin-density distribution then gives

(**XX**)   (**XXI**)   (**XXII**)

an idea about the extent of spin delocalization. This approach has led to the conclusion that 1/10, 1/13, 1/22, and 1/30 of an α spin is delocalized into each chelate ring of structures **XVII**, **XX**, **XXI**, and **XXII** respectively. The significant inference is that an N_4 donor set is more effective than an N_2O_2 set in causing delocalization of π-spin density.

The important question of metal–ligand (M—L) bonding will now be taken up. In a strictly tetrahedral nickel(II) complex, the two unpaired electrons are in a t_2 orbital. The actual symmetry of Ni(ATI)$_2$ is D_{2d}. In D_{2d}, t_2 splits into $b_2 + e$. The e orbital is presumably higher in energy and contains the two unpaired electrons (32). The highest filled bonding π-MO (HFMO) of the ligand also has symmetry e. As a result there is Ni—L π bonding and an electron is partially donated from the HFMO to Ni (L $\rightarrow$ M charge transfer). In presence of the magnetic field the metal atom has net α spin. As a result delocalization of α spin occurs in ligand e orbital. The observed spin densities are in good agreement (32) with this mechanism. A similar mechanism is undoubtedly operative in the other nickel chelates described above.

Subject to symmetry limitations, there are three paths by which unpaired π spin can be placed on the ligand system: (1) α-spin transfer from HFMO of L to an M orbital; (2) β-spin transfer involving the same path as in (1); (3) α-spin transfer from M to lowest unoccupied MO (LUMO) of L. Paths (1) and (2) correspond to L $\rightarrow$ M and path (3) to M $\rightarrow$ L charge transfer. Path (2) was used in the last paragraph. If L $\rightarrow$ M charge transfer is involved, the choice between paths (1) and (2) should be straightforward. Thus only if the π-bonding metal orbital is less than half filled, path (1) will be involved.

An important factor determining the choice between L $\rightarrow$ M and M $\rightarrow$ L paths is the relative energies of metal d- and ligand π-orbitals. As one moves across the first transitional series, the d orbitals become more and more stable. Thus if toward the end of the series, HFMO matches d_π orbitals in energy, it is conceivable that toward the beginning d_π–LUMO interaction will become energetically favorable. The ATI complexes provide a case. In Ni(ATI)$_2$ the path is practically pure L $\rightarrow$ M involving HFMO. The same appears to be more or less true of Co(ATI)$_2$ (d^7), Fe(ATI)$_2$ (d^6), Fe(ATI)$_3$ (d^5), and Mn(ATI)$_3$ (d^4). However in Cr(ATI)$_3$ and V(ATI)$_3$ there are definite indications of M $\rightarrow$ L participation (30). Similarly in tris(salicylaldimino) vanadium(III) (119), in contrast to the nickel(II) analog, M $\rightarrow$ L charge transfer appears to be the predominant path. The acetylacetonates of the first transitional series follow the same general pattern (28). In Ti(acac)$_3$ and V(acac)$_3$ practically pure M $\rightarrow$ L process is involved, while toward the end of the series the L $\rightarrow$ M path makes an additional contribution.

Unsymmetrical mixed complexes (i.e., two chelate rings that are of the same type but are not identical) have been thoroughly investigated for aminotroponeimineate (32) and salicylaldimine (16) chelates of nickel(II).

In both cases each chelate ring is found to give rise to its characteristic signals. The better π-bonding ligand gets the larger share of delocalized spin. This provides an elegant method for studying subtle effects of various substituents on π-bonding ability of the ligand system.

C. Diastereoisomeric Complexes

Pseudotetrahedral $M(A\!-\!B)_2$ complexes are dissymmetric at the metal. The two optical isomers can be designated as Δ (right-handed helicity along C_2 axis) and Λ (left-handed helicity). If in addition the ligand $A\!-\!B$ has an optically active center, the following diastereoisomers shown together with their enantiomers (connected by $\equiv\!\!\equiv$) are possible

$$\Delta(+, +)\!\equiv\!\!\equiv\!\Lambda(-, -)$$
$$\Lambda(+, +)\!\equiv\!\!\equiv\!\Delta(-, -)$$
$$\Delta(+, -)\!\equiv\!\!\equiv\!\Lambda(+, -).$$

If all three active sites are stable on the NMR time scale, it should be possible, in principle, to observe separate signals from the three diastereoisomers.

Holm *et al.* have observed (*35*) that complexes of the type **XXI** and **XXII** with optically active R groups do show distinct signals for two diastereoisomeric species, viz., active $[(+, +)$ or $(-, -)]$ and meso $[(+, -)]$. The systems **XXI** and **XXII** exhibit planar–tetrahedral equilibria in solution. The difference $\Delta\Delta f_i$ in the contact shift of the two forms is primarily attributed to a difference $(\Delta\Delta F)$ in ΔF of Eq. (22), although a difference of a_i may also make some contribution. Generally $\Delta F_{\text{active}} < \Delta F_{\text{meso}}$. Some representative results for the β-ketoamine chelates are shown in Table IX.

The absence of splitting of the active peak into signals associated with Δ and Λ configurations has led to the conclusion (*107*) that either the splitting is too small to be resolved or more likely that the configuration at the metal is rapidly racemized by the interconversion shown in the diagram.

$$\Delta \qquad\qquad \textit{trans} - \text{Planar} \qquad\qquad \Lambda$$

TABLE IX

Some Data on Diastereoisomeric β-Ketoamine Complexes in $CDCl_3$ at 100 MHz (30°)

Compound		Percentage of tetrahedral form in equilibrium	Position	$\Delta\Delta f_i$ (Hz)[a]	a_i (gauss)	
R	R'				Active	Meso
CH(Me)(Et)	Me	~100	α-Me	+32	−0.303	−0.300
			β-H	0	−0.831	−0.831
CH(Me)(Et)	H	~80	β-H	+117	−0.917	−0.917
CH(Me)(Ph)	Me	~100	α-Me	+335	−0.306	−0.276
			β-H	+565	−0.834	−0.783
CH(Me)(Ph)	H	~40	β-H	−980	−1.08	−1.08

[a] Δf_i(active) − Δf_i(meso).

An interesting situation arises in the complex shown in Fig. 8, which is about 10% tetrahedral at 25°C in $CDCl_3$ (*107*). In this case $\Delta \rightleftharpoons \Lambda$ interconversion will occur only with simultaneous racemization of the biphenylyl group. Because of the unfavorable activation energy of the latter process the Δ and Λ configurations are found to be quite stable in practice. Three distinct azomethine signals can indeed be seen for this complex (Fig. 8). The basis for the assignment shown in Fig. 8 is discussed in detail in the original paper (*107*).

Recently Pignolet and Horrocks (*111*) have reported that the fully tetrahedral complexes of the type $(R'R''R'''P)_2MI_2$ (M = Ni or Co) show doubling of contact-shifted ^1H NMR signals of the various R groups. In this case the doubling is almost certainly due to the inequality a_i(meso) $\neq a_i$(active)*. When R is an aryl group, spin is delocalized into the π orbital of the former (*78*). The behavior of the cobalt complexes is complicated by pseudocontact shifts (*78*).

* More recently, the same authors have stated that doubling is due to contamination by phosphine oxide complexes and not due to presence of diastereoisomers: L. H. Pignolet and W. Dew Horrocks, Jr., *J. Am. Chem. Soc.* **91**, 3979 (1969).

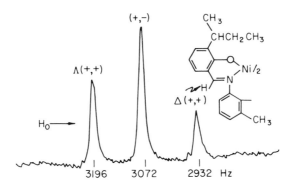

FIG. 8. ^{1}H NMR spectrum in azomethine region of a salicylaldimine complex in CDCl$_3$ at 60 MHz and $\sim 27°$C. The shifts are downfield from tetramethylsilane (*107*).

D. Contact Shift Due to Spin Density in Sigma Orbitals

Until now we have considered spin delocalization in ligand π orbitals only. NMR studies of spin-density distributions in σ orbitals is still at a very early stage. One type of problem arises when a saturated system is connected to a center containing spin density in a π orbital. An example is the Ni(ATI)$_2$ complex (*29*) with R = *n*-Bu. The protons on the alkyl chain show the follow-ing contact shifts (60 MHz, 30°C, CDCl$_3$): α-CH$_2$, $-10,603$; β-CH$_2$, -697; γ-CH$_2$, -431; δ-CH$_3$, -212 Hz. Since in Ni(ATI)$_2$ the primary process of spin delocalization involves π orbitals, the observation of contact shifts in the sigma frame requires $\pi-\sigma$ interaction.

One possible mechanism is the direct overlap of hydrogen 1s orbitals with the spin-containing p_π orbital. For the α-protons this is essentially a hyper-conjugative process. Once spin has reached the α-protons, σ delocalization can lead to appreciable spin density at remote atoms. When the sigma-bonded group is long and flexible enough, as is true for the example we are discussing, it is also possible that the free end of the chain will come close enough to the spin-bearing position to make overlap and spin delocalization possible from this end. Whatever the details are, a direct $\pi-\sigma$ spin transfer leads to spin density of the same sign as that of the π center. The sign of the contact shifts in the *n*-Bu protons is in agreement with such a mechanism. The attenuation of the shifts along the chain may be crudely attributed to "localization" of σ electrons in bonds.

A second mechanism is spin polarization. The unpaired electron (say, of α spin) in the π orbital polarizes the electron pair in the adjacent σ bond(s), favoring an excess of β spin around the neighbor. This is essentially the mechanism implied in the McConnell equation (18). The important point here

(XXIII)

is that the process can be carried further along the sigma frame **XXIII**. The sign of spin densities and contact shifts will alternate along the chain in this model. This mechanism apparently makes appreciable contributions to ^{1}H NMR shifts of, for example, the cyclohexyl group of some Ni(ATI)$_2$ chelates.

A somewhat different situation obtains when spin density is released directly from σ orbitals of the metal into σ orbitals of ligands *via* M—L sigma bonds. A case at hand is the octahedral complex Ni(n-PrNH$_2$)$_6^{2+}$ (*44*). Contact-shift data in dichloromethane at 60 MHz and 28°C are as follows: NH$_2$, +8250; α-CH$_2$, −2067; β-CH$_2$, −406; γ-CH$_3$, −15 Hz. In this complex there are two unpaired electrons in the metal e_g orbital. As a result, M—L σ interaction places positive spin density in the antibonding σ^* orbital, which is primarily a metal e_g orbital. In other words, positive spin density is placed on all atoms that make substantial contribution to the σ^* orbital. Extended Hückel calculations (*44*) on the basis of this model is in good agreement with the observed relative shifts of alkyl protons. The large positive shift of amine protons, also observed in Ni(NH$_3$)$_6^{2+}$ and other amine complexes (*95*), is generally ascribed to spin polarization of N—H bonds by the "lone pairs." Proton contact shifts of M(NH$_3$)$_6^{2+}$ show an interesting trend (*133*). The approximate shifts are: Cu(II), +4800; Ni(II), +7000, Co(II), +2500, and Mn(II), −9800 Hz (in liquid ammonia at 60 MHz). In the Co(II) and Mn(II) ammoniates there are unpaired electrons in both e_g and t_{2g} orbitals. The latter can undergo direct hyperconjugative interaction in the L → M sense with the hydrogen 1s orbitals. As a result positive spin density reaches the protons directly. This interaction is apparently much more important than the e_g interaction in the case of Mn(II), leading to negative shifts.

The work of Luz *et al.* (*83*) on ^{17}O NMR contact shifts of Mn(acac)$_3$ brings out an important point. It was already noted that the ^{1}H NMR contact shifts of M(acac)$_3$ can be explained on the basis of π delocalization. However, strict π–σ separation in this and many of the chelates dealt with in previous subsections is *not* symmetry-allowed. Even then, it is possible to identify predominant π and σ delocalization modes from experimental observation of the pattern of shifts. The protons of the acetylacetone ligand are separated from the metal ion by several bonds, and effects due to spin delocalization directly through the σ frame are expected to be small. On the other hand, the oxygen atoms of the ligand are directly bonded to the metal ion and can receive unpaired spin from the metal via σ bonds as well as via π bonds. The ^{17}O NMR signal of enriched Mn(acac)$_3$ is at -8598 ppm from the external $H_2^{17}O$ standard (25°C). The corresponding shifts of diamagnetic Co(acac)$_3$ and Al(acac)$_3$ are -174 and -274 ppm respectively, This large contact shift ($a_0 = 1.88$ G) of the Mn(III) chelate is compatible (83) only with a predominant σ delocalization.

An observation (75) of biochemical interest is the ^1H NMR shifts of hemin derivatives. All protons are shifted downfield, and on this basis σ delocalization is suggested. However, no detailed analysis is available.

E. Pseudocontact Shifts

Happe and Ward (55) showed that shifts of pyridine (py) protons in Co(acac)$_2$(py)$_2$, unlike those of Ni(acac)$_2$(py)$_2$, have a sizable contribution from pseudocontact interaction. The higher anisotropy of the g factor in Co(II) (d^7) in comparison to that of Ni(II) (d^8) is expected and is of general occurrence.

Pure pseudocontact interactions can be seen in ion pairs. For example (76), the shifts of the butyl protons of [n-Bu$_4$N][Ph$_3$PCoI$_3$] in CDCl$_3$ at 25° from the diamagnetic reference [n-Bu$_4$N] I are: α-CH$_2$, $+6.0 \pm 0.1$; β-CH$_2$, $+3 \pm 0.3$; γ-CH$_2$, $+1.7 \pm 0.3$; and δ-CH$_3$, $+0.5 \pm 0.3$ ppm. In this complex, $g_{\parallel} < g_{\perp}$ and the shifts are positive. In the corresponding Ni(II) complex, the shifts are negative, requiring $g_{\parallel} > g_{\perp}$. These shifts can arise only by way of formation of *oriented* cation–anion complexes, presumably of the ion-pair type. Lack of orientation due to random tumbling would lead to zero average shifts [averaging of the geometric factor in Eq. (20)]. By a computerized procedure it is possible to fit the observed ratio of shifts with a set of geometric factors [Eq. (21)]. In this way the probable shape and dimension of the ion pair can be obtained. However, the flexibility of the butyl chain makes such a fit difficult to achieve in the above example. This difficulty diminishes in [Ph$_4$As][Ph$_3$PCoI$_3$]. The arsonium phenyl protons show (77) the following shifts (34°C; reference, [Ph$_4$As]I): *ortho*, $+2.98 \pm 0.01$; *meta* $+2.25 \pm 0.01$ and *para*, $+1.39 \pm 0.01$ ppm. Agreement between calculated and observed shifts (ratios) was obtained with a model in which the cation lies along the C_3 axis of the pseudotetrahedral anion. This axis is collinear with a C_2 axis of the cation. The separation of the two ions is calculated to be 9.0 ± 0.8 Å.

Interesting examples of pseudocontact shifts (in ^1H and ^{11}B spectra) are provided by some pyrazolylborate chelates (66, 67), which have unusually large anisotropy of the g tensor.

F. Kinetics of Ligand Exchange

Brief reference has previously been made to isotropic shifts in complexes of the type $(Ar_3P)_2MX_2$ (Section VII.C). In the solution phase they undergo exchange reactions of the type $(L = Ar_3P; M = Fe, Co, Ni)$

$$L_2MX_2 + L^* \rightarrow LL^*MX_2 + L$$

However, little was known about the mechanism of this reaction (and of ligand-exchange reactions in tetrahedral complexes in general) before the NMR technique was applied ($110, 112$). The chemical shifts of bound and free L are quite different because of the isotropic shift in the former. It is therefore possible to determine kinetic data by analysis of lineshapes and/or linewidths. Some results are shown in Table X. In every case the average residence time

TABLE X

SOME DATA ON LIGAND EXCHANGE OF L_2MX_2 AT 25° IN $CDCl_3$

Compound	k_2 ($\pm 0.5 \times 10^n$)	$\Delta H^{\ddagger}$ (kcal/mole)	$\Delta S^{\ddagger}$ (eu)
$(Ph_3P)_2FeBr_2$	2.0×10^5	3.8 ± 0.5	-22 ± 3
$(Ph_3P)_2NiBr_2$	6.9×10^3	4.7 ± 0.4	-25 ± 3
$(Ph_3P)_2CoBr_2$	8.7×10^2	7.7 ± 0.5	-19 ± 3

of the ligand in the complex was found to be of the form $k_2[L]$, indicating a second-order process (rate constant k_2). The second-order kinetics requires an associative mechanism with a five-coordinate transition state. The observed trend in lability, viz., Fe > Ni > Co, is mainly dictated by the enthalpy of activation.

The system (2-picoline)$_2$CoCl$_2$ behaves similarly (136). Interestingly the exchange of hexamethylphosphoramide, $OP(NMe_2)_3$, with dihalobis(hexamethylphosphoramide)cobalt(II) involves both a first-order dissociative (tricoordinated intermediate) and the usual second-order associative mechanisms (135). The complex is probably forced to the former path by the bulkiness of the ligand.

VIII. Double Resonance

A. Introduction

The NMR phenomenon described in Section II.A and illustrated in the various other sections is of the single-resonance type. In a nuclear magnetic double-resonance experiment, transitions among spin levels are examined in presence of *two* oscillatory rf fields, both polarized in the xy plane (H_0 along the z direction). Of these the relatively strong field H_2 of frequency v_2 is applied at or near the resonance frequency of one type of nucleus, say X. Simultaneously, the transitions of some other nucleus, say A, in the molecule are investigated, employing the usual weak rf field of frequency v_1. A convenient notation (6) for this double-resonance experiment is A—{X}. Depending on whether A and X are the same or different isotopic species, the experiment is called homonuclear or heteronuclear double resonance respectively. The experimental set-ups used in the two cases are usually quite different.

Normally H_2 is strong enough to satisfy the relation $\gamma_X H_2{}^2 (T_1 T_2)_X \gg 1$ and H_1 is weak in the sense that $\gamma_A H_1{}^2 (T_1 T_2)_A \ll 1$. The H_2 field perturbs the spin system in a predictable way, and the observed spectra can thus be modified in a controlled manner. For a given system the kind of spectrum obtained depends on the strength of the H_2 field. Two limiting situations, viz., *spin decoupling* and *tickling* will be described below. Phenomena (generalized Overhauser effect) related to redistribution of spin-level populations due to the presence of H_2 will not be discussed. Two excellent reviews (6, 59) on the various aspects of double- (and multiple-) resonance studies are available. The short article of McFarlane will also be found to be quite useful (93).

Double-resonance experiments can be carried out either in the field-sweep or in the frequency-sweep mode. Unlike the single-resonance experiment, these two modes are, in general, not equivalent. In a field-sweep double-resonance experiment, the radio frequencies v_1 and v_2 are kept at preset values while the H_0 field is swept to attain resonance. In the experiment A—{X}, the resonance frequency of X is proportional to H_0 [Eq. (7)] and it thus varies with time. Since v_2 is fixed, the different parts of the A spectrum show the effects of a slowly varying perturbation. In the frequency-sweep mode, v_2 and H_0 remain constant while v_1 is varied. The perturbation due to a preset value of v_2 on the whole of the A spectrum can be examined in one scan. Frequency-sweep experiments require a very good stability of the field–frequency ratio (H_0/v_2). A third mode is the INDOR (internuclear double-resonance) mode. Here v_2 is swept for preset values of v_1 and H_0. In this type of experiment v_1 is usually set on an A line. The variation in the signal height at this particular frequency is examined as a function of v_2. Extremely good stability of the H_0/v_1 ratio is required for INDOR experiments.

B. Spin Decoupling

This is the most popular form of double resonance and refers to the collapse of spin multiplets. In the experiment A—{X} the spin multiplet due to A collapses to a single line when H_2 becomes large enough to satisfy the relation $\gamma_X H_2/2\pi \gg J_{AX}$. Under this condition I_X is essentially quantized along H_2 while I_A is quantized along H_0 [in the rotating frame $(6, 59)$]. Since H_2 and H_0 are at right angles to each other, the expectation value of the spin–spin interaction vanishes. This happens because the spin–spin coupling energy E_{AX} is of the form

$$E_{AX} = h J_{AX} I_A \cdot I_X$$

where I_A and I_X are angular-momentum vectors in units of h. Obviously spin decoupling will lead to simplification of otherwise complex spectra. Homonuclear proton-spin decoupling finds many applications in studies of complicated organic molecules.

In Fig. 9 is shown an illustration of spin decoupling (106). The single resonance ^{31}P spectrum of methyl dimethylphosphonate is quite complex. When the Me and OMe ^1H resonances are selectively irradiated, the ^{31}P spectrum reduces to a first-order septet and a quartet respectively (Fig. 9b and c). By using sophisticated electronic circuitry it is possible to distribute the H_2 field over a frequency range (36). Such incoherent irradiation fields can be used to decouple an entire range of absorption (Fig. 9d). However, such experiments are not easy to perform.

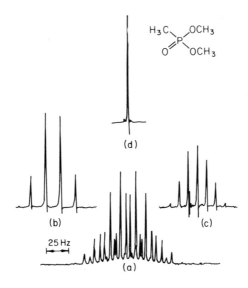

FIG. 9. ^{31}P NMR spectra of methyl dimethylphosphonate: (a) normal spectrum, (b) OMe protons decoupled, (c) Me protons decoupled, and (d) complete decoupling by incoherent irradiation (106).

^{11}B—{^{1}H} decoupling has often been very useful in structural elucidation of boron hydrides and their derivatives. For example (49), the case of μ-Me$_3$GeB$_5$H$_8$ is shown in Fig. 10. It is normal with ^{11}B NMR of boron hydrides that couplings with terminal protons alone are resolvably large. The low-field tripletlike group of the normal spectrum actually consists of two overlapping doublets, since on ^{1}H decoupling two equally intense signals representing the two types of basal boron atoms are observed. The fact that there are only two types of basal boron atoms suggests that the Me$_3$Ge group resides in a bridging position between two equivalent basal boron atoms (Fig. 10). The behavior of the apical ^{11}B signal at higher field is straightforward.

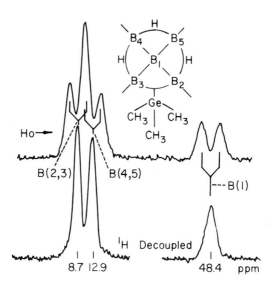

FIG. 10. Normal and ^{1}H-decoupled ^{11}B NMR spectra of μ-Me$_3$GeB$_5$H$_8$. ^{11}B chemical shifts in ppm are from BF$_3$·OEt$_2$ taken as the standard (49).

The ^{19}F—{^{14}N} double-resonance study (5) on cis-N$_2$F$_2$ brings out some important points. The single-resonance ^{19}F spectrum of this molecule consists of a symmetric five-line pattern. Decoupling of ^{14}N occurred at a characteristic value of H_0 (field-sweep experiment). The spin system in cis-N$_2$F$_2$ is of the type AA'XX', since $^1J_{NF} \neq {}^3J_{NF}$. The ^{19}F nuclei (A and A') are symmetrically equivalent and have identical chemical shifts, but are not magnetically equivalent. The same applies to the two ^{14}N nuclei (X and X'). A property of field-sweep double-resonance spectra of the AA'XX' spin system (among several other systems) is that they are mirror images when v_2 is set at

$$
\begin{array}{cc}
\overset{F}{\underset{\diagdown}{}}\quad\overset{F}{\underset{\diagup}{}} & \overset{\hphantom{+}}{\underset{-}{N}}{=}\overset{+}{N}\overset{\diagup F}{\underset{\diagdown F}{}} \\
N{=}N & \\
\textbf{(XXIV)} & \textbf{(XXV)}
\end{array}
$$

equal frequency intervals above and below v_X. The mirror-image rule applies to cis-N_2F_2 (Fig. 11). This establishes structure **XXIV** and further disproves the isomeric structure **XXV**. The mirror-image rule can be used for accurate determination of the chemical shift of the irradiated nucleus. The value of v_2 at which the A spectrum is symmetric is determined. This is the resonance frequency of the X nucleus. The procedure can be checked by shifting v_2 and noticing whether mirror images are produced. In this way the ^{14}N chemical shifts of cis- and trans-N_2F_2 are found to be 13.3 and 75.2 ppm respectively (down-field from NF_3) (5).

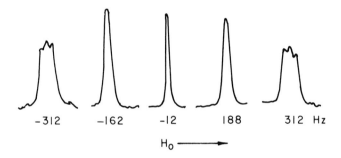

FIG. 11. ^{19}F-$\{^{14}N\}$ double-resonance spectra of cis-N_2F_2, illustrating the mirror-image rule. Numerical figures are a measure of how far v_2 is from the ^{14}N frequency. The normal spectrum is similar in structure to the ±312 spectra (5).

A nucleus with $I = \frac{1}{2}$ may give rise to a very broad resonance when it is coupled to a quadrupolar nucleus. Decoupling of the latter nucleus by double irradiation may dramatically sharpen such signals. For example, the extremely broad NH proton signal of pyrrole becomes quite sharp on ^1H—$\{^{14}N\}$ decoupling.

C. Tickling

At this limit the amplitude of H_2 is such that $\gamma_X H_2/2\pi$, while being comparable to observed linewidths, is considerably smaller than J_{AX}. Under these circumstances the spin–spin couplings are not removed, but they are slightly perturbed. This leads to new features in the spectrum. Tickling experiments are best (but not necessarily) carried out in the frequency-sweep mode so that the effect of a given H_2 on each line can be examined in one scan.

The theory of the tickling phenomenon was developed by Freeman and Anderson (48). This theory predicts that any transition that has an energy level in common (Fig. 12) with a nondegenerate transition excited by H_2 will be split into a doublet. The magnitude of the splitting is proportional to the square root of the intensity of the irradiated line.

Tickling provides a powerful method for indirect determination of the detailed NMR spectrum of the irradiated nucleus X. The principle involved is simple. In an A—{X} tickling, those irradiation frequencies that split A lines symmetrically are the frequencies of the X nucleus. A small shift in the tickling frequency, originally centered on X, leads to a considerable deviation from symmetrical splitting of A signals. Consequently the X frequencies can be determined with considerable precision. In many cases a direct single-resonance experiment, apart from being difficult because of unfavorable (due to low abundance of nuclei or due to intrinsic weakness of signals) signal-to-noise ratio, will not be able to provide the X frequencies with such precision.

The ^{129}Xe spectrum of $XeOF_4$ was determined (12) by observing ^{19}F spectrum. A field-sweep mode was used in this case. The normal spectrum consists of a symmetrical three-line pattern. The outer weak lines are satellites due to ^{129}Xe—^{19}F coupling ($J = 1163$ Hz). The strong central line represents ^{19}F bonded to other Xe isotopes, which are either nonmagnetic or are too quadrupolar to show observable coupling. The ^{129}Xe frequencies were tick-

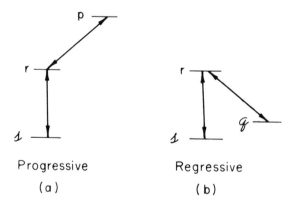

Progressive   Regressive

(a)   (b)

FIG. 12. Two types of transitions with common energy levels: (a) the common energy level r is intermediate between the energies of the other two levels; (b) the common energy level falls outside the energy range of the other two levels. In each case, when one of the transitions is tickled, the other splits into a doublet. The doublet is much better resolved in the (b) than in the (a) case (48).

led around 16.7 MHz while splitting of the high-field fluorine satellite was recorded. The results are shown in Fig. 13.

Five effective frequencies were observed. This established the quintuplet structure for the ^{129}Xe resonance. The separation between any two adjacent frequencies is 1163 Hz. This is clearly in accord with a square pyramidal structure with four equivalent fluorine atoms. It remains to comment on the tripletlike structure of each line in Fig. 13. Each fluorine transition involves degenerate levels, all of which are not connected with any particular xenon transition. As a result, when a xenon transition is tickled, only the connected transitions split into a doublet; the unconnected transitions are not affected and appear as the central line.

The ^{195}Pt chemical shifts of various *trans*-PtHL(PEt$_3$)$_2$ shown in Table IV were determined by tickling ^{195}Pt frequencies while the splitting of the hydridic proton resonance was observed (*25*). ^{1}H—{^{13}C} tickling holds considerable promise for investigation of ^{13}C NMR at natural abundance in organic compounds.

$$\nu_1 = 56.441370 \text{ MHz}$$

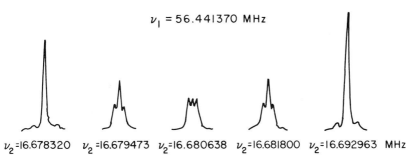

ν_2=16.678320 ν_2=16.679473 ν_2=16.680638 ν_2=16.681800 ν_2=16.692963 MHz

FIG. 13. Field-sweep double-resonance ^{19}F spectra of the high-field ^{129}Xe^{19}F satellite in XeOF$_4$. The tickling frequencies are indicated at the bottom of spectra (*12*).

The INDOR mode can be similarly used to determine the spectrum of the irradiated nucleus X. ν_1 is fixed exactly on a satellite line and the tickling ν_2 frequency is swept. As lines of the X nucleus are traversed, the intensity *at* ν_1 decreases, since the line becomes split. ^{119}Sn NMR spectrum in, for example, Me$_3$SnEt can be determined in this way (*93*).

D. Other Applications

Tickling and other forms of double resonance are the commonly employed tool for determination of the relative signs of spin–spin coupling constants. For a simple example the article of Cunliffe *et al.* (*21*) is recommended. Double resonance finds application in studies of processes in which very slow exchange is involved. One mode of study is to saturate the NMR signals due to the exchanging nuclei in one or more sites, the effect being detected as an intensity change (population transfer) of the NMR signal at some other site. For further details the article of Hoffman and Forsén (59) should be consulted.

Acknowledgements

The author thanks the Council of Scientific and Industrial Research, India, for financial assistance.

References

1. Abragma, A., "The Principles of Nuclear Magnetism." Oxford Univ. Press, London and New York, 1961.
2. Adams, A. C., and Larsen, E. M., *Inorg. Chem.* **5**, 228, 814 (1966).
3. Alei, M., and Jackson, J. A., *J. Chem. Phys.* **41**, 3402 (1964).
4. Atkins, P. W., Green, J. C., and Green, M. L. H., *J. Chem. Soc.* (A), 2275 (1968).
5. Baldeschwieler, J. D., Noggle, J. H., and Colburn, C. B., *J. Chem. Phys.* **37**, 182 (1962).
6. Baldeschwieler, J. D., and Randall, E. W., *Chem. Rev.* **63**, 81 (1963).
7. Barfield, M., and Grant, D. M., *Advan. Magn. Resonance* **1**, 149 (1965).
8. Becconsal, J. K., and McIvor, M. C., *Chem. Brit.* **5**, 147 (1969).
9. Bradley, R. C., and Holloway, C. E., *J. Chem. Soc.* (A), 219 (1968).
10. Bramley, R., Figgis, B. N., and Nyholm, R. S., *J. Chem. Soc.* (A), 861 (1967).
11. Brown, M. P., Haseltine, R. W., and Sutcliffe, L. H., *J. Chem. Soc.* (A), 612 (1968).
12. Brown, T. H., Whipple, E. B., and Verdier, P. H., *J. Chem. Phys.* **38**, 3029 (1963).
13. Buckingham, A. D., and McLauchlan, K. A., *in* "Progress in Nuclear Magnetic Resonance Spectroscopy," (J. W. Emsley, J. Feeney, and L. H. Sutcliffe, eds.), Vol. 2. Pergamon Press, Oxford, 1967.
14. Buckingham, A. D., and Stephens, P. J., *J. Chem. Soc.* (A), 2747, 4583 (1964).
15. Calderazzo, F., Lucken, E. A. C., and Williams, D. F., *J. Chem. Soc.* (A), 154 (1967).
15a. Calderon, J. L., Cotton, F. A., and Legzdino, P., *J. Am. Chem. Soc.* **91**, 2528 (1969).
16. Chakravorty, A., and Holm, R. H., *J. Am. Chem. Soc.* **86**, 3999 (1964).
17. Cooke, D. W., *Inorg. Chem.* **5**, 1141 (1966).
18. Corio, P. L., *Chem. Rev.* **60**, 363 (1960).
19. Cotton, F. A., *Acc. Chem. Res.* **1**, 257 (1968).
20. Cotton, F. A., and Legzdino, P., *J. Am. Chem. Soc.* **90**, 6233 (1968).
21. Cunliffe, A. V., Finer, E. G., Harris, R. K., and McFarlane, W., *Mol. Phys.* **12**, 497 (1967).
22. Davis, D. G., and Kurland, R. J., *J. Chem. Phys.* **46**, 388 (1967).
23. Davison, A., and Rakita, P. E., *J. Am. Chem. Soc.* **90**, 4479 (1968).
24. Dean, P. A. W., and Evans, D. F., *J. Chem. Soc.* (A), 698 (1967).
25. Dean, R. R., and Green, J. C., *J. Chem. Soc.* (A), 3047 (1968).

26. Douglas, A. W., *J. Chem. Phys.* **45**, 3465 (1966).
27. Dybvig, D. H., *Inorg. Chem.* **5**, 1795 (1966).
28. Eaton, D. R., *J. Am. Chem. Soc.* **87**, 3097 (1965).
29. Eaton, D. R., Josey, A. D., and Benson, R. E., *J. Am. Chem. Soc.* **89** 4040 (1967).
30. Eaton, D. R., McClellan, W. R., and Weiher, F., *Inorg. Chem.* **7**, 2040 (1968).
31. Eaton, D. R., and Phillips, W. D., *Advan. Magn. Resonance* **1**, 103 (1965).
32. Eaton, D. R., and Phillips, W. D., *J. Chem. Phys.* **43**, 392 (1965).
33. Eaton, D. R., Phillips, W. D., and Caldwell, D. J., *J. Am. Chem. Soc.* **85**, 397 (1963).
34. Emsley, J. W., Feeney, J., and Sutcliffe, L. H., "High Resolution Nuclear Magnetic Resonance Spectroscopy," Vol. I, 1965; Vol. II, 1966. Pergamon Press, Oxford.
35. Ernst, R. E., O'Connor, M. J., and Holm, R. H., *J. Am. Chem. Soc.* **89**, 6104 (1967).
36. Ernst, R. R., *J. Chem. Phys.* **45**, 3846 (1966).
37. Everett, G. W., Jr., and Holm, R. H., *J. Am. Chem. Soc.* **87**, 2117 (1965).
38. Fay, R. C., and Lowry, R. N., *Inorg. Chem.* **6**, 1512 (1967).
39. Fay, R. C., and Piper, T. S., *Inorg. Chem.* **3**, 348 (1964).
40. Fay, R. C., and Piper, T. S., *J. Am. Chem. Soc.* **84**, 2303 (1962).
41. Fiat, D., and Connick, R. E., *J. Am. Chem. Soc.* **88**, 4754 (1966).
42. Fiat, D., and Connick, R. E., *J. Am. Chem. Soc.* **90**, 608 (1968).
43. Figgis, B. N., Kidd, R. G., and Nyholm, R. S., *Proc. Roy. Soc.* (*London*) **A269**, 469 (1962).
44. Fitzgerald, R. J., and Drago, R. S., *J. Am. Chem. Soc.* **90**, 2523 (1968).
45. Fluck, E., "Der Kernmagnetische Resonanz und ihre Anwendung im der Anorganischen Chemie." Springer, Berlin, 1963.
46. Fratiello, A., Lee, R. E., Miller, D. P., and Nishida, V. M., *Mol. Phys.* **13**, 349 (1967).
47. Fratiello, A., and Schuster, R. E., *Inorg. Chem.* **7**, 1581 (1968).
47a. Fratiello, A., Lee, R. E., Nishida, V. M., and Schuster, R. E., *J. Chem. Phys.* **48**, 3705 (1968).
48. Freeman, R., and Anderson, W. A., *J. Chem. Phys.* **37**, 2053 (1962).
49. Gaines, D. F., and Iorns, T. V., *J. Am. Chem. Soc.* **90**, 6617 (1968).
50. Glass, G. E., Schwabacher, W. B., and Tobias, R. S., *Inorg. Chem.* **7**, 2471 (1968).
51. Grant, D. M., and Litchmann, W. M., *J. Am. Chem. Soc.* **87**, 3994 (1965).
52. Green, M. L. H., and Jones, D. J., *Advan. Inorg. Chem. Radiochem.* **7**, 115 (1965).
53. Greenwood, N. N., and Srivastava, T. S., *J. Chem. Soc.* (A), 703 (1966).
54. Griffith, J. S., and Orgel, L. E., *Trans. Faraday Soc.* **53**, 601 (1957).
55. Happe, J. A., and Ward, R. L., *J. Chem. Phys.* **39**, 1211 (1963).
56. Harris, R. K., *Can. J. Chem.* **42**, 2275 (1964).
57. Herbison-Evans, D., and Richards, R. E., *Mol. Phys.* **8**, 19 (1964).
58. Hinton, J. F., and Amis, E. S., *Chem. Rev.* **67**, 367 (1967).
59. Hoffman, R. A., and Forsén, S., *in* "Progress in Nuclear Magnetic Resonance Spectroscopy" (J. W. Emsley, J. Feeney, and L. H. Sutcliffe, eds.), Vol. 1. Pergamon Press, Oxford, 1966.
60. Holm, R. H., Chakravorty, A., and Dudek, G. O., *J. Am. Chem. Soc.* **86**, 379 (1964)
61. Holmes, R. R., and Dieters, R. M., *J. Am. Chem. Soc.* **90**, 5021 (1968).
62. Hunt, J. P., "Metal Ions in Aqueous Solutions." Benjamin, New York, 1963.
63. Jackman, L. M., "Applications of Nuclear Magnetic Resonance in Organic Chemistry." Pergamon Press, Oxford, 1959.
64. Jackson, J. A., Lemons, J. F., and Taube, H., *J. Chem. Phys.* **32**, 553 (1960).
65. Jenkins, J. M., and Shaw, B. L., *J. Chem. Soc.* (A), 770 (1966).
66. Jesson, J. P., *J. Chem. Phys.* **47**, 583 (1967).
67. Jesson, J. P., Trofimenko, S., and Eaton, D. R., *J. Am. Chem. Soc.* **89**, 3149 (1967).

68. Johnson, C. S., Jr., *Advan. Magn. Resonance* **1**, 33 (1965).
69. Jørgensen, C. K., *Prog. Inorg. Chem.* **4**, 73 (1962).
70. Kalia, K. C., and Chakravorty, A., *Inorg. Chem.* **7**, 2016 (1968).
71. Kalia, K. C., and Chakravorty, A., *Inorg. Chem.* **8**, 2586 (1969).
72. Kidd, R. G., and Traux, D. R., *J. Am. Chem. Soc.* **90**, 6867 (1968).
73. Klanberg, F., and Muetterties, E. L., *Inorg. Chem.* **7**, 155 (1968).
74. Kotz, J. C., Schaeffer, R., and Clouse, A., *Inorg. Chem.* **6**, 620 (1967).
75. Kurland, R. J., Davis, D. G., and Chen Ho, *J. Am. Chem. Soc.* **90**, 2700 (1968).
76. La Mar, G. N., *J. Chem. Phys.* **41**, 2992 (1964).
77. La Mar, G. N., Fischer, R. H., and Horrocks, W. DeW., Jr., *Inorg. Chem.* **6**, 1798 (1967).
78. La Mar, G. N., Horrocks, W. DeW., Jr., and Allen, L. C., *J. Chem. Phys.* **41**, 2126 (1964).
79. Lipscomb, W. N., *Advan. Magn. Resonance* **2**, 138 (1966).
80. Lowenstein, A., Shporer, M., Lauterbur, P. C., and Ramirez, J. E., *Chem. Commun.* 214 (1968).
81. Lucken, E. A. C., Noack, K., and Williams, D. F., *J. Chem. Soc.* (A), 148 (1967).
82. Luz, Z., and Meiboom, S., *J. Chem. Phys.* **40**, 1058 (1964).
83. Luz, Z., Silver, B. L., and Fiat, D., *J. Chem. Phys.* **46**, 469 (1967).
84. Lynden-Bell, R. M., and Sheppard, N., *Proc. Roy. Soc.* (*London*) **A269**, 385 (1962).
85. Maher, J. P., and Evans, D. F., *J. Chem. Soc.* (A), 637 (1965).
86. Malinowski, E. R., and Knapp, P. S., *J. Chem. Phys.* **48**, 4989 (1968).
87. Maslowsky, E., and Nakamoto, K., *Inorg. Chem.* **8**, 1108 (1969).
88. Matwiyoff, N. A., *Inorg. Chem.* **5**, 788 (1966).
88a. Matwiyoff, N. A., Asprey, L. B., Wageman, W. E., Reisfeld, M. J., and Fukushima, F., *Inorg. Chem.* **8**, 750 (1969).
89. Matwiyoff, N. A., and Taube, H., *J. Am. Chem. Soc.* **90**, 2796 (1968).
90. Mavel, G., *in* "Progress in Nuclear Magnetic Resonance Spectroscopy" (J. W. Emsley, J. Feeney, and L. H. Sutcliffe, eds.), Vol. 1. Pergamon Press, Oxford, 1966.
91. McCarthy, S. J., *in* "Spectroscopy and Structure of Metal Chelate Compounds" (K. Nakamoto and S. J. McCarthy, eds.). Wiley, New York, 1968.
92. McConnell, H. M., and Dearman, H. H., *J. Chem. Phys.* **28**, 51 (1958).
93. McFarlane, W., *Chem. Brit.* **5**, 142 (1969).
94. McLachlan, A. D., *Mol. Phys.* **3**, 233 (1960).
95. Milner, R. S., and Pratt, L., *Discussions Faraday Soc.* **34**, 88 (1962).
96. Misono, A. Uchida, Y., Hidai, M., and Araki, M., *Chem. Commun.*, 1044 (1968).
97. Moedritzer, K., *Advan. Organometal. Chem.* **6**, 171 (1968).
98. Moedritzer, K., and Van Wazer, J. R., *Inorg. Chem.* **5**, 547 (1966).
99. Movius, W. G., and Matwiyoff, N. A., *Inorg. Chem.* **6**, 847 (1967).
100. Muetterties, E. L., *Inorg. Chem.* **4**, 769 (1965).
101. Muetterties, E. L., *J. Am. Chem. Soc.* **90**, 5097 (1968).
102. Muetterties, E. L., and Phillips, W. D., *Advan. Inorg. Chem. Radiochem.* **4**, 231 (1962).
103. Muetterties, E. L., and Phillips, W. D., *J. Am. Chem. Soc.* **81**, 1084 (1959).
104. Muetterties, E. L., and Schunn, R. A., *Quart. Rev.* **20**, 245 (1966).
105. Nakamura, S., and Meiboom, S., *J. Am. Chem. Soc.* **89**, 1765 (1967).
106. "NMR at Work," No. 99, Varian Associates, Palo Alto, California.
107. O'Connor, M. J., Ernst, R. E., and Holm, R. H., *J. Am. Chem. Soc.* **90**, 4561 (1968).
108. Parks, J. E., and Holm, R. H., *Inorg. Chem.* **7**, 1408 (1968).
109. Pidcock, A., Richards, R. E., and Venanzi, L. M., *J. Chem. Soc.* (A), 1970 (1968).
110. Pignolet, L. H., Forster, D., and Horrocks, W. DeW., Jr., *Inorg. Chem.* **7**, 828 (1968).

111. Pignolet, L. H., and Horrocks, W., DeW., Jr., *Chem. Commun.*, 1012 (1968).
112. Pignolet, L. H., and Horrocks, W. DeW., Jr., *J. Am. Chem. Soc.* **90**, 922 (1968).
113. Pinnavaia, T. J., and Fay, R. C., *Inorg. Chem.* **5**, 233 (1966).
114. Pople, J. A., and Santry, D. P., *Mol. Phys.* **8**, 1 (1964).
115. Pople, J. A., Schneider, W. G., and Bernstein, H. J., "High-Resolution Nuclear Magnetic Resonance." McGraw-Hill, New York, 1959.
116. Reeves, L. W., *J. Chem. Phys.* **40**, 2423 (1964).
117. Richards, R. E., *in* "Determination of Organic Structures by Physical Methods," (F. C. Nachod and W. D. Phillips, eds.), Vol. 2. Academic Press, New York, 1962.
118. Roberts, J. D., "Nuclear Magnetic Resonance." McGraw-Hill, New York, 1959.
119. Röhrscheid, F., Ernst, R. E., and Holm, R. H., *J. Am. Chem. Soc.* **89**, 6472 (1967).
120. Saika, A., and Slichter, C. P., *J. Chem. Phys.* **22**, 26 (1954).
121. Sargeson, A. M., *in* "Transition Metal Chemistry" (R. L. Carlin, ed.), Vol. 3. Dekker, New York, 1966.
122. Saxena, K. M. S. and Narasimhan, P. T., *Intern. J. Quantum Chem.* **1**, 731 (1967).
123. Schreiner, A. F., and Brown, T. L., *J. Am. Chem. Soc.* **90**, 336 (1968).
124. Serpone, N., and Fay, R. C., *Inorg. Chem.* **6**, 1835 (1967).
125. Shaw, B. L., and Smithies, A. C., *J. Chem. Soc.* (A), 2784 (1968).
126. Slichter, C. P., "Principles of Magnetic Resonance." Harper and Row, New York, 1963.
127. Smith, B. B., and Sawyer, D. T., *Inorg. Chem.* **7**, 922 (1968).
128. Smith, B. B., and Sawyer, D. T., *Inorg. Chem.* **7**, 2020 (1968).
129. Storr, A., *J. Chem. Soc.* (A), 2605 (1968).
130. Supran, L. D., and Sheppard, N., *Chem. Commun.*, 832 (1967).
131. Thomas, S., and Reynolds, W. L., *J. Chem. Phys.* **44**, 3148 (1966).
132. Wawro, R. G., and Swift, T. J., *J. Am. Chem. Soc.* **90**, 2792 (1968).
133. Wayland, B. B., and Rice, W. L., *Inorg. Chem.* **6**, 2270 (1967).
133a. Wiedenheft, C., *Inorg. Chem.* **8**, 1174 (1969).
134. Witanowski, M., *J. Am. Chem. Soc.* **90**, 5683 (1968).
135. Zumdahl, S. S., and Drago, R. S., *Inorg. Chem.* **7**, 2162 (1968).
136. Zumdahl, S. S., and Drago, R. S., *J. Am. Chem. Soc.* **89**, 4319 (1967).

NUCLEAR QUADRUPOLE RESONANCE

Harry D. Schultz

U.S. DEPARTMENT OF THE INTERIOR, BUREAU OF MINES
MORGANTOWN ENERGY RESEARCH CENTER
MORGANTOWN, WEST VIRGINIA

I. Introduction

Nuclear quadrupole resonance (NQR) spectroscopy depends upon the interaction between the nuclear quadrupole moment of the nucleus and the electric-field gradients at the nucleus due to the charge distribution in a solid.

The nucleus can be represented as a rigid particle with a specific magnetic-dipole moment and electric-quadrupole moment: higher-order moments may exist, but they are small enough to be neglected in most NQR chemical applications. Therefore, many problems concerned with the electronic structure in crystalline solids can be studied by this method. Dehmelt and Krüger (16) obtained the first NQR signal in 1949. Much of the research to date has been on compounds where the results can be interpreted in terms of the electronic structure of atoms or molecules and compared easily with results from other methods such as x-ray diffraction, nuclear-magnetic-resonance (NMR) spectrometry, etc.

NQR pertains only to samples in the solid state because molecular motion averages the electric-field gradients in gases and liquids so that quadrupolar splitting does not occur. Theoretically it is applicable to all atoms with nuclear spin greater than $\frac{1}{2}$, since all such nuclei have a quadrupole moment greater than zero. In practice the nucleus under study must have reasonably large values for the nuclear quadrupole moment and natural abundance, and the chemical bonds associated with it must have an appreciable measure of p-orbital character to give a sufficiently large field gradient.

NQR frequencies corresponding to the energy of coupling between nuclear quadrupole moments and the gradients of molecular electric fields lie in the range 1–1000 MHz. NQR is similar to NMR spectrometry in that both use an applied rf field to effect nuclear transitions. However, NQR does not require a high-field magnet to observe the frequency of the nuclear spin transitions, but instead relies on the electric-field gradients of the crystal. Unlike NMR, which often exhibits large line-width broadening with lowering of temperature, NQR signals normally show only small line-width effects with variation in temperature (73). However, they frequently shift with temperature. Shifts of as much as 0.5–0.7 MHz can be observed for ^{35}Cl resonances between ambient temperature and 77°K with line widths remaining constant at about 0.002–0.003 MHz (2–3 KHz). Consequently a temperature must be specified in order to correlate line position with the determination of structure of compounds.

Of the 130 different isotopes whose nuclei can be detected by the NQR technique, ^{35}Cl and ^{14}N have been investigated more than any of the others. The reason for the extensive investigation of ^{35}Cl is that standard radio-frequency oscillators and detectors can be built to function at maximum sensitivity in the range of 10–70 MHz (^{35}Cl resonances generally fall in the 20–40-MHz range), and chlorine resonance signals are not easily saturated by high radio-frequency power levels. On the other hand, the investigation of ^{14}N is of interest because of the wide variety of different bonding types that exist in nitrogen-containing compounds. Table I lists the active NQR nuclides with the most frequently investigated isotopes marked with an asterisk.

II. Theory

A. Electric-Quadrupole Moment

The electric-quadrupole moment is a measure of the deviation from spherical symmetry of a nucleus and is defined by the equation

$$Q = \frac{1}{e} \int \rho(3z_m^2 - r^2)\, dx\, dy\, dz \tag{1}$$

where e is the electronic charge, ρ is the density distribution of the charge in the nucleus, $r^2 = x^2 + y^2 + z^2$, and z_m is an axis coincident with the direction of the nuclear-spin vector I. Only nuclei with spins greater than $\frac{1}{2}$ have electric-quadrupole moments. Q is positive if the nucleus is elongated along the spin vector and is negative if the nucleus is shortened in that direction (Fig. 1).

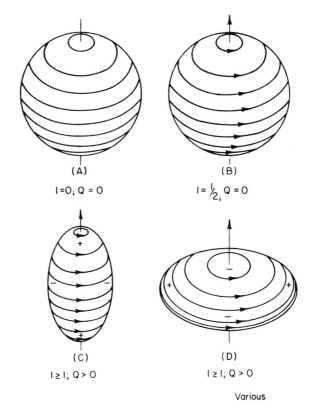

(A)

$I = 0;\ Q = 0$

(B)

$I = \frac{1}{2};\ Q = 0$

(C)

$I \geqslant 1;\ Q > 0$

(D)

$I \geqslant 1;\ Q > 0$

Various

FIG. 1. Representations of nuclei for quadrupole configurations: (a) $I = 0$, $Q = 0$; (b) $\frac{1}{2} = I$, $Q = 0$; (c) $I \geqslant 1$, $Q > 0$; (d) $I \geqslant 1$, $Q < 0$.

TABLE I

CURRENT NQR VALUES[a]

Nuclide	Natural abundance (%)	Nuclear spin, I	Quadrupole moment (in units of $Qe \times 10^{-24}$ cm^2)
^{2}H$_1$	1.4×10^{-2}	1	2.7×10^{-3}
^{6}Li$_3$	7.43	1	4.6×10^{-4} (5×10^{-4})
^{7}Li$_3$	92.57	$\frac{3}{2}$	-4.2×10^{-2} (-1.2×10^{-2})
*^{9}Be$_4$ [b]	100.0	$\frac{3}{2}$	2×10^{-2}
*^{10}B$_5$	18.83	3	11.1×10^{-2} (8.6×10^{-2})
*^{11}B$_5$	81.17	$\frac{3}{2}$	3.55×10^{-2} (3.6×10^{-2})
*^{14}N$_7$	99.63	1	2×10^{-2}
^{17}O^8	4×10^{-2}	$\frac{5}{2}$	-4×10^{-3} (-5×10^{-3})
^{21}Ne$_{10}$	—	$\frac{5}{2}$	—
^{23}Na$_{11}$	100.0	$\frac{3}{2}$	0.1 (0.02)
^{12}Mg$_{25}$	10.1	$\frac{5}{2}$	—
^{27}Al$_{13}$	100.0	$\frac{5}{2}$	0.149 (0.150)
^{33}S$_{16}$	0.74	$\frac{3}{2}$	-6.4×10^{-2} (-6.7×10^{-2})
*^{35}Cl$_{17}$	75.4	$\frac{3}{2}$	-7.97×10^{-2} (-8.5×10^{-2})
^{36}Cl$_{17}$	—	2	-1.68×10^{-2}
*^{37}Cl$_{17}$	24.6	$\frac{3}{2}$	-6.21×10^{-2} (-6.7×10^{-2})
^{39}K$_{19}$	93.1	$\frac{3}{2}$	(0.14) 0.09
^{41}K$_{19}$	6.9	$\frac{3}{2}$	0.11
^{43}Ca$_{20}$	0.13	$\frac{7}{2}$	—
^{45}Sc$_{21}$	100.0	$\frac{7}{2}$	(−0.22)
^{49}Ti$_{22}$	5.51	$\frac{7}{2}$	—
^{50}V$_{23}$	0.24	6	—
^{51}V$_{23}$	98.78	$\frac{7}{2}$	(0.3) 0.2
^{53}Cr$_{24}$	9.5	$\frac{3}{2}$	—
^{55}Mn$_{25}$	100.0	$\frac{5}{2}$	0.5 (0.4)
*^{59}Co$_{27}$	100.0	$\frac{7}{2}$	0.5
*^{63}Cu$_{29}$	69.1	$\frac{3}{2}$	−0.15 (−0.16)
*^{65}Cu$_{29}$	30.9	$\frac{3}{2}$	−0.14
^{67}Zn$_{30}$	4.12	$\frac{5}{2}$	0.18
*^{69}Ga$_{31}$	60.2	$\frac{3}{2}$	0.2318 (0.24)
*^{71}Ga$_{31}$	39.8	$\frac{3}{2}$	0.1461 (0.15)
^{73}Ge$_{32}$	7.7	$\frac{9}{2}$	−0.22
*^{75}As$_{33}$	100.0	$\frac{3}{2}$	(−0.3) 0.27
^{79}Se$_{34}$		$\frac{7}{2}$	0.9
*^{79}Br$_{35}$	50.57	$\frac{3}{2}$	0.33
*^{81}Br$_{35}$	49.43	$\frac{3}{2}$	0.28
^{83}Kr$_{36}$	11.5	$\frac{9}{2}$	0.15 (0.16)
*^{85}Rb$_{37}$	72.8	$\frac{5}{2}$	0.31 (0.3) (0.28)
*^{87}Rb$_{37}$	27.2	$\frac{3}{2}$	0.15 (0.14)
^{87}Sr$_{38}$	7.1	$\frac{9}{2}$	0.136
^{91}Zr$_{40}$	11.2	$\frac{5}{2}$	—
*^{93}Nb$_{41}$	100.0	$\frac{9}{2}$	−0.4
^{95}Mo$_{42}$	15.8	$\frac{5}{2}$	—
^{97}Mo$_{42}$	9.6	$\frac{5}{2}$	—

[a] Values are taken from a table prepared by Wilks Scientific Corporation, 1969.
[b] Frequently investigated nuclei.

TABLE I (continued)

Nuclide	Natural abundance (%)	Nuclear spin, I	Quadrupole moment (in units of $Qe \times 10^{-24}$ cm^2)
^{99}Ru$_{44}$	12.9	$\frac{5}{2}$	0.3
^{101}Ru$_{44}$	16.9	$\frac{5}{2}$	—
^{105}Pd$_{46}$	22.2	$\frac{5}{2}$	—
^{113}In$_{49}$	4.2	$\frac{9}{2}$	1.144 (1.18) (0.75)
^{115}In$_{49}$	95.8	$\frac{9}{2}$	1.161 (1.20) (0.76)
*^{121}Sb$_{51}$	57.25	$\frac{5}{2}$	-0.8 (-0.5)(-1.2)
*^{123}Sb$_{51}$	42.75	$\frac{7}{2}$	-1.0 (-0.7)(-1.5)
*^{127}I$_{53}$	100.0	$\frac{5}{2}$	-0.75 (-0.6) (-0.61)
^{129}I$_{53}$	—	$\frac{7}{2}$	-0.78
^{131}Xe$_{54}$	21.2	$\frac{3}{2}$	-0.12
^{133}Cs$_{55}$	100.0	$\frac{7}{2}$	(-0.003) (-0.03)
*^{135}Ba$_{56}$	6.59	$\frac{3}{2}$	—
*^{137}Ba$_{56}$	11.32	$\frac{3}{2}$	—
^{138}La$_{57}$	0.089	5	1.5 (3.0)
^{139}La$_{57}$	99.9	$\frac{7}{2}$	0.9 (0.3)
^{141}Pr$_{59}$	100.0	$\frac{5}{2}$	5.4×10^{-2} (5×10^{-2})
^{143}Nd$_{60}$	12.2	$\frac{7}{2}$	1.2
^{145}Nd$_{60}$	8.3	$\frac{7}{2}$	1.2
^{147}Sm$_{62}$	15.07	$\frac{7}{2}$	0.72
^{149}Sm$_{62}$	13.84	$\frac{7}{2}$	0.72
^{151}Eu$_{63}$	47.8	$\frac{5}{2}$	(1.2)
^{153}Eu$_{63}$	52.2	$\frac{5}{2}$	2.5 (2.6)
^{155}Gd$_{64}$	14.7	$\frac{7}{2}$	1.1
^{157}Gd$_{64}$	15.6	$\frac{7}{2}$	1.0
^{159}Tb$_{65}$	100.0	$\frac{3}{2}$	—
^{161}Dy$_{66}$	18.7	$\frac{7}{2}$	—
^{163}Dy$_{66}$	25.0	$\frac{7}{2}$	—
^{165}Ho$_{67}$	100.0	$\frac{7}{2}$	—
^{167}Er$_{68}$	22.8	$\frac{7}{2}$	(10.0)
^{173}Yb$_{70}$	16.1	$\frac{5}{2}$	3.9 (0.4)
^{175}Lu$_{71}$	97.4	$\frac{7}{2}$	5.9 (6.5)
^{176}Lu$_{71}$	2.6	7	(8.0)
^{181}Ta$_{73}$	100.0	$\frac{7}{2}$	6.5 (7.0)
^{185}Re$_{75}$	37.1	$\frac{5}{2}$	2.8 (2.9)
^{187}Re$_{75}$	62.9	$\frac{5}{2}$	2.6 (2.7)
^{189}Os$_{76}$	16.1	$\frac{3}{2}$	2.0 (2.0)
^{191}Ir$_{77}$	38.5	$\frac{3}{2}$	1.2 (1.5)
^{193}Ir$_{77}$	61.5	$\frac{3}{2}$	1.0 (1.5)
^{197}Au$_{79}$	100.0	$\frac{3}{2}$	0.56 (0.6)
*^{201}Hg$_{80}$	13.2	$\frac{3}{2}$	0.5 (0.65)
*^{209}Bi$_{83}$	100.0	$\frac{9}{2}$	-0.4
^{235}U$_{92}$	0.72	$\frac{5}{2}$	—
^{237}Np$_{93}$		$\frac{5}{2}$	—
^{241}Pu$_{94}$		$\frac{5}{2}$	—
Am$_{95}$	—		—
Cm$_{96}$	—		—
Bk$_{97}$	—		—
Cf$_{98}$	—		—

B. Electric-Field Gradient

Two atoms bonded together possess an inhomogeneous electric field produced from asymmetry in the electron distribution in the molecule, which is defined by the electric-field gradient q at the nucleus. The electric-field gradient is the second derivative of the electrostatic potential V at the nucleus produced by nearby charges. This gradient is a tensor quantity having nine components in Cartesian coordinates, where

$$V_{ij} = \frac{\partial^2 V}{\partial X_i \, \partial X_j} \qquad (X_i X_j = X, Y, Z) \qquad (2)$$

The components of the field gradient are determined by the structure and symmetry properties of the solid.

If q is axially symmetric with respect to Z—that is,

$$\frac{\partial^2 V}{\partial X^2} = \frac{\partial^2 V}{\partial Y^2} \qquad (3)$$

and the coordinate axes are selected such that the tensor for q takes a diagonal form (Fig. 2), then we can represent the tensor by six irreducible components (14). The tensor components about the X, Y, and Z axes are q_{zz}, q_{yy}, and q_{xx}, with q_{zz} referring to the largest field gradient, q_{yy} the next largest, and q_{xx} the smallest. The sum of the tensor components equals zero (Laplace's equation), so that only two parameters are required to specify the field-gradient tensor. By convention these two quantities are q_{zz} and the asymmetry parameter η.

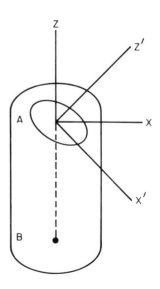

FIG. 2. Nonspherical nucleus in an inhomogeneous electric field of a molecule.

If $q_{zz} = q_{xx} = q_{yy}$, the field gradient is spherical, and the interaction of the quadrupole moment with the electronic charge vanishes, leading to degenerate quadrupole levels. If $q_{zz} \neq q_{xx} = q_{yy}$, the maximum field gradient q_{zz} lies along the highest-fold symmetry axis Z and there is axial symmetry. When $q_{xx} \neq q_{yy} \neq q_{zz}$, the field gradients are termed nonsymmetric and specify the asymmetry parameter where

$$\eta = (q_{xx} - q_{yy})/q_{zz} \tag{4}$$

A nucleus with a quadrupole moment situated in an inhomogeneous electric field possesses a potential energy that depends upon the orientation of the quadrupole moment with respect to the electric field. The quadrupole coupling between different energy states is directly proportional to the quantity eQq, called the nuclear quadrupole coupling constant.

C. Nuclear Quadrupole Resonance

In an axially symmetric field ($\eta = 0$), quantum mechanics (14) shows that the energy arising from Q in the field of q takes the form

$$E_m = \frac{eqQ[3m^2 - I(I + 1)]}{4I(2I - 1)} \tag{5}$$

where m is the magnetic quantum number, which takes $2I + 1$ values between I and $-I$. The energy transitions can be stimulated by an electromagnetic field whose quanta have the energy

$$h\nu = E_m - E_{m-1} \tag{6}$$

The Boltzmann distribution states that the E_{m-1} level will be the more densely populated. Although very small at room temperature, the difference in the population densities makes it possible to measure the absorption. The tendency for the field in accordance with Eq. (6) to produce transitions upward is greater than the tendency for the upper level to produce emission. Although the energy levels of Eq. (5) are electrical in origin, the transitions are caused by the interaction between the magnetic component of the rf field and the magnetic moment of the nucleus.

From Eqs. (5) and (6) we can derive the resonance frequency as

$$\nu_r = \frac{3eQq}{4hI(2I - 1)}(2|m_I| - 1) \tag{7}$$

therefore, $\nu_r = 3eQq/4h$ for $I = 1$, $eQq/2h$ for $I = \frac{3}{2}$, etc.

In a nonsymmetric environment ($\eta \neq 0$) the energies of the different quadrupole states are not defined by Eq. (5). If $I = \frac{3}{2}$ the value for eQq

cannot be obtained from a measured frequency because both η and q are unknown; however, the quantities can be obtained from the Zeeman effect (Section II.D.2) on a single crystal of the compound. For I values other than $\frac{3}{2}$, where more than one line is usually observed, both η and eQq can be obtained from the spectrum. From Table II it can be seen that the spectrum becomes more complicated as the nuclear spin increases. The resonance intensities depend upon the angle between the axis of the electric-field gradient and the applied rf field, and the population of the states in different energy levels. The two quantities q_{zz} and eQq are very sensitive to details of the structure of molecules and crystals. q_{zz} is sensitive to the structure of the electron shells, and eQq gives information concerning the ionic character and hybridization of covalent bonds. The asymmetry parameter η can be used to estimate the π-electron concentration near the atom.

D. Methods for Measuring Quadrupole Coupling Constants

1. PURE QUADRUPOLE RESONANCE

Pure NQR measures the quadrupole-resonance frequency in solids at zero magnetic field as a direct transition between levels corresponding to different nuclear orientations (energy states).

2. ZEEMAN EFFECT

The Zeeman splittings (11, 14) of the NQR frequency in a small magnetic field H may be used to determine the z direction of the field-gradient tensor in a single crystal. The symmetry axis of the cone formed by the components of H coincides with the z axis of the field-gradient tensor. In the case where ($\eta \neq 0$, $I = \frac{3}{2}$), a lengthy calculation gives

$$\sin^2 \theta_0 = \frac{2}{3 - \eta \cos 2\phi} \tag{8}$$

where θ_0 is the angle between H and the axis of q as a function of the azimuthal angle ϕ. The asymmetry parameter η, and the directions of x and y may be determined from the above equation.

3. NUCLEAR MAGNETIC RESONANCE

In many crystals it is more convenient to measure the quadrupole splittings of the NMR lines (27) rather than of the pure NQR, which may be difficult at low frequencies (small coupling constants). If the magnetic interaction is much larger than the quadrupole interaction, the latter can be treated as a

TABLE II

QUADRUPOLE RESONANCE FREQUENCIES

I	Secular energy equation	$\nu\,(\eta = 0)$	$\nu\,(\eta \neq 0)$
1	—	$3eQq_{zz}/4$	$3eQ(q_{zz})/4 \cdot (1 \pm \eta/3)$
$\tfrac{3}{2}$	$E^2 - 3\eta^2 - 9 = 0$	$eQq_{zz}/4I(2I-1)$	$eQ(q_{zz})/2 \cdot (1 + \eta^2/3)^{1/2}$
$\tfrac{5}{2}$	$E^3 - 7(3+\eta^2)E^2$ $-20(1-\eta^2) = 0$	$eQq_{zz}/2I(2I-1)$	$\nu_1 = 3eQq_{zz}/20 \cdot (1 + 1.09259\eta^2 - 0.63403\eta^4)$ $\nu_2 = 3eQ(q_{zz})/10 \cdot (1 - 0.20370\eta^2 + 0.16215\eta^4)$
$\tfrac{7}{2}$	$E^4 - 42(1+\eta^3/3)E^2$ $-64(1-\eta^2)E$ $+105(1+\eta^2/3)^2 = 0$	$3eQq_{zz}/4I(2I-1)$	$\nu_1 = eQ(q_{zz})/14 \cdot (1 + 3.63333\eta^2 - 7.26070\eta^4)$ $\nu_2 = 2eQ(q_{zz})/14 \cdot (1 - 0.56667\eta^2 + 1.85952\eta^4)$ $\nu_3 = 3eQ(q_{zz})/14 \cdot (1 - 0.1000\eta^2 - 0.01804\eta^4)$
$\tfrac{9}{2}$	$E^5 - 11(3+\eta^2)E^3$ $-44(1-\eta^2)E^2$ $+44/3(3+\eta^2)^2E$ $+48(3+\eta^3)(1-\eta^2) = 0$	$3eQq_{zz}/2I(2I-1)$	$\nu_1 = eQ(q_{zz})/24 \cdot (1 + 9.03333\eta^2 - 45.69070\eta^4)$ $\nu_2 = 2eQ(q_{zz})/24 \cdot (1 - 1.338095\eta^2 + 11.72240\eta^4)$ $\nu_3 = 3eQ(q_{zz})/24 \cdot (1 - 0.1857\eta^2 - 0.12329\eta^4)$ $\nu_4 = 4eQ(q_{zz})/24 \cdot (1 - 0.08095\eta^2 - 0.004258\eta^4)$

small perturbation, which gives a fine structure to the magnetic resonance frequency. The quadrupole coupling constant can be determined from the fine structure.

4. MICROWAVE SPECTRUM

In the gaseous state a molecule can rotate freely, and the transition between rotational levels can be observed in the microwave spectrum (71). The nuclear spin couples with the angular momentum of the molecule, and the quadrupole coupling energy can be obtained from the fine structure of the spectrum.

5. MÖSSBAUER SPECTROSCOPY

Mössbauer resonance measures the resonant absorption of nuclear gamma rays in crystals (76). Gamma rays are emitted or resonantly absorbed via transitions between a nuclear excited state and the ground state. Even in the case of the most favorable Mössbauer isotopes (^{57}Fe, ^{119}Sn, ^{121}Sb), nuclear quadrupole coupling constants, local magnetic fields, etc. cannot be measured so precisely as with NQR or NMR. However, while the NQR/NMR lines are often so narrow that they are multiplied in real crystals and disappear in the noise, there is always something to see in the Mössbauer spectrum because of the natural line width of resonant absorption.

6. OTHER TECHNIQUES

One may also determine the quadrupole coupling constants from the hyperfine structure of electron-spin-resonance techniques (38). The values of eQq in atoms can be obtained from atomic-beam spectra (58).

III. Factors Affecting the Resonance Signal

A. Line Shape

The NQR line is often broadened by static and dynamic influences to such an extent that it may be difficult to detect the resonance. One may minimize this problem by cooling the sample, allowing for a more favorable population density, according to Eq. (6). However, NQR instrumentation has not yet advanced to the level where sensitivity loss due to the factors discussed below can be completely compensated or avoided. Some problems that are inherent characteristics of the molecule such as dipole–dipole broadening cannot be avoided.

1. Static Influences on the Line Width

The neighboring nuclei around the species of interest may possess magnetic moments, in which case local magnetic fields occur, resulting in either splitting or, more often, broadening, depending upon the strength and direction of the fields. Dipole–dipole interaction occurring between adjacent nuclei causes a fine structure. In this case the influence of the more distant nuclei is suppressed and causes only a small broadening of the fine-structure components. If no dipole–dipole interaction is possible between nearest neighbors, then more distant nuclei with dipole moments will produce broadening of the main resonance line, but fine-structure splitting will not be observed (21). If the natural line width is that caused by the lifetime of the quadrupole states, then the ratio of the line widths for two isotopes of an element is equal to the ratio of the squares of the quadrupole moments. Theoretical and experimental studies on $NaClO_3$ and $NaBrO_3$ (37) show that the dipole–dipole interactions are responsible for the major increase of the observed line width over the true line width, although in this case both values are close. Often the effect is so pronounced that the splitting of the molecule is masked. For example, in Br_2, theory predicts a spectrum with 18 lines. The width of each line exceeds 10 kHz; thus the spectrum is not resolved because the individual line separation is only 3 kHz (34). In this case one actually observes a broad band with three weak peaks.

Terrestrial fields can affect the line width in some cases, such as for $NaBrO_3$ and $KClO_3$, and the line becomes narrower when this field is eliminated (37).

Another static factor influencing the resonance signal width arises from defects and stresses in the crystal lattice produced by rapid growth, annealing, quenching, grinding, etc., which result in a random variation of q that in turn leads to broadening. In the case of $KBrO_3$ a comparison of the observed and calculated (including the contribution of dipole–dipole interactions) line width leads to the conclusion that defects in the lattice system account for a great deal of the broadening. An exact theory of this effect cannot be formulated until we know the laws governing the molecular interaction and their subsequent effect on the electric-field gradient. Although Das and Hahn (14) and others have made some interesting proposals, the problem has not yet been solved.

The study of structural imperfections is important in NQR investigations, especially the effect of impurities. For many NQR experiments purification of the material by recrystallization or zone refinement may be necessary before usable signals can be obtained. The size and charge of the impurity atom affects the distribution of the electric-field gradients in the host crystal. The gradients are affected only by interactions confined within the crystal.

Consequently impurities outside the crystal lattice system do not affect the line shape. Indications are that the size effect of the impurity atom is the most important, at least in molecular crystals. The line width increases and the peak intensity decreases as the amount of impurity concentration increases, but the integrated intensity remains constant. Also it has been observed that neither the spin–lattice relaxation time T_1 nor the spin–spin relaxation time T_2 is affected by impurity concentration. Line broadening is attributable to strain due to the substitution of an atom of different size into the crystal lattice, which subsequently causes a random distribution of the electric-field gradients at different sites in the sample. Deformation of the lattice around each impurity atom diminishes with increasing distance from that molecule.

When the impurity concentration is less than 0.2%, the line width $\Delta\omega$, peak amplitude A, and concentration of impurity c are exponentially related by

$$A = A_0 e^{-Nc} \tag{9}$$

and

$$Nc = (\Delta\omega - \Delta\omega_0)/\Delta\omega \tag{10}$$

where the subscript 0 corresponds to the pure sample.

One can derive numerical relationships for the deformation of the molecule based on dimensions of the impurity and host molecule. If the assumption is made that the size of the atoms involved in the intermolecular contacts is proportional to the corresponding interatomic distances, then we have

$$\Delta v/v = -3\,\Delta r/r \tag{11}$$

where $\Delta v/v$ is the relative change of frequency and $\Delta r/r$ is the relative scatter of the intermolecular distances in the crystal (*19*).

The effect of gamma radiation is generally similar to that of impurities in that the frequency does not change, but the line width increases and the peak height decreases. The rate of the signal amplitude decrease with radiation dose measures the radiation stability of the specimen. The radiation stability of a specimen to γ-radiation is the dose of rays that reduces the intensity of the signal by 40%. The peak amplitude A and dosage D are exponentially related by

$$A = A_0 e^{-KD} \tag{12}$$

However, this relationship is not valid at high doses of radiation, because irreversible chemical changes take place in the defective regions of the crystal lattice. When all the voids in the molecular crystal system are filled with radiolysis products, the mobility of the molecular fragments decreases and further radiation dose has little effect on the amplitude of the NQR signal.

Calculations based on this mechanism show that a single crystal has higher radiation resistance than a polycrystalline specimen (*33*). Thus geometric factors are most decisive in explaining the influence of ionizing radiation on quadrupole resonance; polar factors are of minor importance. Randall *et al.* (*59*) observed that for sodium chlorate and potassium chlorate the integrated intensity of the NQR signal decreased rather than remaining constant with increased irradiation. Although paramagnetic centers are produced in these compounds, the NQR lines recover their original amplitude in several days, whereas the paramagnetic resonance spectrum is unchanged after several months. The cause of line broadening is due to a size effect and formation of relatively short-life reaction products, rather than because of paramagnetic centers.

2. DYNAMIC INFLUENCES ON THE LINE WIDTH

The dynamic effects on the width of the NQR line are usually smaller than the static factors. Nevertheless, thermal oscillations may cause a spread in q values and thus broaden the line. Spin-echo techniques provide the most accurate method for measuring these effects. Quanta of energy from a radio-frequency field tend to equalize the population densities of the nuclear energy states. If the rf field is momentarily turned off, then the population-density difference Δn produced by the field tends to return to the original difference Δn_0 in accordance with the following rule:

$$\Delta n = \Delta n_0 (1 - e^{-t/T_1}) \tag{13}$$

where t is the elapsed time, and T_1 is the spin–lattice relaxation time.

Since T_1 is independent of concentration and temperature at low concentrations, the lattice dynamics remain unchanged. This is substantiated by the absence of frequency shifts in the main resonant signal. If the excited state has a long lifetime, then T_1 is large and saturation may result—that is, the amount of energy absorbed in unit time equals the amount emitted in that time. Although it is more difficult to produce saturation in quadrupole resonance as compared to NMR, the line width may be increased even if the signal does not vanish (*24, 74*).

If the spread of resonance frequencies $g(v)$ is written so that

$$\int_{-\infty}^{\infty} g(v)\, dv = 1 \tag{14}$$

then

$$T_2 = 2g(v_r) \tag{15}$$

where T_2 is the spin–spin relaxation time, which is used to specify the line shape. The application of spin-echo techniques to mineral investigations will be discussed in Section VI.C.

Another aspect of molecular vibrations has been observed for the NQR signal from Cl in BCl_3 (36), which shows a doublet with splitting of 2.8 kHz, even though the crystal structure excludes the possibility of two nonequivalent positions in the unit cell. This observation is best explained in terms of internal vibrations of the molecule. The $^{10}BCl_3$ and $^{11}BCl_3$ differ in their vibration of the B atom, giving an effect on q_{zz} equal to 2 kHz. This model correctly predicts both the position and intensity ratio of the ^{10}B and ^{11}B atoms. Since ^{10}B is the less abundant isotope, the low-frequency component of the doublet is the weaker one.

B. Position of the Resonance Signal

For small thermal vibrations the effective q is a function of temperature (35); consequently the frequency is also related to temperature, and it may be expressed by

$$v = v_0(1 + aT + b/T) \tag{16}$$

where v_0 is the frequency for a static molecule. In most cases the b/T term is small compared to the aT term, and thus the frequency decreases with increasing temperature. Here a and b are coefficients independent of temperature. When the aT term is small, the resonance frequency will decrease with increasing temperature. Positive temperature coefficients have been reported for many compounds, such as AsI_3, WCl_6, and $TiBr_4$.

However, Barnes and Engardt (3) have found that the frequency for $TiBr_4$ has a negative temperature coefficient below $-50°C$. This can be explained by pressure produced within the lattice when the temperature is lowered. The frequency change with pressure at constant volume is given by the thermodynamic relationship

$$\left(\frac{\partial v}{\partial T}\right)_P = \frac{\alpha}{\chi}\left(\frac{\partial v}{\partial P}\right)_T + \left(\frac{\partial v}{\partial T}\right)_\Delta \tag{17}$$

where α is the thermal expansion coefficient and χ is the compressibility. Consequently the observed frequency can be related to thermal vibration in the crystal. The frequency change with change in pressures at various temperatures was observed for ^{35}Cl in $KClO_3$, ^{63}Cu in Cu_2O (35), and for SnI_4, As_4O_6, and $KBrO_3$ (20).

IV. Instrumentation

In most NQR spectrometers the signal is produced by a regenerative or superregenerative oscillating detector.

In a regenerative oscillating detector the sample is inserted into the oscillator coil and becomes an element of the resonant circuit. The frequency of the oscillations is varied by a variable capacitor and induced into the circuit by feedback. When the frequency of the oscillations in the circuit corresponds to the frequency between the quadrupole energy levels, governed by Eq. (6), the sample in the coil absorbs energy from the circuit, thus varying the conductance of the circuit, and changing the high-frequency amplitude at resonance. The modulation of the conductance γ is defined by (2)

$$\gamma = -4\pi\eta\chi''Q_0 \tag{18}$$

where η is the filling factor, χ'' is the component of the dynamic magnetic susceptibility of the specimen, and Q_0 is the figure of merit of the circuit. Spectrometer design calculation must be based on the values of χ of the order of 10^{-6}–10^{-7}. The relative change ϵ of the circuit voltage is directly proportional to γ, $\epsilon \sim 10\gamma \sim 10^{-5}$–$10^{-6}$; thus high radio-frequency voltages are not used in the circuit (66). The ϵ/γ ratio and the signal shape are dependent on the time constant of the detecting circuit; at maximum sensitivity the signal produced by the detector is not the true shape of the absorption line. Many regenerative NQR spectrometers are described in the literature, but all are essentially modifications of the Hopkins (28) circuit or the Pound (57) circuit, having approximately the same sensitivity. The Pound circuit has been much reproduced because it allows stabilization of the generator and detector with large frequency changes. Therefore it is suitable for automatic searches over a broader frequency region than most other regenerators. Wang (74) increased the sensitivity of the Hopkins circuit by feedback from the resonant signal, permitting work at high rf levels, but this led to marked distortion of signal shape. Thus the sensitivity of regenerative quadrupole spectrometers over a broad frequency range often proves to be inadequate.

In a superregenerative oscillator the radio-frequency oscillations are not continuous but are quenched at a frequency γq by means of an external source or by production of an internal voltage, permitting high radio-frequency voltages of up to 40 V in the circuit. Although a five- to tenfold sensitivity gain results from using a superregenerator, sideband responses are generated ($2n + 1$ NQR signals, $n = 1, 2, 3, \ldots$) that often hinder accurate measurements of frequencies in overlapping multiplet spectra. Dean and Pollak (15) proposed a method for eliminating the superregenerative sideband response of the signal by slow modulation of the quenching frequency V_q. Since the circuit tuning frequency V_0 does not depend on V_q and variations of V_q suppress the various sideband components, the true NQR signal is observed without appreciable loss of intensity. The theory of operation of superregenerative receiver circuits has been developed by Whitehead (77). Also, Graybeal and Cornwell (23) have developed the superregenerative oscillator–detector and explained the line-shape effects in terms of the components of the complex magnetic susceptibility. Because of the intrinsic sensitivity and because of the power-level allowances, the superregenerative systems are good for search purposes, but less satisfactory for line-shape, multiple-structure, and relaxation-time studies as compared to regenerative spectrometers. Peterson (56) describes a superregenerator in which the bandwidth can be increased while spectrometer still maintains a high degree of sensitivity and coherence in a 15–90 MHz range. Dean (15) used purely electronic feedback, whereas Peterson uses a servo motor to adjust the oscillator grid bias. The resonance signal is rejected from the feedback loop by using a rejection slot at 30 Hz. Later Peterson, working with automatic coherence control, extended the range to cover the 5–15 MHz and 270–327 MHz ranges (7). In all the Peterson systems the noise level was essentially constant throughout the scan.

Smith (68) proposes a frequency-modulated superregenerative spectrometer operating in the 10–50 MHz range as a means of solving the interrelated problems of automatic gain control, sideband suppression, and frequency measurement. Since the gain of the circuit depends mainly on the length of the damping time, gain change is negligible when the quench frequency is varied to suppress the sidebands if the damping time is made independent of quench frequency. This is accomplished with a feedback loop that maintains constant noise output. Frequency measurements are made by a separate marker pen on the recorder chart. A small delay in the marker channel ensures that sideband zero-beats are not printed when sideband suppression is in operation.

Another method for the observation of NQR signals is afforded by the quadrupole spin-echo technique (8, 14, 67). If two rf pulses separated by time τ are applied to the sample, then 2τ seconds after the first pulse an echo signal is produced in the specimen, caused by interaction of the damping signals of nuclear induction from the first and second pulses. Damping of the echo with known τ makes it possible to determine the T_2 of spin–spin relaxation. The spin–lattice relaxation time (T_1) can be determined by measurement of the echo envelope and amplitude of the free induction signal. The spin-echo method can be used with double-resonance techniques, where it offers promise for NQR studies at low frequencies because the double-resonance method does not require a Boltzmann population difference for the second spin system.

The problem of automatic searching over a large frequency range and obtaining spectra of high sensitivity and good coherence cannot be regarded as completely solved. Oscillator circuits employing field-effect transistors have performance characteristics comparable to circuits using vacuum tubes (72). In addition, the noise figure of field-effect transistors is a factor of 4 better than that of their vacuum-tube counterparts. Also, these circuits can be operated at low voltage, thus avoiding the problems in filtering to reduce hum and pick-up associated with high-voltage power supplies. This low-voltage operation maintains a nearly constant sensitivity over a wide range of oscillation levels and frequencies. Field-effect transistors have noise figures as low as 0.5 dB. Use of these devices in the oscillator circuit, audio amplifier, and filter circuit yields better signal-to-noise ratios. NQR spectrometers employing solid-state circuitry entered the commercial market in 1967. More commercial activity is imminent and indicates that nuclear quadrupole resonance will become a common and useful research technique.

V. Correlation of Nuclear Quadrupole Coupling Constants with Molecular Structure

The electric-field gradient q is composed of three components, $q_{xx} = q_{yy} = \frac{1}{2}q_{zz} = \frac{1}{2} \partial^2 V / \partial Z^2$, and since $V = -e/r$, then $q_{zz} = e/r^3$. Thus q is directly proportional to $1/r^3$ for a charge at a distance r from the quadrupolar nucleus. The principal axis coordinate system is defined so that the Z axis coincides with the line of centers between the two charges along r, and hence q_{zz} is the component most sensitive to field-gradient fluctuations caused by variations in chemical bonding. Since q_{zz} is inversely proportional to r^3, it is expected that only those electrons that are located close to the nucleus will have an effect on q_{zz}. The magnitude of q_{zz} at the nucleus is most dependent upon the electronic distribution in the valence p electronic shell. Any s orbital or any closed valence electronic shell is spherically symmetrical and therefore gives no contribution to q_{zz}, and the valence d and f orbitals do not penetrate sufficiently close to the nucleus to have an effect on q_{zz}. To evaluate q at a particular nucleus in a covalent molecule it is necessary only to consider contributions of valence electrons of the nucleus in question.

The first attempt to interpret observed nuclear quadrupole coupling constants and to correlate them with electronic structure was made by Townes and Dailey (70). Since a neutral Cl atom has five valence p electrons and lacks one electron of acquiring a symmetrical closed-shell configuration, it would be expected to have a rather large quadrupole coupling constant (Cl eQq for Cl_2 molecule = -110.4). In the case where a Cl atom is covalently bonded, a first approximation would be to assume a defect of one valence p electron, and hence the Cl should have a quadrupole coupling constant almost as great as that for a neutral Cl (ClCN $eQq = -82.5$ MHz). On the other hand, a Cl atom chemically bonded with a highly electropositive atom such as Na or K has acquired a good share in another p electron and consequently has a very nearly symmetrical closed valence shell, and eQq would have a value close to zero (NaCl < 1). The very low value of eQq for NaCl indicates that its bond is almost completely ionic in character, and agrees with the cubic configuration of NaCl. eQq values for covalently bonded Cl are approximately 25% lower than the value for atomic Cl. This is attributed to s–p hybridization of the covalent bond and not to fractional ionic character.

In a subsequent publication (13), Townes and Dailey revised their estimates of s–p hybridization in halogens to include Br and I in addition to Cl. They proposed a rule for the estimation of the amount of s character in a hybrid halide bond: Cl, Br, and I bonds are taken to have 15% s character whenever the halogen is bonded to an atom that is more electropositive than the halogen by 0.25 of a unit; otherwise no hybridization is allowed. This rule is based on observed quadrupole coupling data and on promotional energy for halogens. They also propose the amount of d character to be 5% or less and in most cases consider it to be negligible. The following equations can be formulated, based on the above arguments:

$$(eQq)_{mol} = (1 - s + d)(1 - i)(eQq)_{at} \tag{19}$$

$$(eQq)_{mol} = (1 - s + d)(1 - i) + 2(1 + c)(eQq)_{at} \tag{20}$$

where i is the ionic character, s and d are the amount of s and d character in the halide bond, c is a constant (usually approximated to be 0.25 to account for the fact that positive ionization increases q by pulling all of the electrons closer to the nucleus), and $(eQq)_{at}$ is the coupling constant of an isolated atom and has been determined for the halogens by atomic-beam measurements (71). Equation (19) is specific for correlation of quadrupolar coupling constants for chlorine compounds, and Eq. (20) is used for the other halogens (Br, I). Employing Eqs. (19) and (20), Townes and Dailey made a plot of ionicity versus electronegativity difference for a series of diatomic halides (13). The curve revealed that, for electronegativity differences greater than 2, molecules are essentially 100% ionic.

Gordy (22) has proposed that pure p orbitals are involved in Br, Cl, and I bonds, except for the two first-row elements C and F, based upon observed coupling constants and dipole moments. Gordy defines the quantity

$$P_q = \frac{eQq_{mol}}{eQq_{at}} \tag{21}$$

$$1 - P_q = \frac{X_{hal} - X_M}{2} \quad \text{for} \quad X_{hal} - X_M < 2 \tag{22}$$

where $1 - P_q$ is the ionicity and $X_{hal} - X_M$ is the electronegativity difference between the halogen and the atom bonded to the halogen. Quadrupole measurements can be used to evaluate effective electronegativities of atomic bonds, since ionic character is a sensitive function of electronegativity difference between atoms. Gordy plotted $1 - P_q$ versus $X_{hal} - X_M$ and concluded, as did Townes and Dailey, that for electronegativity differences greater than 2, bonding is essentially 100% ionic. The Townes–Dailey curve and the Gordy curve diverge at lower electronegativity differences; this divergence can be attributed to the different hybridization rules that were used by the different groups. Pauling derived an ionicity versus electronegativity curve from dipole-moment calculations made on hydrohalides (49). Both the Townes–Dailey and Gordy curves differ markedly from Pauling's curve. This divergence can be explained by the fact that in his dipole moment calculations, Pauling did not correct for a large overlap moment, which in the hydrohalides opposes the primary moment; hence he obtained low values of ionicity. Dailey (12) studied the halogen-bond character of 24 alkyl halides using a modified form of Eq. (19) $(eQq(\text{mol}) = I - s + d - i - II)$. Assuming negligible double-bond character II and making an independent calculation of the ionicity i, he determined the sum total of s and d hybridization. The average values were 13.6% for C—Cl bonds, 8.6% for C—Br bonds, and 1.8% for C—I bonds, in good support of the Townes–Dailey hybridization rule.

When correlating molecular structure of a given family of compounds, it is convenient to define a quantity

$$U_p = \frac{(eQq)_{mol}}{(eQq)_{at}} \tag{23}$$

called the "number of unbalanced p electrons directed along the bond axis." With this quantity one is able to compare coupling constants obtained from different quadrupolar nuclei. It should be noted that U_p has the same meaning as P_q as defined in Eq. (21). From Eq. (19) one can see that

$$U_p = \frac{(eQq)_{mol}}{(eQq)_{at}} = (1 - s + d - i - II) \tag{24}$$

Therefore s hybridization, ionicity, and double-bond character tend to decrease U_p, but d hybridization tends to increase U_p. Considerable insight can be gained by studying families of compounds where some of the parameters remain constant or are zero, and utilizing the formulated hybridization rules and ionicity curve.

However, in ionic crystals contributions from distortion of the closed-shell electrons around the nucleus and from charge distributions associated with adjacent atoms or ions should be taken into account. The field gradient calculated from a point-charge model is often in poor agreement with experiment. The disagreement can be reduced by correction for the polarizability of the ions—i.e., by multiplying q_{ion} by the antishielding factor $(1 - \gamma_\infty)$ (69), which usually increases the field gradient by a large value (57.6 for Cl^-). The large antishielding correction is responsible for difficulties encountered in the chemical interpretation of quadrupole coupling constants for ionic halogen compounds. For nonhalogen ionic compounds the point-charge-model calculations often give results that are in agreement with experimental observations. However, even in the latter case, agreement is sometimes poor. Examples are given in the next section.

VI. Applications

A. Halides

It is evident from the discussion on chemical bonding (Section V) that charge distributions in solid molecules can be studied in detail by quadrupole resonance. In addition, NQR is highly sensitive to crystal and structural effects due to (1) the thermal motion in the crystal, (2) line broadening because of crystal-lattice imperfections, and (3) nonequivalence of the positions of the molecule in the crystal lattice. It is sometimes difficult to establish whether observed frequency shifts are chemical in origin or are due to crystal non-equivalence. However, chemical nonequivalence usually causes a greater frequency shift then nonequivalent molecules in the crystal lattice. Section VI will present the NQR data that have been used to elucidate molecular and crystal structure, demonstrate how NQR data relate to the orientation of molecules when phase transitions occur, and show the influence of pi and sigma bonding on absorption frequencies. Although the halogen nuclei have been the most extensively investigated, significant results from nonhalogen nuclei are included. Recent quantitative analysis investigations by NQR are discussed in Section VI.D.

1. GROUP-III HALIDES

Barnes and Segel (4) found that the halogen spectrum of Group-III trihalides ($AlBr_3$, $GaCl_3$, InI_3, etc.) consisted of three resonances, two being closely spaced and lying well above the third. The difference in the high-field frequencies is assigned to a physical crystalline inequivalence around a set of chemically identical atoms. NQR and x-ray data support a dimerized molecule with the two metal ions bonded by two halogen bridge bonds (Fig. 3).

For Al_2Br_2 and In_2I_6 the quadrupole resonance of both the cation and halogen nuclei were observed and U_p and η were calculated (10). The four

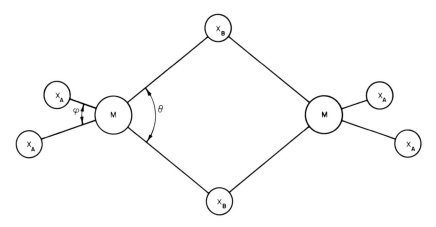

FIG. 3. Structure of Group-III halide molecules.

bonds around the metallic atom M, which defines the valence angles θ and ϕ, can be represented by the s–p hybridized orbits:

$$\psi_1 = (s + \alpha p_x + \beta p_y)(1 + \alpha^2 + \beta^2)^{-1/2} \tag{25}$$

$$\psi_2 = (s + \alpha p_x - \beta p_y)(1 + \alpha^2 + \beta^2)^{-1/2} \tag{26}$$

$$\psi_3 = (s - \gamma p_x + \delta p_z)(1 + \gamma^2 + \delta^2)^{-1/2} \tag{27}$$

$$\psi_4 = (s - \gamma p_x - \delta p_z)(1 + \gamma^2 + \delta^2)^{-1/2} \tag{28}$$

with

$$\tan \theta/2 = \beta/\alpha \quad \text{and} \quad \tan \phi/2 = \delta/\gamma \tag{29}$$

The field gradients can then be derived as a function of θ, and the resultant asymmetry parameter has the form $\eta_M = -3 \cos \theta$ (2); the valence angles θ and ϕ can be calculated from η and compared with the experimental x-ray data. In the case of Al_2Br_6 the values of ϕ calculated from quadrupole data ($\phi = 116°$) compared well with the x-ray measurement ($\phi = 114°$), but the NQR value for θ ($104°$) is larger than the x-ray value ($\theta = 98°$). This suggests that the bonding orbits do not lie along the straight lines connecting the nuclei. Formation of two hybrid orbits with an angle of less than $90°$ using only s and p wave function is impossible; therefore, the angles between the orbits of the aluminum and the bromine atoms must be larger than the angles between the lines connecting the nuclei; thus the wave functions overlap outside the line of centers to form a bent bond.

U_p is determined by the relative proportion of electrons in each different orbital, as shown by Eq. (24). The electron distribution in all the bonds can be calculated, since U_p is known for all the atoms in the trihalide molecule. These results showed that the bonds connecting the metal atoms with the bridged halogens were weaker than the others. This agrees with the x-ray and electron-diffraction analyses, which shows that the halogen bridge bonds are the longer by 0.1 Å and 0.2 Å in Al_2Br_6 and Al_2Cl_6, respectively, as compared to the nonbridged halogen bonds.

2. GROUP-IV HALIDES

The quadrupole coupling constants of a series of Group-IV halides of the general formula MH_3X, where M = C, Si, Ge, and X = Br, Cl, have been interpreted in terms of the percent ionic character and percent single-bond and double-bond character of the M–X bond (39). The double-bond character can be calculated from

$$\gamma = r_1 - r_0/r_1 + 2r_0 - 3r_2 \qquad (30)$$

where γ is the fractional double-bond character, r_1 and r_2 are actual M–X single and double bond distances, respectively, and r_0 is the observed internuclear distance of the molecule. NQR revealed that the double-bond character is greatest for SiH_3Cl. Consequently it was argued that the dipole moment of GeH_3Cl is considerably greater than that of SiH_3Cl (2.12 versus 1.30), despite the fact that the electronegativity difference between Ge and Si is very small. It is rationalized that increased π character in Si compounds tends to shorten the Si–X internuclear distance and hence decrease the dipole moment.

The halogen quadrupole coupling has been interpreted for a series of Group-IV tetrahalides using Townes and Dailey's hybridization rule (63). The results show increasing ionic character with increasing electronegativity difference between the Group-IV metal and the halogen. It was observed that U_p was less for the Si-halides than for any of the other M-halides. This is interpreted not as an ionic effect, but rather as indicative of the greater π character of the Si–X bonds.

Assuming negligible s and d character in halide bonding orbitals, the molecular parameters of the pure quadrupole spectra obtained by several groups of workers have been calculated (22). The results show the same trend in ionic character as noted above for Schawlow's data, but in general show a significantly smaller percent of double-bond character. Gordy's calculation of the effective electronegativity X_M of the central metal ion shows that Si is less electronegative than Ge, contrary to accepted belief.

The electronegativity of the central-metal Group-IV atom has been evaluated from observed quadrupole coupling constants (1) for compounds of the type MX_4, MH_3X when M is C, Si, Ge, or Sn and X is Cl, Br, or I. For each series, eQq versus electronegativity difference was plotted and an excellent linear relationship was obtained. From this it can be concluded that, since

SiX compounds show the smallest halogen coupling constants, Si should be most electropositive of the four elements under investigation. Thus the following electronegativity order is postulated as accounting for the observed linear relationship between eQq and $[X_X - X_M]$: C > Ge > Sn > Si.

Drago (18) argues against this order of Group-IV electronegativities, stating that the quadrupole coupling constants of Si halides are lower than those observed for the corresponding Ge and Sn halides because additional π character and s hybridization exist in the Si–X bond. The amount of s character in the M–X bond decreases as one goes from C to Sn, and hence the quadrupole coupling constants of Si compounds should be lower than those for the corresponding Ge and Sn compounds, as observed.

3. TRANSITION-METAL HALIDES

A comparison of covalency determined by the magnetic transferred hyperfine interaction (THFI) at the ligands and by quadrupole coupling in transition-metal halides (5) is discussed in terms of antibonding molecular orbitals of the form

$$\psi = N^{-1/2}(\phi_M - \gamma_{\chi L}) \tag{31}$$

Since the bonding state contains two electrons and therefore does not contribute to the spin density, the field is calculated more conveniently by considering the contributions of the empty or partially occupied antibonding state to the p-charge density.

Using the Townes–Dailey approach and assuming that the contributions from p electrons to the quadrupolar interaction dominate, it can be shown that

$$f_Q = \frac{(eQq)_{\text{crystal}}}{n_h(eQq)_{\text{atom}}} \tag{32}$$

where f_Q is the excess hole density in the ligand p orbital along the direction of maximum field gradient and $n_h = 1$ or 2 is the number of holes in the state ϕ_M prior to bonding. It must be emphasized that $(eQq)_{\text{crystal}}$ is a measure of the hole excess in one of the p orbitals as compared to the average of the hole excess in the other two. However, the anisotropic THFI measures only the differences of spin density in two p orbitals with respect to the third. In a similar fashion the fractional amount of spin density in the chlorine $3p$ orbitals is

$$f_\sigma - f_\pi = 2S(A_\sigma - A_\pi)/A_{3p} \tag{33}$$

in which $(A_\sigma - A_\pi)$ is the anisotropic THFI. A_s, the isotropic THFI, is determined by a spin Hamiltonian of the form

$$A_s \mathbf{I} \cdot \mathbf{S} + (A_\sigma - A_\pi)(\tfrac{3}{2}I_z S_z - \tfrac{1}{2}\mathbf{I} \cdot \mathbf{S}) \tag{34}$$

If symmetry does not allow spin density to be transferred by covalent interactions to the π orbitals (as in d^8 or d^9 configurations in an octahedral environment) then $f_\sigma - f_\pi$ reduces to f_σ. In the complementary cases where σ-spin transfer does not occur (d^1, d^2, or d^3), ($A_\sigma - A_\pi$) determines f_π. Once f_σ or f_π is known, the coefficients λ of the LCAO–MO (Linear Combination of Atomic Orbitals–Molecular Orbitals) of Eq. (30) can be determined. Then f_Q can be related to the antibonding wavefunction by

$$f_Q = q/q_{\text{atom}} = N_\sigma^{-1}\eta_\sigma \tfrac{1}{3}\lambda_\sigma{}^2 - N_\pi^{-1}\eta_\pi \tfrac{1}{4}\lambda_\pi{}^2 \tag{35}$$

Table III summarizes the eQq and THFI results obtained by Bersohn and others on transition-metal halides. Bersohn drew several general conclusions from these data:

(1) The fluoride f_s values are consistent with the small amount of hyperfine interaction introduced by the orthogonalization of the metal-atom $3d$

TABLE III

COVALENCY DETERMINED BY THFI AND NQR IN TRANSITION-METAL HALIDES

Compound	A_s $(10^{-4}\,\text{cm}^{-1})$	f_s (%)	$A_\sigma - A_\pi$ $(10^{-4}\,\text{cm}^{-1})$	$f_\sigma - f_\pi$ (%)	$\nu Q = (eqQ/2h)$ $\times (1+\tfrac{1}{3}\eta^2)^{1/2}$ (MHz)	fQ (%)
$CuCl_2\cdot2H_2O$	7.8	0.50	5.0	9.8	9.0	15.8
$CdCl_2$:Cu^{++}	9.5	0.61	4.5	8.8	—	—
$CuF_2\cdot2H_2O$	86	0.55	24	5.5	—	—
$CoCl_2$	—	—	—	—	2.56	4.7
$CoCl_2\cdot2H_2O$	5.6	1.08	0.8	4.7	5.2	9.4
$CoCl_2\cdot6H_2O$	—	—	—	—	5.5	10.0
$KCoF_3$	26	—	8	—	—	—
$KMgF_3$:Co^{2+}	23	0.48	6	2.9	—	—
$FeCl_2$	—	—	—	—	2.37	4.3
$AgCl$:Fe^{3+}	2.8	0.90	0.5	4.9	—	—
$FeCl_3$	—	—	—	—	10.12	18.3
$KMgF_3$:Fe^{3+}	24.0	0.76	2.9	3.3	—	—
$CrCl_2$	—	—	—	—	8.52	15.5
$CrCl_3$	—	—	—	—	12.92	23.4
K_2NaCrF_6	−1.1	−0.02	−7.2	−4.9	—	—
VCl_3	—	—	—	—	9.40	17.0
$TiCl_2$	—	—	—	—	4.17	—
$TiCl_3$	—	—	—	—	7.39	13.4
$(NH_4)_2PtCl_6$:Ir^{4+}	—	—	8.8	−4.3	—	—
K_2PtCl_6	—	—	—	—	26.02	47.0

orbitals to the fluorine $2s$ function. The f_s values for chlorine are also small, but larger than fluorine; consequently, in both cases the halogen-atom bonding functions are not appreciably hybridized with s functions.

(2) The value f_Q measures the holes found in both spin-paired and unpaired orbitals, whereas $f_\sigma - f_\pi$ measures the holes associated with the unpaired orbitals. However, both measure the excess p holes in the bond direction as an average in the two orthogonal directions and roughly increase or decrease simultaneously as a measure of covalency. It would be helpful if more abundant data were available to offer more support to this conclusion.

(3) A comparison of the eQq data on $CrCl_2$, $CrCl_3$; $FeCl_2$, $FeCl_3$; and $TiCl_2$, $TiCl_3$ demonstrates that the higher halides are more covalent than the lower ones.

(4) The values of f_σ are larger for Cl than for F, expressing quantitatively the generalization that fluorides are more ionic than chlorides.

The only compound where a direct comparison between THFI and eQq becomes possible is $CuCl_2 \cdot 2H_2O$; here, Cu^{2+} ion with its $3d^9$ configuration can only transfer spin density to the σ orbitals of the ligands, and THFI would directly measure the σ-bond spin density instead of the σ–π imbalance. The value of $f_\sigma = 9.8\%$ for $CuCl_2 \cdot 2H_2O$ is close to the value of $f_Q = 15.8\%$ calculated from the observed quadrupole splitting with $\eta = 0.42$. There are two factors that could make $f_Q > f_\sigma$. The first is the point-charge contribution to q, which reduces the covalency contribution to f_Q by 3% and would reduce it further if antishielding γ_∞ is taken into consideration. A second reason is that the metal-ion $4s$ and $4p$ orbitals bonding with the ligand will contribute to f_Q but not to the spin density measured by f_σ.

Another compound for which both THFI and eQq measurements have been made is $CoCl_2 \cdot 2H_2O$ (46). However, the crystal-structure characteristics of $CoCl_2 \cdot 2H_2O$ (each chlorine has two equidistant neighbors with a bond angle $\sim 90°$, which allows cancellation of the anisotropic interactions) and the complex electronic structure of Co^{2+}, where the σ–π cancellation is uncertain, prevent a meaningful comparison of f_σ and f_Q. On the other hand, in $CoCl_2 \cdot 6H_2O$ each Cl^- has one Co^{2+} neighbor. Hence ν_Q is about the same as in $CoCl_2 \cdot 2H_2O$, except that in the hexahydrate the principal axis of the largest fixed gradient lies along the Co–Cl bonds rather than perpendicular to them as in the dihydrate.

In $FeCl_2$ each chlorine has three equidistant Fe^{2+} neighbors lying approximately along three Cartesian coordinates, and both f_Q and f_σ are small, as would be expected. The value for f_Q for $FeCl_3$ approximately equals f_Q for $CuCl_2 \cdot 2H_2O$ (just as $f_\sigma - f_\pi$ for $[FeF_6]^{-3}$ is $\sim f_\sigma$ for $CuF_2 \cdot 2H_2O$) because even though the σ bonding in the ferric complex is larger than in the cupric complex ($+3$ metal-ion compared to $+2$ metal-ion complexes), the σ–π

cancellation brings the imbalance down to a value slightly less than that of the cupric complex, where π bonding cannot contribute.

As the number of π electrons decreases in the series $CrCl_3$, VCl_3, $TiCl_3$, the t_{2g} orbitals of the metal empty, the π cancellation becomes more important, and eQq decreases, as demonstrated in Table III.

The above considerations can be used to help our understanding of the relative roles of σ and π bonding in the low-spin transition-element complexes. Measurements by optical absorption and ESR determine only the unpaired spins in the π orbitals. From correlations by NQR we can evaluate the relative importance of σ and π bonding in the low-spin compound K_2PtCl_6. Quadrupole resonance (45) studies of $PtCl_6$ give f_Q as approximately 50%, interpreted in terms of covalency. Hence, in contrast to high-spin complexes such as $CuCl_2 \cdot 2H_2O$, where f_σ is almost as large as f_Q, in these low-spin complexes $f_\sigma \ll f_Q$. In other words, the metals and p functions account for most of the σ bonding in these low-spin complexes.

Bersohn argues against the point-charge-model interpretation employed by several other workers, stating that if one treats all the ions as point charges and computes the necessary lattice sums to determine the field gradients, one obtains results that are too small, usually by one order of magnitude. The model of a $^1S_0$ ion in an electric field gradient coming from external point charges is inadequate. Bersohn also rejects the extension of the point-charge model so as to include the polarization of negative and positive ions. This results in a large increase in the magnitude of the field gradient by a factor $(1 + |\gamma_\infty|)$ that may be as much as 100, where γ_∞ is the Sternheimer antishielding factor. However, no calculated value of γ_∞ can be completely satisfactory because different values of γ_∞ are often used to fit the experimental field gradients. If the correct wave functions of the crystal were available, the correct field gradient could be obtained without any antishielding correction. Dipoles make a substantial contribution to the total field gradient and can be computed from the electric field at the ions by using their electric polarizability. Another approach is to recompute the total field gradient at each ion and, by using their electric quadrupole polarizabilities, to determine the induced quadrupole moment. If one considered only the cations in a metal, with a fairly uniform conduction-electron density, this procedure would be satisfactory because cations have small polarizabilities and the terms following the first are only small corrections. However, in ionic crystals the anions have large and often poorly known polarizabilities and are highly sensitive to the crystal-structure parameters. The final result would be a distortion of the ion in question, caused by the nearest neighbors, which can be described in terms of a small amount of σ and π covalent character of the p electrons.

Other research groups have also raised serious questions about the adequacy of the point-charge model. Nicholson and Burns (48) studied ^{57}Fe

in several rare-earth garnets in which Fe^{3+} occupies both octahedral and tetrahedral sites. The field gradients calculated for the octahedral sites with a point-charge model were in agreement with experiment, but at the tetrahedral sites, where the isomer shifts indicate a greater degree of covalent character, the breakdown of the point-charge model became evident. Also, in ionic CdI_2, which contains two inequivalent iodide sites, the two frequencies differ by only 2%, although the point-charge field gradients at the two sites differ by 50%. Also, for the compounds BeO and Al_2O_3, poor agreement has been reported (65), using the point-charge model. From these and other experiments it can be concluded that the point-charge model with antishielding corrections is inadequate and the experimental results might better be interpreted in terms of covalency, where experimental observations and theoretical calculations are more compatible.

4. Hexahalides in Heavy Atoms

Nakamura and colleagues (30, 31, 41–45) interpreted the NQR spectra of a series of hexahalides of heavier atoms in terms of the covalency of the metal–ligand bonds. Using the Townes–Dailey method (12, 63), the covalent character $(1 - i)$ can be calculated from

$$eQq = (1 - i)(i - s)(eQq)_{atom} \qquad (36)$$

where $(eQq)_{atom}$ is the atomic quadrupole coupling constant, i is the ionic character of the metal–ligand bond, and s is the s character in the bonding orbital of the halogen. The assumption is made that the ionic contribution to the field gradient at the nucleus is negligible. That is, the quadrupole coupling constants in Table IV, calculated from the NQR frequencies measured at 77°K, are based on the covalent contribution of the M–X bonds, assuming 15% s character and negligible d character in halide bonding orbitals. The charges migrating toward the central metal ion of metal–ligand bonds partially neutralize the formal charge and reduce it to a fraction of an electronic charge. Consequently the net charge ρ on the metal ion may be calculated from

$$\rho = \text{formal charge} - \text{coordination number} (1 - i) \qquad (37)$$

The covalent character of metal–ligand bonds increases with decreasing electronegativity of the halogen or with decreasing difference between the electronegativity of atoms forming the bond. This rule of linearity permits one to estimate the extent of covalent character of M–X bonds for complexes of analogous structures. For example, hexahaloselenates and hexahalotellurates with the same outer electronic configuration of the central atoms yield a single straight line, while hexahaloplatinates and hexahalopalladates lie on

TABLE IV

NQR Data on Hexahalides

Hexahalide	X = ^{35}Cl				X = ^{79}Br				X = ^{127}I			
	ν^a	eQq (MHz)	$1-i$	ρ	ν	eQq (MHz)	$1-i$	ρ	ν	eQq (MHz)	$1-i$	ρ
K_2PtX_6	26.021	52	56	0.64	202	406	62	0.28	407	1360	70	−0.20
K_2PdX_6	26.75	53.2	57	0.58	205.34	411	63	0.22	—	—	—	—
K_2SeX_6	20.58	41.2	44	1.36	173	346.2	53	0.82	—	—	—	—
NH_4SeX_6	20.877	41.8	45	1.30	172.623	345.2	53	0.82	—	—	—	—
Cs_2SeBr_6	—	—	—	—	177.44	354	54	0.76	—	—	—	—
K_2TeX_6	—	—	—	—	135.670	271	42	—	154, 302	1022, 1011	52	—
NH_4TeCl_6	15.137	30.2	32	2.08	—	—	—	—	—	—	—	—
$M_2TeX_6{}^b$	—	—	—	—	129.946	273.5	42	1.48	154	1017	52	0.88
Cs_2TeBr_6	—	—	—	—	135.962	271.92	—	—	—	—	—	—
$N(Me)_4TeBr_6$	—	—	—	—	142.58	285.16	—	—	—	—	—	—
K_2SnX_6	15.73	31.5	34	1.96	130.5	261	40	1.60	—	884	45	—
$(NH_4)_2SnBr_6$	15.65	31.3	34	1.96	126.9 @ −72°C	253	39	1.66	—	—	—	—
K_2WX_6	10.22 @22.5°C	20.4	0.22 (0.57)	0.58	—	—	—	—	—	—	—	—
K_2ReX_6	13.9	27.9	0.28 (0.55)	0.70	115	231.34	34 (61)	.34	123, 246	822.21	57 (68)	0.08
K_2OsX_6	16.897	33.8	0.38 (0.53)	0.80	—	—	—	—	—	—	—	—
K_2IrX_6	20.841	41.7	0.45 (0.53)	0.80	—	—	—	—	—	—	—	—
$(NH_4)_2PbX_6$	17.26	34.5	0.37	1.78	—	—	—	—	—	—	—	—

^{a} Temperature is liquid-nitrogen temperature unless otherwise specified.
^{b} M = an average of K^+, Cs^+, NH_4^+, and $N(CH_3)_4^+$.

another straight line, when percentage covalent character is plotted against electronegativity difference ($\Delta_\chi = \chi_X - \chi_M$) (43). In accordance with the increase in electronegativity of selenium (2.4) over that of tellurium (2.1), the covalency of Se–X bonds is greater than that of Te–X bonds in the corresponding complexes.

Nakamura *et al.* have extended the Townes–Dailey equation (19) to include π bonding (30). The π character of the Ir–Cl bond was obtained from the hyperfine structure of EPR spectra in hexachloroiridates, and the total ionic character was calculated from

$$I = 1 - \sigma - \pi \tag{38}$$

and also

$$eQq = [(1 - s)(1 - i - \pi) - \pi/2](eQq)_{\text{atom}} \tag{39}$$

The result is $I = 0.47$ when $\pi = 0.054$. In estimating the π bonding in other $5d^n$ halides, Nakamura assumed that the number of electrons migrating from the halogen ion to the central metal ion through bonding is proportional to the number of electronic vacancies in the t_{2g} orbitals of the central metal ion. According to molecular theory, this assumption is valid if the overlap integrals between the central ion and the ligands vanish. π is proportional to $6 - n$, where n is the number of electrons in the antibonding t_{2g} orbital. Since $n = 2, 3, 4,$ and 5 in W^4, Re^4, Os^4, and Ir^4 complexes, therefore $\pi = 0.22, 0.16, 0.11,$ and 0.05, respectively. If we use these π values to calculate covalency, we obtain the numbers shown in parentheses in Table IV, and they are appreciably higher than the values calculated when only pure σ bonding is considered.

The halogen resonance frequencies of the complexes studied by Nakamura were temperature dependent, with Δv varying approximately from 1 to 4 MHz as the temperature was changed from room temperature to liquid-nitrogen temperature. To locate transition points, one plots resonance frequency versus temperature, the transition point being obtained as a break in the curve. For example, K_2ReBr_6 shows the existence of three transition points at -4, -16, and $-27°C$, using the ^{79}Br resonance frequency. The temperature coefficient

$$(\partial v/\partial T)_p = (\partial v/\partial T)_v + (\partial v/\partial V)_T(\partial v/\partial T)_p \tag{40}$$

of quadrupole frequencies is usually negative; however, positive temperature coefficients were obtained for K_2ReCl_6 ($dv/dT = 0.13$ kHz/deg), K_2WCl_6 (0.44), and K_2ReBr_6 (2.8) in the temperature range for which these complexes show a single resonance line. The resonance frequency of the hexachloro complexes decreases, while the temperature coefficient increases progressively with decreasing atomic number of the central metal atoms or with increasing electron deficiency in the $d\epsilon$ orbitals. This indicates that the frequency decrease and the positive temperature coefficient are closely related to the vacancy in the $d\epsilon$ orbitals or the $d\pi$–$p\pi$ bond character of the metal–ligand bonds. As the π-bond character increases, electrons in the p_x and p_y orbitals of chlorine migrate toward the central metal ion; the resulting electron deficiency causes the decrease in the quadrupole resonance frequency of chlorine. The fact that the observed temperature coefficient $(\partial V/\partial T)$ is normally negative indicates that the major term $(\partial V/\partial T)_v$ is negative. Theoretically this term takes into account the thermal vibration of complex anions, but is free from the thermal expansion of the lattice. As the thermal vibration of the complex ion increases, the overlap of the σ orbital of the central metal ion with the p_z orbital of chlorine decreases, leading to decreased covalent character of the metal–ligand bond. Therefore, the field gradient q and consequently the quadrupole resonance frequency decrease with increasing temperature. However, when $d\pi$–$p\pi$ bonds are involved, as in paramagnetic complexes, the decrease of overlap due to thermal vibration causes a decrease of π-bond character. This means that electrons in the p_x and p_y orbitals of a halogen atom migrate toward the $d\pi$ orbitals to a smaller extent, and hence the field gradient increases with increasing temperature. A positive temperature coefficient is obtained when this effect and the second term of Eq. (39) predominate over the normally negative temperature coefficient, as is the case with $K_2Re_2Cl_6$, K_2WCl_6, and K_2ReBr_6, where the metal–ligand bonds involve high π-bond character.

For the series K_2PtCl_6, K_2IrCl_6, K_2OsCl_6 one observes increasing π-bond character of the metal–ligand bonds, while the σ-bond character decreases as the ionic character remains almost constant, in conformity with Pauling's electroneutrality principle. In this series the net charge on the central atom is a positive fraction of the electronic charge; the increase of net charge on the central atom due to the decrease in the σ-bond character is compensated by the decrease of the charge due to the π-bond formation. The data on this series conform to the linear relationship $(1 - i$ versus $\Delta\chi_X - \Delta\chi_M)$; however, if the π-bond character is disregarded, they deviate from the straight-line relationship and the net charges on iridium and osmium are increased in contradiction to Pauling's rule. This establishes the importance of π character in Ir–Cl and Os–Cl bonds.

NQR experiments can throw considerable light on the nature of metal–halogen bonds in complex compounds, and there appears to be need for more data in this area. In polyatomic molecules it is difficult to develop an unambiguous concept of ionicity from nuclear quadrupole coupling constants, as contrasted to diatomic molecules (Section V) where ionicity calculated from nuclear quadrupole coupling constants agree with ionicity developed from dipole moments, bond lengths, force constants, etc. Since the ionic character calculated from NQR is only an approximation within the molecular orbital theory, and molecular orbital theory itself is only an approximation to the wave function of the molecule, it is not surprising that observables calculated from this approximate wave function are often in poor agreement with experiment. The uncertainty of the Sternheimer polarization further hampers the interpretation of NQR data. Nevertheless, a linear relation exists between the ionic characters and the electronegativity differences of the central atoms and the ligand halogen atoms if the central atoms belong to the same family of the periodic table. The concept of ionicity, as represented by the electronegativities and NQR data, is significant in spite of the difficulties inherent in a theoretical treatment of polyatomic molecules.

5. OTHER HALIDES

Ionicity values of 87% and 77% were calculated for $TiCl_4$ and WCl_6 (26) respectively, using the Townes–Dailey procedure. A curve plotted from these ionicity values against electronegativity difference fell appreciably above Townes and Dailey's curve, indicating the existence of π character in the M–Cl bonds. Since the spectrum of $TiCl_4$ contains four resonance lines (60), it is postulated that $TiCl_4$ has a crystal structure similar to the other Group-IV tetrahalides (Section VI.A.2.). An 87% ionicity value for $TiCl_4$ was determined to be high because of admixture of some π character in the M–X bonds. The NQR spectra of $NbCl_5$ and $TaCl_2$ were discussed in terms of crystal structures, but with difficulties caused by inconsistent observed frequencies. The covalent character of paramagnetic halides $TiCl_3$, VCl_3, and $CrBr_3$ calculated from that observed from the pure quadrupole resonance frequencies (4) agreed with those based upon electronegativities.

Many researchers have found that the rare-earth chlorides isomorphic to $LaCl_3$ are especially suitable for quadrupole studies because of simple structure and high symmetry. The simplest point-charge model fails to fit the data, and a calculation that includes the shielding effects of the $5s$ and $5p$ shells has not been made. An alternate hypothesis is that the observed splittings are due to overlap of the rare-earth wavefunctions with the surrounding ligands (32). Support for this argument is offered by a prediction of the relative spacing of the split energy levels and by a determination of covalent bonding in the ligand ions.

The Zeeman effect has been used extensively to study NQR spectra of single crystals. Recently the method has been developed to obtain the asymmetry parameter from the Zeeman NQR spectra of powders for $I = \frac{3}{2}$ (17, 40). When a weak dc magnetic field H is applied parallel to the radio-frequency field, a pair of minima will appear on the spectrum at frequencies shifted by $\pm\gamma H/2\pi$ from the unperturbed resonance line, where γ is the gyromagnetic ratio of the resonating nuclei. Also, two pairs of maxima will be produced at

$\pm(I - \eta)\gamma H/2\pi$ and at $\pm(I + \eta)\gamma H/2\pi$. The asymmetry parameter η is obtained as a function of the frequency separation Δv between the observed maxima and minima on the spectral trace ($\Delta v = 2\eta v H$). Results on polycrystalline HgCl$_2$ reveal that $\eta = 0.30$ for ^{35}Cl at the high-frequency site, whereas $\eta = 0.28$ for the ^{35}Cl at the low-frequency site.

Quadrupole studies on iodine in ammonium, rubidium, and cesium triiodides (62) reveal three v_1 and two v_2 frequencies ($I = \frac{5}{2}$, the third v_2 frequency being beyond the range of the spectrometer), indicating that all three iodine atoms are nonequivalent. The linear configuration of the I_3^- ion is postulated to be in resonance among three electronic structures:

$$I_a^- - I_c - I_b \qquad I_a^- - I_c - I_b^- \qquad I_a^- - I_c^+ - I_b^-$$

$$\text{(I)} \qquad\qquad \text{(II)} \qquad\qquad \text{(III)}$$

Structure **III** takes into account the polarization of an iodine molecule by a nearby negative ion. For I_a and I_b the calculated asymmetry parameter decreases in the order $NH_4I_3 > RbI_3 > CsI_3$, suggesting that the field asymmetry originates from the field gradient of external charges in the crystal rather than to charge distribution in the I_3^- ion. A large positive temperature dependence is characteristic of the quadrupole coupling constants of I_a and I_b in ammonium triiodide, whereas I_c in NH_4I_3 and all iodide atoms in RbI_3 and CsI_3 show normal temperature behavior (a small negative temperature coefficient). This indicates the existence of hydrogen bonding between ammonium ions and terminal iodine atoms in NH_4I_3.

B. Nitrogen Quadrupole Resonance

In contrast to halogen resonances, many failures to observe ^{14}N resonances have been reported because of the low range of frequencies (2–10 MHz) and the resultant low signal-to-noise ratio. Watkins and Pound (75) first observed ^{14}N resonances in 1952, using a spectrometer that employed magnetic-field modulation. However, this technique is ineffective for polycrystalline specimens when the asymmetry parameter is large. Theoretically, ^{14}N holds high interest because of the large number of different bonding structures encountered in inorganic, organic, and biological compounds for this isotope. Table V gives the values of the asymmetry parameter and coupling constants for some inorganic nitrogen compounds. Since the most abundant isotope of nitrogen, ^{14}N, has a nuclear spin of unity, one can determine the asymmetry parameter from quadrupole frequencies in polycrystalline samples, provided that η is finite. Two resonance frequencies can be detected for each inequivalent crystalline site:

$$v^I = \tfrac{1}{4}heQq(3 + \eta) \tag{41}$$

$$v^{II} = \tfrac{1}{4}heQq(3 - \eta) \tag{42}$$

TABLE V

^{14}N QUADRUPOLE RESONANCE

Compound	eQq	η
N_2	4.65	0
N_2 in quinone clathrate	4.6	—
HCN	4.0183	0.85
ClCN	3.219	0.0157
BrCN	3.37	0.006
ICN	3.40	0
NH_3	3.5705	0
ND_3	3.2308	0
$CO(NH_2)_2$	3.507	32.3
$C_3N_3Cl_3$	4.083	1.7
CCl_3CN	4.0521	0.53
$K_2Zn(CN)_4$	4.139	0.00
$K_2Cd(CN)_4$	4.1988	0.00
$K_2Hg(CN)_4$	4.0475	0.00
$K_3Cu(CN)_4$	3.9639	2.94
$K_2Pt(CN)_4 \cdot 3H_2O$	3.467	3.2 (av.)
$K_3CO(CN)_6$	3.684	3.0 (av.)
$Hg(CN)_2$	3.9513	—
$NaNO_2$	5.792	40.5 (av.)

If $\eta = 0$, a single resonance peak is observed and the quadrupole coupling constant is calculated from

$$v^0 = \tfrac{3}{4}heQq \tag{43}$$

This is the case for N_2 with a frequency of 3.6 MHz. The magnitude of eQq and η can be calculated directly from the observed frequencies by application of Eqs. (41–43). If two or more resonance lines are observed, this may be due to a single nitrogen site with a nonzero asymmetry parameter or a symmetrical field gradient with some crystalline field splitting. These two cases can be distinguished by observing the behavior of the resonance line in a weak magnetic field.

Ikeda et al. (29) interpreted the quadrupole data of ^{14}N in metal cyano complexes in terms of π bonding due to the migration of electrons of the central metal atom into the vacant antibonding π molecular orbitals of the cyano groups. Since i_π is positive, the quadrupole resonance frequency decreases with increased π bonding and the observed difference in quadrupole coupling constants is largely due to the extent of i_π. For example, metal–ligand bonds in the platinum and cobalt complexes ($eQq = 3.5$–3.7) have a greater extent of $d\pi$–$p\pi$ bond character than in the zinc, cadmium, mercury, and copper complexes ($eQq = 4.0$–4.2). This is in agreement with the published results for the simple cyanides. A quadrupole coupling constant of 3.8–4.0 MHz is observed for ^{14}N in hydrogen cyanide (47) and alkyl cyanides (9, 47), where a CN group is bonded to the rest of the molecule by a σ bond. In contrast, nitrogen, in having a halogen (XC≡N) or sulfur (RSC≡N) atom bonded to the carbon atom of the cyano group yields a quadrupole coupling constant in a range of 3.2–3.6 MHz, indicating the contribution from resonance structures X^+≡C≡N^- and RS^+≡C≡N^-.

Calculations based upon a point-charge model for $(eQq)_{ext}$ gives values less than 1% of the observed coupling constant; consequently the field gradient comes largely from the charge distribution within the cyano group. The $[Pt(CN)_4]^{2-}$ ion has a square planar configuration. Since the $d\pi$–$p\pi$ bonding is different for parallel and perpendicular directions to the plane of the complex ion, a fairly large asymmetry parameter is expected. Since the observed η is only 3.2%, it can be concluded that Pt–CN bonds have no appreciable intrinsic asymmetry, and the small η is due to the effect of the external ions.

C. Minerals

1. Chemical and Structural Properties

NQR has been used to study structure and chemical properties of minerals that are difficult to determine by more conventional methods. For example, it is possible to analyze NQR spectra in minerals of arsenic, antimony, and bismuth, whose structures contain the pyramidal coordination complexes RX_3 where R is As, Sb, or Bi, and X is S or O. Qualitative correlation between the value of eQq and the electron state of the atom R yields information on the distribution of electric charges within the coordination complex RX_3 that represents the basic functional grouping of atoms in R_2X_3 minerals.

In antimonite and orpiment the RX_3 complexes are bound together by means of common atoms X; in the more complex compounds, by means of metallic atoms Cu, Ag, Pb, etc. The eQq value can be related to the geometry of these complexes. Bond-angle determinations (49, 50) indicate that s hybridization changes by only about 10% in RX_3-type minerals. Therefore eQq shifts are brought about mostly by changes in the effective charge of the atoms R and X. Donor–acceptor bonds, induction, and steric effects are the causes of changes in the effective charge of atoms in R_2X_3-type minerals.

a. Donor–Acceptor Bonds. When the atom R is the acceptor and is carrying a positive charge, polarity of the principal bonds is decreased (46), which leads to a lowering of the field gradient at the nucleus of R. The field gradient depends substantially on the ion component of the R–X bond. Also, eQq is sensitive to small gradient shifts brought about by weak (long-range) bonds of the donor–acceptor type. For instance, in realgar (As_4S_4) the values of eQq for the four nonequivalent positions of As atoms are 178.44, 182.10, 184.10, and 185.92 MHz (50). Their fields of force differ somewhat because of the effect of donor–acceptor and van der Waals bonds, thus bringing about the above-mentioned lack of equivalence. In the As_4S_4 molecule, the As–S bonds are accompanied by two homoatomic bonds As–As. The polarities of the pair of As–As bonds differ somewhat in the molecule, because of different effective charges on these atoms. In orpiment (As_2S_3, the product of realgar alteration), weak As–As bonds are identifiable by means of pulsed NQR echo signals (54) and are assigned to the existence of dative bonding between As atoms of two adjacent chains. Had this effect been caused by a direct dipole–dipole interaction between magnetic moments of As nuclei in the same chain, an identical spectrum would have been observed for all As nuclei. The absence of the latter suggests dative-bond formation between As atoms of two adjacent chains.

Donor–acceptor bonds often lead to formation of coordination polymer compounds, as demonstrated by antimonite and orpiment. The structure of antimonite contains macromolecules $(Sb_4S_6)_n$ regarded as dimers of $(Sb_2S_3)_n$ linked by a pair of equidistant bridge bonds, so that Sb atoms in the dimer are equivalent in pairs (61). The polymerization is accompanied by an appreciable redistribution of electron densities within the (Sb_2S_3) molecules. Because of this, two SbS_3 polyhedrons (out of four) within the dimer are strongly distorted ($\eta = 38.10\%$), and the corresponding Sb atoms acquire additional bonds. Accounting for the latter, the effective number of atoms reaches five (the coordination polyhedron now becoming a semioctahedron).

Additionally, realgar exhibits features of the coordination polymer of a donor–acceptor nature. However, even the lowest of the four values of eQq in realgar (178.44 MHz) is considerably higher than the maximum eQq value for orpiment (143.88 MHz). This difference is caused by the different bonding possibilities for orpiment atoms, and polymerization proceeds here in a different way, as compared with antimonite. The S—As—S—As chains in orpiment may be regarded as condensates of realgar half-molecules As_2S_2 linked by means of the bridging atom S. With such a construction the number of bonds in each atom remains unchanged, as in realgar. However, the number of sulfur atoms increases from two to three, with AsS_3 polyhedrons formed.

b. Induction Effect. The induction (polarization) effect is realized in compounds with two or more "electropositive" atoms of various kinds, such as mineral sulfosalts. Sulfosalts are structurally similar to certain simple sulfides. For example, bournonite ($CuPbSbS_3$) (65) is an analog of the antimonite structure, derived from the latter by replacing the Sb atoms, in turn, by those of Pb and placing Cu atoms in the tetrahedral voids. eQq correlations are valid for structures with different patterns, where the basic element is represented by the coordination complex RX_3. For example, the series antimonite, Sb_2S_3; pyrargyrite, Ag_3SbS_3; bournonite, $CuPbSbS_3$ shows that the eQq values of ^{121}Sb increase progressively from 318.0, 332.3, 376.2 MHz for the series. This can be explained by the growing number of rival "electropositive" atoms of the copper type that influence the bond properties of atom R, and thus affect the quadrupole coupling constant. As electropositive atoms grow more numerous in the lattice, the probability of forming supplementary bonds at the expense of the undivided s-electron pair for R atoms would be decreased. This is why the undivided s-electron pair for R atoms does not participate in bond formation in the sulfosalts, as contrasted to the simple sulfides (Sb_2S_3, etc.) where the undivided s-electron pair is more active in forming donor–acceptor bonds.

Another example of the induction effect is demonstrated by atoms of boron emplaced in the monoclinic lattice of bismuth oxide, α-Bi_2O_3 (52). The effect of these atoms causes a small but discernible frequency shift for the ^{209}Bi nuclei. Boron atoms (with electrophilic properties) tend to arrange the oxygen atoms (donors) about themselves in the same way as in the B_2O_3 structure. This process is accompanied by distortion of the Bi_2O_3 structure and thus by a growth of the asymmetry parameter. Furthermore, the additive boron atoms lower the ion nature of Bi–O bonds. These factors are responsible for resonance shift to lower frequencies.

c. Steric Effect. The steric effect is closely related to induction. For instance, the lattice of pyrargyrite, Ag_3SbS_3, contains an addition of arsenic atoms (55) in the Ag_3SbS_3 matrix. The NQR frequency shift for the additive of As–S bonds in pyrargyrite indicates more ionic nature than for those in a specimen of proustite, Ag_3AsS_3, with the additive atoms forced to accommodate themselves to the position of the matrix atoms.

The NQR data demonstrate that the usual valence assignments to atoms in minerals are by no means a complete representation of the real structure. A representation of Sb^{3+} and As^{3+} in antimonite, orpiment, etc. is inaccurate, to say the least.

2. ORDER–DISORDER PHENOMENA

The high sensitivity of NQR frequency shifts due to different structures lends itself to the study of fine features of minerals of order–disorder phenomena such as degree of distortion in the coordination polyhedrons, the nature of flaws and their distribution throughout the lattice, the nature of structural additions and their interaction with the lattice, the dynamics of the lattice, etc. These parameters are not always determined with the desired accuracy by x-ray diffraction because the data are usually an average for both standard and distorted unit cells. For instance, x-ray identifies the spatial symmetry group in bournonite, but indicates that all of the polyhedrons are distorted to the same extent as implied by the interatomic distances. However, according to NQR data (51), one of the polyhedrons is almost ideally symmetrical ($\eta < 0.2\%$), while another is appreciably distorted ($\eta = 22.8\%$).

The high sensitivity of NQR for crystallochemical effects has been used to establish the structure of arsenic selenide (As_2Se_3) (53). Arsenic atoms occupy four nonequivalent positions in the As_2Se_3 cell. The two principal frequencies (79.80 and 82.38 MHz) turned out to be close to that for orpiment, with two supplementary lines 0.34 and 0.73 MHz away from the second principal line. Consequently arsenic selenide and orpiment are not strictly isostructural, as indicated by x-ray data. As revealed by NQR, symmetry of the spatial group for As_2Se_3 is lower (Pc) than for As_2S_3 ($P2_1/c$). The NQR spectrum of stephanite (Ag_5SbS_4) shows that all Sb atoms in the lattice of this mineral are crystallochemically nonequivalent. Three S atoms are situated in the immediate vicinity of the Sb atoms, with none of the Sb–S distances shorter than the usual equivalent distances attributed to the substantially covalent structures. The SbS_3 polyhedron is appreciably distorted ($\eta = 10.7\%$). The Sb–S bonds in stephanite appear to be more polar than in antimonite and pyrargyrite, and are close to those in bournonite. The higher eQq value for stephanite, as contrasted with pyrargyrite, is associated with a higher coordination number of the Ag atom.

In recent years structural states of the order–disorder type have been intensively studied in mineralogy and crystallochemistry by nuclear magnetic resonance (25). Fewer but similar results have been obtained by NQR. A widening of the resonance line is a manifestation of disordered elements in the minerals. The degree of perfection for the crystal structure determines the width of the resonance line. Flaws in the lattice cause widening of the resonance peak, dislocations being the most effective because deformations caused by them are more conspicuous than, for example, point flaws. In low-symmetry crystals, characterized by the presence of several nonequivalent positions of R atoms, dislocation is confined chiefly to the vicinity of nuclei with a higher value of the asymmetry parameter (52, 54).

The possibility of observing NQR spectra of additive atoms in the corresponding matrices depends on their concentration. The latter directly determines the intensity of the signals, as discussed in Section VI.D on quantitative analysis. Detecting the resonance of nonstructural impurities such as mechanical inclusions is not difficult, as long as such impurities are present in sufficient amounts. Obviously the NQR spectrum of nonstructural impurities would be identical with that of a standard specimen of the compound under investigation. The detection of structural impurities is much more complicated. These distort the matrix lattice system, either wholly or locally. Two examples are isomorphous mixtures and phases of emplacement. Proustite (Ag_3AsS_3) and pyrargyrite ($AgSbS_3$) exhibit restricted isomorphism, and the natural specimens usually contain a small number of isomorphous atoms. The NQR of a specimen of natural pyrargyrite containing 2% As gave a ^{75}As signal at 67.58 MHz (at 77°K) as compared to a ^{75}As signal of 67.311 MHz for proustite (55). The slight frequency shift indicates the similar symmetry for the closest environment of the ^{75}As nuclei—a phenomenon characteristic of isomorphous systems. With a symmetry deviation the frequency difference would have been considerably greater. Difference in the size of As and Sb atoms must be reflected in the structure of Ag_3SbS_3 matrix itself. This is obvious from the fact that the width of resonance lines obtained from additive nuclei ^{75}As is over 300 KHz compared to a ^{75}As line width of 24 KHz in a standard specimen of proustite. α-Bi_2O_3 alloyed with boron (52) demonstrates the effect of atom emplacement on the width of the NQR signal. Boron atoms bring about a marked widening of the lines, mostly for a single position of the Bi atoms, corresponding to the higher value of the asymmetry parameter ($\eta = 38\%$). This demonstrates the selective nature of impurity distribution within the α-Bi_2O_3 matrix. Because dislocations interact with impurities and create concentration clouds, a definite correlation exists between the asymmetry parameter and the distribution of the impurity.

Temperature relationships between the spin–lattice relaxation (T_1) and the resonance frequencies yield information on the dynamics of individual functional groups of atoms in mineral structures. For example, in antimonite, one of the R–X bonds in SbS_5 ($\eta = 38.1\%$) is 2.49 Å long, appreciably shorter than the ordinary bond. The value of T_1 for this type of complex is approximately three times that for the symmetrical complexes ($\eta = 0.8\%$) (61).

It follows from the above discussion that the NQR method opens up new approaches to the study of fine features of the chemistry and structure of minerals. In its chemical aspect this method affords a means for evaluating the distribution of electron density within a functional grouping of atoms such as in the molecule and coordination complex; studying chemical shifts due to chemical and structural differences, including the very slight ones from the effect of crystalline impurities; studying the trend of changes in chemical bonds; and determining the effect of long-range coordination bonds on the distribution of charges within the region of the nucleus. In its structural aspect the NQR method is successful in determining the degree of perfection for the structure; the presence of impurity elements; disorder and especially dislocations and the associated structural distortions; the nature of structural additions (isomorphous and emplacement phases); phase transformations; and dynamics of the lattice. Also, it should be noted that NQR can be used as a rapid method for mineral identification, including quantitative analysis.

D. *Quantitative Analysis*

Correlation charts are valid for relating NQR frequencies with the structure of compounds (*6, 14*). For the structure determination of unknown compounds the NQR spectra can be interpreted by their frequency position and ratios of the observed peak.

The techniques for quantitative spectroscopic analysis by NQR have recently been established (*64*). The position of the sample in the rf coil is important for obtaining the maximum sensitivity on a given nuclide, the optimum position being the center of the rf oscillator coil, although a weak signal can be obtained on that portion of the sample along the axis immediately outside of the coil area (Fig. 4). It should be noted that only the region corresponding to the volume of the rf coil gives a nearly linear response to the isotope under investigation. In any case, for a given quantitative analysis all samples must occupy a given volume within the same portion of the rf coil,

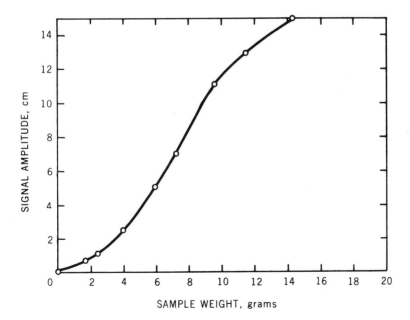

FIG. 4. Effect of quantity of sample in rf coil. [Reprinted from *Anal. Chem.* **41**, 661 (1969). Copyright by the American Chemical Society. Reprinted by permission of the copyright owner.]

which may require blending of the sample under investigation with a diluent that does not react chemically under these conditions nor have an NQR signal in that portion of the spectrum.

By plotting concentration versus signal amplitude of a compound containing an active NQR isotope one obtains linear calibration curves comparable to other spectroscopic techniques in linearity, reproducibility, etc. Figure 5 shows a plot for various concentrations of Cu_2O ([65]Cu at 26.5 MHz) using $HgCl_2$ as the diluent. For each sample representing a point on the calibration line the height of the total sample corresponded to the height of the rf coil. The signal peak amplitude was taken as the height of the center line. Similarly calibration lines have been constructed for five different isotopes ([75]As in As_2O_3, [63]Cu and [65]Cu in Cu_2O, [35]Cl in $HgCl_2$, [209]Bi in $BiCl_3$). Two different isotopes are in equivalent environments ([63]Cu and [65]Cu), and one isotope is in two nonequivalent environments ([35]Cl in $HgCl_2$). Area measurements obtained on perimeter tracings of the hand-drawn envelope enclosing the center line and sideband responses of the resonance signal gave calibration lines of less reproducibility and linearity than peak-height measurements.

Synthetic mixtures of $BiCl_3$ ([209]Bi at 37.5 MHz) in KCl and of $HgCl_2$ ([35]Cl at 22.4 and 22.5 MHz) in NaCl were analyzed according to the condi-

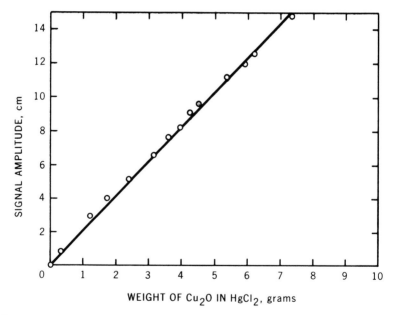

FIG. 5. Calibration curve for [65]Cu (24.6 MHz) in Cu_2O. [Reprinted from *Anal. Chem.* **41**, 661 (1969). Copyright by the American Chemical Society. Reprinted by permission of the copyright owner.]

tions stated above. In each case the diluent was a different compound than that used to plot the calibration line. Since the amplitude of the resonance signal corresponds to the weight of the compound in the sample, one can calculate total weight percentage of the compound under investigation if the total weight of the sample is known. Table VI summarizes the results on synthetic mixtures.

Several different samples of the minerals cuprite and arsenolite were analyzed for Cu_2O and As_2O_3 content, respectively. A sample vial was filled with small pieces of an ore sample to a line representing the calibration volume, and the height of the center line of the resonance peak was measured as a determination of weight Cu_2O or As_2O_3 in the cuprite ore. Fifteen analyses of five different cuprite ores were performed for Cu_2O content by this method using the 26.5-MHz resonance of ^{63}Cu. Duplicate analyses on these samples gave agreement ranging from 1.1 to 6.7%. In seven representative samples the 24.6-MHz values were also determined and gave good agreement with the 26.5-MHz values (within 5%). Similarly six analyses were performed on three different arsenolite minerals. The results are listed in Table VII.

As part of the procedure to establish sampling techniques for quantitative NQR analysis a detailed grind study of inorganics and minerals was undertaken. There is a decrease of the ^{63}Cu signal at 26.5 MHz in both cuprite ore and synthetic Cu_2O as the result of grinding in a tungsten carbide vial. The NQR signal is completely destroyed after 20 min of grinding (Fig. 6).

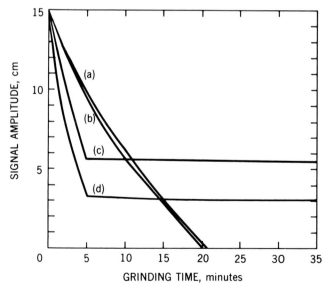

FIG. 6. Effect of tungsten carbide grinding on NQR signal: (a) cuprite ^{63}Cu at 26.5 MHz; (b) Cu_2O ^{63}Cu at 26.5 MHz; (c) $HgCl_2$ ^{35}Cl at 22.5 MHz; (d) $HgCl_2$ ^{35}Cl at 22.4 MHz.

TABLE VI

QUANTITATIVE NQR ANALYSIS OF SYNTHETIC MIXTURES

Wt of $BiCl_3$ in KCl (g)	Total wt of mixture (g)	$BiCl_3$ (wt%)	Wt of $BiCl_3$ (g) from 37.5-MHz calibration curve	$BiCl_3$, (wt%) from 37.5-MHz calibration curve	Difference (wt%)
1.20	4.23	28.4	1.05	24.8	+3.6
2.56	4.86	52.7	2.63	54.1	−1.9
3.77	5.20	72.5	3.82	73.5	−1.0

Wt of $HgCl_2$ in NaCl (g)	Total wt of mixture (g)	$HgCl_2$ (wt%)	Wt of $HgCl_2$ (g) from calibration curves		$HgCl_2$ (wt%) from calibration curves		Difference (wt%)	
			22.4 MHz	22.5 MHz	22.4 MHz	22.5 MHz	22.4 MHz	22.5 MHz
1.04	3.55	29.3	1.10	1.00	31.0	28.2	−1.7	+1.1
2.41	4.73	51.0	2.45	2.30	51.8	48.6	−0.8	+2.4
3.92	5.07	77.3	3.90	3.87	76.9	76.3	+0.4	+1.0

TABLE VII

QUANTITATIVE ANALYSIS OF MINERALS BY NQR

Weight of	Weight of Cu_2O	Weight-percent Cu_2O in cuprite	Weight of As_2O_3	Weight-percent As_2O_3

Sample	Source	sample (g)	in cuprite (g) (from 26.5 MHz)	26.5 MHz	24.6 MHz	in arsenolite (from 116 MHz)	in arsenolite from 116 MHz
Cuprite	Butte, Montana No. 1	5.01	1.00	20.0	—	—	—
Cuprite	Butte, Montana No. 1	4.58	1.15	25.1	29.5	—	—
Cuprite	Butte, Montana No. 1	3.07	1.00	32.6	33.9	—	—
Cuprite	Butte, Montana No. 2	4.57	1.70	37.1	—	—	—
Cuprite	Bisbee, Arizona No. 1	4.03	0.90	20.9	21.2	—	—
Cuprite	Bisbee, Arizona No. 1	4.53	0.65	14.2	—	—	—
Cuprite	Bisbee, Arizona No. 2	6.52	6.50	99.7	99.5	—	—
Cuprite	Bisbee, Arizona No. 2	6.38	6.40	100.3	—	—	—
Cuprite	Ajo, Arizona	6.89	7.00	101.6	99.1	—	—
Cuprite	Ajo, Arizona	6.89	7.05	102.3	—	—	—
Cuprite	Ajo, Arizona	6.79	6.95	102.4	—	—	—
Cuprite	Ajo, Arizona	6.79	7.05	103.8	—	—	—
Cuprite	Ajo, Arizona	5.70	5.60	98.2	99.4	—	—
Cuprite	Burra, Australia	5.63	1.45	25.6	—	—	—
Cuprite	Burra, Australia	6.21	1.50	24.2	24.8	—	—
Arsenolite	Nevada, No. 1	4.43	—	—	—	1.011	22.8
Arsenolite	Nevada, No. 1	5.17	—	—	—	1.20	23.3
Arsenolite	Nevada, No. 2	5.25	—	—	—	2.804	53.4
Arsenolite	Nevada, No. 2	6.28	—	—	—	3.28	52.2
Arsenolite	Nevada, No. 3	4.57	—	—	—	1.109	21.9
Arsenolite	Nevada, No. 3	5.62	—	—	—	1.26	22.5

This is attributed to the alteration of the crystal lattice system by introduction of dislocations into the lattice system, which causes a random distribution of the electric-field gradient q at the ^{63}Cu site in the unit cell. In support of this proposition x-ray diffraction patterns of these same samples showed broadening in the back-reflection region, in contrast to the diffraction pattern of unground cuprite and synthetic Cu_2O. However, if hand grinding with an agate mortar and pestle or agate-vial grinding is employed, no reduction in signal intensity is observed. When $HgCl_2$ is subjected to tungsten carbide grinding, the signal intensity is reduced by approximately 67%, as compared to the original unground signal, in 5 min, and no further loss in signal stength is detected after 35 min of grinding. Apparently, an equilibrium condition between the ordered structure and the disordered system is reached in 5 min, and not enough energy is generated to further change this equilibrium after 35 min of tungsten carbide grinding. Again, agate-vial and hand grinding result in no loss of signal intensity. If grinding is to be part of the analytical procedure, then the effect of the method of grinding on the peak amplitude of the specific compound under investigation should be established before quantitative analysis is attempted. Some ore samples must not be subjected to vigorous grinding before analysis. This presents no analytical problems because an average value can be determined from several different selections of small pieces of ore, or a mild grinding method can be employed. The former is demonstrated for five different selections of small pieces of high-purity cuprite from Ajo, Arizona. All samplings consistently gave values close to 100% Cu_2O. In this regard, NQR analysis is faster and much simpler than infrared analysis, in which samples must be ground until reproducible absorptivities are obtained. Although additional data are badly needed, the results summarized here demonstrate that quantitative NQR analysis of inorganic compounds and minerals gives reproducible results that are comparable to those of other spectroscopic methods in accuracy, and has considerable promise for future development.

References

1. Allred, A. L., and Rochow, E. G., *J. Inorg. Nucl. Chem.* **5**, 269 (1958).
2. Andrew, E. R., "Nuclear Magnetic Resonance." Cambridge Univ. Press, London and New York, 1955.
3. Barnes, R. G., and Engardt, R., *J. Chem. Phys.* **28**, 731 (1958).
4. Barnes, R. G., and Segel, S. L., *Phys. Rev. Letters* **3**, 462 (1959).
5. Bersohn, R., and Shulman, R. G., *J. Chem. Phys.* **45**, 2298 (1966).
6. Brame, E. G., *Anal. Chem.* **39**, 918 (1967).
7. Bridenbaugh, P. M., and Peterson, G. E., *Rev. Sci. Instr.* **36**, 702 (1965).
8. Buchta, J., *Rev. Sci. Instr.* **29**, 55 (1958).

9. Casabella, P. A., and Bray, P. J., *J. Chem. Phys.* **28**, 1182 (1958).
10. Casabella, P. A., Bray, P. J., and Barnes, R. G., *J. Chem. Phys.* **30**, 1393 (1959).
11. Cohen, M., *Phys. Rev.* **96**, 1278 (1954).
12. Dailey, B. P., *J. Chem. Phys.* **33**, 1641 (1960).
13. Dailey, B. P., and Townes, C. H., *J. Chem. Phys.* **23**, 118 (1955).
14. Das, T. P., and Hahn, E. L., "Nuclear Quadrupole Resonance Spectroscopy." Academic Press, New York, 1958.
15. Dean, C., and Pollak, M., *Rev. Sci. Instr.* **29**, 630 (1958).
16. Dehmelt, H., and Krüger, H., *Naturwissenschaften* **37**, 11 (1950).
17. Dinish, and Narasimhan, P. T., *J. Chem. Phys.* **45**, 2170 (1966).
18. Drago, R. S., *J. Inorg. Nucl. Chem.* **15**, 237 (1960).
19. Fedin, E. I., and Kitaigorodskii, A. I., *Kristallografiya* **6**, 3 (1961).
20. Fuke, T., *J. Phys. Soc. Japan* **18**, 1154 (1963).
21. Fuke, T., and Koi, V., *J. Chem. Phys.* **29**, 973 (1958).
22. Gordy, W., *Discussions Faraday Soc.* **19**, 14 (1955).
23. Graybeal, J. D., and Cornwell, C. D., *J. Phys. Chem.* **62**, 483 (1958).
24. Gutowsky, H., and Woessner, D., *J. Chem. Phys.* **27**, 1072 (1957).
25. Hafner, S. S., and Brinkmann, D., Applications of NMR/NQR in Mineralogy, "Resonance Spectroscopy of Mineralogy," pp. HB1–12. Am. Geological Inst., Washington, D.C., 1968.
26. Hamlen, R. P., and Koski, W. S., *J. Phys. Chem.* **25**, 360 (1956).
27. Hartmann, H., and Sillescu, H., *Theoret. Chim. Acta* **2**, 371 (1964).
28. Hopkins, N., *Rev. Sci. Instr.* **20**, 401 (1949).
29. Ikeda, R., Nakamura, D., and Kubo, M., *J. Phys. Chem.* **72**, 2982 (1968).
30. Ito, K., Nakamura, D., Ito, I., and Kubo, M., *Inorg. Chem.* **2**, 690 (1963).
31. Ito, K., Nakamura, D., Kurita, Y., Ito, K., and Kubo, M., *J. Am. Chem. Soc.* **83**, 4526 (1961).
32. Jorgensen, C. K., Pappalardo, R., and Schmidtke, H. H., *J. Chem. Phys.* **39**, 1422 (1963).
33. Kitaigorodskii, A. I., and Fedin, E. I., *Dokl. Akad. Nauk SSSR* **130**, 1005 (1960).
34. Kojima, S., and Tsukada, K., *J. Phys. Soc. Japan* **10**, 591 (1955).
35. Kushida, T., Benedek, G. B., and Bloembergen, N., *Phys. Rev.* **104**, 1364 (1956).
36. Livingston, R., *J. Phys. Chem.* **57**, 496 (1953).
37. Livingston, R., and Zeldes, H., *J. Chem. Phys.* **26**, 351 (1957).
38. Low, W., "Paramagnetic Resonance in Solids". Academic Press, New York, 1960.
39. Mays, J. M., and Dailey, B. P., *J. Chem. Phys.* **20**, 1695 (1952).
40. Morino, V., and Toyama, M., *J. Chem. Phys.* **35**, 1289 (1961).
41. Nakamura, D., *Bull. Chem. Soc. Japan* **36**, 1662 (1963).
42. Nakamura, D., Ito, K., and Kubo, M., *Inorg. Chem.* **1**, 592 (1962),
43. Nakamura, D., Ito, K., and Kubo, M., *Inorg. Chem.* **2**, 61 (1963).
44. Nakamura, D., Ito, K., and Kubo, M., *J. Am. Chem. Soc.* **84**, 163 (1962).
45. Nakamura, D., Kurita, Y., Ito, K., and Kubo, M., *J. Am. Chem. Soc.* **82**, 5783 (1960).
46. Narath, A., *Phys. Rev.* **140**, A552 (1965).
47. Negita, H., Casabella, P. A., and Bray, P. J., *J. Chem. Phys.* **31**, 730 (1959).
48. Nicholson, W. J., and Burns, G., *Phys. Rev.* **133**, 1568 (1964).
49. Pauling, L., "The Nature of the Chemical Bond." Cornell Univ. Press, Ithaca, New York, 1960.
50. Pen'kov, I. N., and Safin, I. A., *Dokl. Akad. Nauk SSSR* **153**, 692 (1963).
51. Pen'kov, I. N., and Safin, I. A., *Dokl. Akad. Nauk SSSR* **161**, 1404 (1965).
52. Pen'kov, I. N., and Safin, I. A., *Fiz. Tverd. Tela* **7**, 190 (1965).

53. Pen'kov, I. N., and Safin, I. A., *Kristallografiya* **168**, 1148 (1966).
54. Pen'kov, I. N., and Safin, I. A., *Proc. Acad. Sci. USSR, Phys. Chem. Sect. (English Transl.)* **156**, 459 (1964).
55. Pen'kov, I. N., and Safin, I. A., *Soviet Phys.—Solid State (English Transl.)* **6**, 1957 (1965).
56. Peterson, G. E., and Bridenbaugh, P. M., *Rev. Sci. Instr.* **35**, 698 (1964).
57. Pound, R. V., and Knight, W., *Rev. Sci. Instr.* **21**, 219 (1950).
58. Ramsey, N. F., "Molecular Beams." Oxford Univ. Press (Clarendon), London and New York, 1956.
59. Randall, I., Molton, W., and Ard, W., *J. Chem. Phys.* **31**, 730 (1959).
60. Reddoch, A. H., *J. Chem. Phys.* **35**, 1085 (1961).
61. Safin, I. A., and Pen'kov, I. N., *Proc. Acad. Sci. USSR, Phys. Chem. Sect. (English Transl.)* **147**, 815 (1962).
62. Sasane, A., Nakamura, D., and Kubo, M., *J. Phys. Chem.* **71**, 3249 (1967).
63. Schawlow, A. L., *J. Chem. Phys.* **22**, 1211 (1954).
64. Schultz, H. D., and Karr, C., Jr., *Anal. Chem.* **41**, 661 (1969).
65. Sharma, R. R., and Das, T. P., *J. Chem. Phys.* **41**, 3581 (1964).
66. Shpigel, I., Raizer, M., and Myae, E., *Zh. Tekhn. Fiz.* **27**, 387 (1957).
67. Smith, G. W., *Phys. Rev.* **149**, 346 (1966).
68. Smith, J. A. S., and Tong, D. A., *J. Sci. Instr.* **1**, 8 (1968).
69. Sternheimer, R. M., *Phys. Rev.* **95**, 736 (1954).
70. Townes, C. H., and Dailey, B. P., *J. Chem. Phys.* **17**, 782 (1949).
71. Townes, C. H., and Schawlow, A. L., "Microwave Spectroscopy." McGraw-Hill, New York, 1955.
72. Viswanathan, T. L., Viswanathan, T. R., and Sane, K. V., *Rev. Sci. Instr.* **39**, 472 (1968).
73. Volpicelli, R. J., Rao, B. D. N., and Baldeschwieler, J. D., *Rev. Sci. Instr.* **36**, 150 (1965).
74. Wang, T. C., *Phys. Rev.* **99**, 566 (1955).
75. Watkins, G. D., and Pound, R. V., *Phys. Rev.* **85**, 1062 (1952).
76. Wertheim, G. K., "Mössbauer Effect: Principles and Applications." Academic Press, New York, 1964.
77. Whitehead, J. R., "Super-Regenerative Receivers." Cambridge Univ. Press, London and New York, 1950.

MÖSSBAUER SPECTROSCOPY AND ITS APPLICATIONS TO INORGANIC CHEMISTRY

H. B. Mathur

NATIONAL CHEMICAL LABORATORY
POONA, INDIA

I. Introduction

The discovery of recoil free emission and absorption of gamma rays by Rudolf Mössbauer (*61*) in 1958 led to a new and rapidly expanding field of scientific investigation that has produced a great deal of information on several scientific disciplines. This phenomenon, which is now called the Mössbauer effect, has proved to be a powerful tool in the hands of the physicist for studies on diverse problems, such as the measurement of red shift in an accelerated system (36), determination of nuclear moments (30, 46), estimation of nuclear isomeric lifetimes (62), estimation of local internal magnetic fields (3, 57), and determination of nuclear quadrupole (10) and Zeeman splittings (20). Soon after 1961 an increasing number of chemists realized the potential of this new technique in the studies on chemical bonding (21), crystal structure (22), ionic states (89), electron density (67), and magnetic and several other properties (27). Mössbauer spectroscopy is now accepted as a complementary tool to other established spectroscopic techniques such as EPR, NMR, and NQR. A number of reviews (9, 19, 63), books (25, 27, 83), and conference proceedings (4, 13) are now available to the interested reader. In the present chapter, therefore, no attempt is made either to review the current literature on the subject or to discuss in detail the basic physics of the Mössbauer effect. We shall start with a discussion of the phenomenon of resonant absorption of photons leading to Mössbauer's discovery and limit ourselves to only a brief description of some of the theoretical aspects that are of relevance to inorganic chemists. Greater attention will be paid to the discussion of the types of problems the inorganic chemists can fruitfully handle by the use of Mössbauer spectroscopy. This will be done by means of selected examples from the literature on the subject.

II. Resonant Absorption of Photons and Mössbauer's Discovery

The phenomenon of resonance in optics was first demonstrated by Wood (88) in 1904 with the help of sodium D lines. The possibility of observing a similar phenomenon in gamma rays was suggested by Kuhn (51) in 1929, but such resonance remained undetected for the next 20 years because the recoil energy associated with the emission and the absorption of gamma rays was in most cases very large as compared to the gamma-ray width. The natural line width Γ and the lifetime τ of an excited state are related by the Heisenberg uncertainty relation:

$$\Gamma\tau = \frac{h}{2\pi} = 1.05 \times 10^{-27} \quad \text{erg sec} \tag{1}$$

where h is Planck's constant. The resonance excitation probability $W(E)$ at any energy E is given by the Breit–Wigner dispersion formula (11)

$$W(E) = \frac{1}{1 + [(E - E_0)/(\Gamma/2)]^2} \tag{2}$$

which has been normalized so that $W(E_0) = 1$, where E_0 is the exact resonance energy or the energy of the excited state. A plot of $W(E)$ versus E has a Lorentzian shape (Fig. 1) and shows that the resonance probability $W(E)$ drops sharply as the difference between W and W_0 increases. At $W_0 - W = \Gamma/2$, $W(E) = \frac{1}{2}$; and at $W_0 - W = \Gamma$, $W(E) = \frac{1}{5}$.

The energy dependence of the cross section σ for resonant absorption can also be found from Eq. (2) by the expression

$$\sigma = \sigma_0 W(E) \tag{3}$$

where

$$\sigma_0 = \frac{2I_e + 1}{2(2I_g + 1)} 4\pi\lambda_0{}^2 \tag{4}$$

I_g and I_e are the spins of the ground and the excited states and λ_0 is the wavelength at resonance.

When an isolated atom of mass m emits gamma radiation, the atom recoils with the conservation of momentum so that the energy E_γ' of the emitted gamma ray is less than E_0 by an amount E_R, the recoil energy, which is given by the expression

$$E_R = \frac{1}{2} mv^2 = \frac{p^2}{2mc^2} = \frac{E'^2}{2mc^2} \tag{5a}$$

where c is the velocity of light, and v and p are the velocity and the momentum of the recoiling nucleus. E_R is very small as compared to E_γ' or E_0. Hence Eq. (5a) may be rewritten as

$$E_R = \frac{E_\gamma'^2}{2mc^2} \simeq \frac{E_0{}^2}{2mc^2} \tag{5b}$$

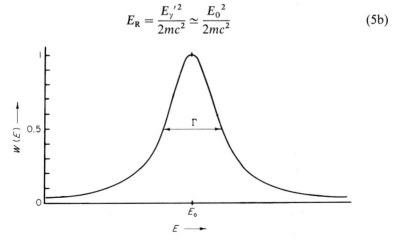

FIG. 1. Excitation probability $W(E)$ as a function of the energy E of resonance radiation.

The energy E_γ'' of the gamma radiation that is absorbed would also have to be greater than E_0 by an amount E_R, since the recoil energy is also involved in the absorption process. The positions of the maxima of the emitted and the absorbed radiations would be separated in energy by $2E_R$. If $2E_R \ll \Gamma$, the two curves would overlap, as indeed happens in optical transitions. One therefore observes resonant absorption. Since E_R is proportional to the square of the transition energy (Eq. 5b), it increases by a very large amount as we go from the optical region (~ 2 eV) to the region of the gamma radiation (~ 10 keV). There is thus no overlap of the emitted and the absorbed gamma-ray lines. Hence no resonant absorption can take place. Table I gives the

TABLE I

COMPARISON OF THE RECOIL ENERGIES IN THE ATOMIC AND
NUCLEAR TRANSITIONS

	Visible sodium D line	Gamma-ray transition in	
		^{57}Fe	^{119}Sn
E_0 (eV)	2.1	14.4×10^3	23.8×10^3
E_R (eV)	1.13×10^{-8}	1.9×10^{-3}	2.6×10^{-3}
Γ (eV)	6.6×10^{-8}	4.55×10^{-9}	2.4×10^{-8}
E_0/Γ	3.18×10^8	3.17×10^{12}	9.9×10^{11}
E_R/Γ	1.71×10^{-1}	4.2×10^5	2.8×10^6

values of the recoil energies for the electromagnetic radiations in the visible and the gamma-ray regions. Figure 2 illustrates the two cases. It may, however, be mentioned that on account of thermal motion the width of the gamma-ray emission and absorption lines can no longer be characterized by the natural line width Γ, but by the Doppler width D, given by the relation (26)

$$D = \frac{1}{c}\left(\frac{2kT}{m}\right)^{1/2} E_0 \tag{6}$$

where k is the Boltzman constant and $D > \Gamma$. Therefore, in order to observe resonance, the requirement now would be $2E_R \ll D$.

Mössbauer (61, 62) showed that when the emitting and the absorbing nuclei are bound in a crystal lattice, Eq. (5), derived for an isolated atom, is no longer valid. The recoil momentum is taken up by the crystal as a whole, and the mass m of the isolated atom should be replaced by the effective mass M of the whole crystal. The crystal lattice is a quantized system and hence cannot be excited in any arbitrary manner. The recoil energy, when taken up

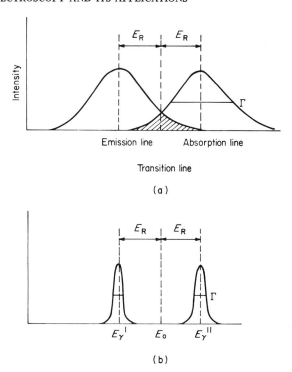

Fig. 2. Schematic representation of (a) overlapping of emission and absorption lines in optical transition, (b) absence of overlapping of the emitted and absorbed gamma-ray lines on account of the recoil loss in free atoms.

by the crystal lattice, can only be reirradiated in the form of collective excitation quanta or phonons. If the recoil energy is smaller than the phonon energy, it is possible than no phonons will be emitted, and the recoil momentum equal to the momentum of the gamma quantum will be taken up by the whole crystal lattice. The recoil energy E_R then becomes much less than D and thus satisfies the necessary condition for resonance absorption.

In terms of the Debye model of solids, the probability f of obtaining no phonon events can be calculated from the expression (83)

$$f = \exp\left[-\frac{E_R}{k\theta_D}\left(\frac{3}{2} + \frac{\pi^2 T^2}{\theta_D^2}\right)\right] \qquad (T \ll \theta_D) \qquad (7)$$

where θ_D is the characteristic Debye temperature of the solid. The values of f for the 14.4-keV radiation of ^{57}Fe and the 129-keV gamma radiation of ^{191}Ir in their metallic lattices have been reported to be 0.91 and 0.06 at 0°K. It follows from Eq. (7) that the value of the f factor would increase with a decrease in E_0 and T but an increase in θ_D.

At this stage the question naturally arises as to what elements can be studied with the aid of the Mössbauer spectroscopy. The Mössbauer effect has already been observed in 44 nuclides of 33 elements (Fig. 3) and has been predicted for a number of other elements, for which a number of tables (*9, 25*) have been compiled on the basis of the following main requirements for a Mössbauer nuclide:

(1) The lifetime of an excited state should be in the range 10^{-6}–10^{-10} sec. A line with a lifetime longer than 10^{-6} sec gives so narrow a line that its experimental detection becomes difficult. On the other hand, a lifetime shorter than 10^{-10} sec would make the linewidth too broad to be useful for study of hyperfine effects.

(2) The energy of the gamma ray should be in the range 5–150 keV. A gamma ray with an energy lower than 5 keV poses experimental difficulties in its detection due to the background of other higher-energy gamma rays and its large electronic absorption cross section. This limits the use of such gamma radiation to absorbers of very small thicknesses. A nuclear transition

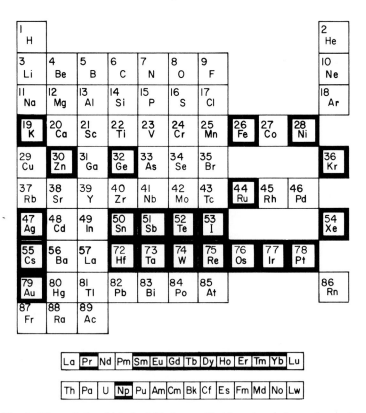

FIG. 3. Elements in which the Mössbauer effect has already been observed.

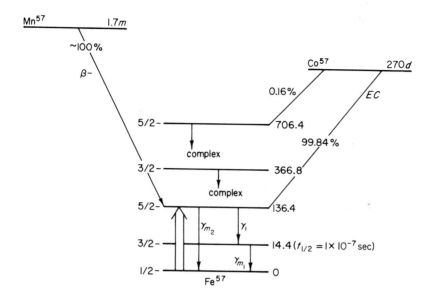

FIG. 4. Decay scheme of 270-day ^{57}Co, showing the energy levels in ^{57}Fe.

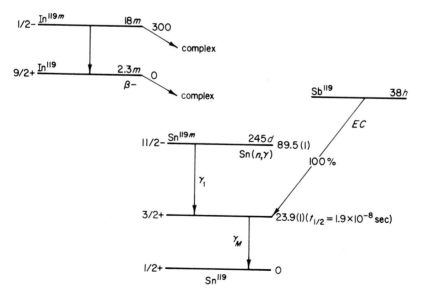

FIG. 5. Decay scheme of 245-day ^{119m}Sn, showing the energy levels in ^{119}Sn.

>150 keV would need to have appreciable recoil effects (Eq. 5) and a very low value of recoilless fraction (Eq. 7) to be useful.

(3) The conversion coefficient for a nuclear transition should be small to ensure that the transition has an appreciable probability of proceeding by photon emission.

(4) The lifetime of the precursor that populates the Mössbauer transition state must be long enough to be used in an experiment lasting several hours. The 270-day ^{57}Co that populates the 14.4-keV Mössbauer transition state in ^{57}Fe is one such example. On the other hand, the 1.7-hr ^{61}Co that populates the 67.4-keV Mössbauer transition (64) in ^{61}Ni would hardly last for one experiment. Some of the nuclides in which Mössbauer-effect experiments have been conducted in detail are ^{197}Au, ^{161}Dy, ^{151}Eu, ^{153}Eu, ^{57}Fe, ^{127}I, ^{129}I, ^{191}Ir, ^{195}Pt, ^{119}Sn, ^{125}Te, ^{169}Tm, ^{182}W, and ^{129}Xe. Of these the largest amount of work has been done on ^{57}Fe and ^{119}Sn, the decay schemes of which are shown in Figs. 4 and 5. We shall therefore confine our discussion to Mössbauer spectroscopic studies using ^{57}Fe and ^{119}Sn, with occasional references to the studies on other nuclides.

III. Experimental Methods

A. Mössbauer Spectrometers

A schematic representation of the Mössbauer spectrometer is shown in Fig. 6. The source is made of a precursor radioactive nuclide that populates the Mössbauer transition state of interest. The absorber is made of the substance to be studied. The detector is a scintillation counter or a proportional counter suitable for measuring the intensity of low-energy gamma rays. The source is moved relative to the stationary absorber when the gamma ray emitted by the source changes in energy by ΔE as a result of Doppler effect so that

$$\Delta E = E \frac{v}{c} \qquad (8)$$

The velocity v of the source is defined by convention to be positive if the source is moved toward the absorber. As noted earlier (see Table I) the gamma-ray linewidths are extremely small, and hence the relative source–absorber velocities required to change the energy of the gamma ray by one linewidth are therefore comparatively small and can be obtained in principle by commonly available mechanical or electromechanical devices. It is sometimes convenient to give a relative velocity to the absorber and keep the

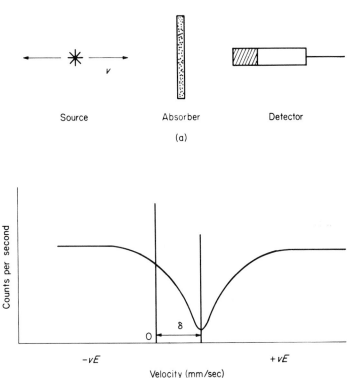

Source   Absorber   Detector

(a)

(b)

FIG. 6. Schematic representation of (a) the main components of a Mössbauer spectrometer, (b) a typical single resonance line Mössbauer spectrum.

source stationary. Most of the Mössbauer experiments have been conducted
with a transmission geometry, but in some cases it is advantageous to measure
the intensity of the gamma rays scattered (*53, 60*) by the absorber. A plot of
the intensity of the gamma rays transmitted through or scattered by the
absorber versus the relative velocity of the source–absorber pair as shown in
Fig. 6b is called the Mössbauer spectrum.

Different types of velocity drives have been used to provide well-defined
relative velocities between the source and the absorber. These may be divided
into two main types.

1. Constant-Velocity Spectrometer

The source or the absorber is moved with a predetermined velocity for a
fixed interval of time. This operation is repeated point by point at different

velocities until the desired velocity interval is covered. Counting time is adjusted to give the desired statistical accuracy. A single-channel analyzer is used to limit the detection to only those pulses that have the required energy. A number of methods of imparting constant velocity, such as eccentric cams, rotating discs, inclined planes, hydraulic pistons, etc., have been described in the literature (13, 25). Such spectrometers, though simple in design and relatively cheap, are subject to the effects of the drift of the electronic system due to the use of a single-channel analyzer.

2. CONSTANT-ACCELERATION SPECTROMETER

In this system a multichannel analyzer is used to cover the entire velocity spectrum, so that any drift in the electronic system is averaged over the whole range of velocities simultaneously. A signal from a function generator produces a linear increase in voltage with time. This voltage is transformed into velocity in the dual voice coil of a loudspeaker, which is used to drive the source relative to the absorber. One voice coil carries the drive signal, and the other is used as the velocity pick-up coil. The signal from the gamma-ray detector as a function of the loudspeaker velocity is measured by using a single-channel analyzer to "command" a multichannel analyzer to measure the velocity at a particular instant. This is done by feeding the amplified output of the velocity pick-up coil to the appropriate place in the analog-to-digital converter in the multichannel analyzer. Therefore, when a gamma ray is detected, it is stored in the channel corresponding to the proper velocity. A number of such constant-acceleration spectrometers have been described in the literature (23, 48, 58, 77, 83).

B. Source and Absorber

A Mössbauer source is prepared by incorporating the precursor radioactive isotope that populates the Mössbauer transition of interest in a suitable crystal lattice such that it gives a large recoil-free fraction and an unsplit line with a width as nearly as possible equal to the natural linewidth. The crystal lattice is not always of the same element as the isotope. For studies of 14.4-keV Mössbauer transition in ^{57}Fe, sources of 270-day ^{57}Co have been prepared by electrodeposition and subsequent diffusion in thin metallic foils, such as 310 stainless steel, Cu, Cr, and Pd, etc. (32, 64). Iron foil is not suitable, as it induces hyperfine splitting. Among the many sources prepared for studies with the 23.8-keV Mössbauer transition in ^{119}Sn, sources made in the form of Pd_3Sn and $BaSnO_3$ have been found most suitable (38).

In a few cases, the low-lying Mössbauer transition states have been induced as a result of nuclear reactions or Coulomb excitation by heavy charged particles. Sources of ^{40}K have been prepared by ^{39}K(n, γ)^{40}K or ^{39}K(d, p)^{40}K reactions (34, 73). The Coulomb excitation of the 14.4-keV excited state in ^{57}Fe and the 67.4-keV excited state in ^{61}Ni has been achieved by irradiation with ^{4}He^{2+} and ^{16}O^{4+} respectively (52, 74).

A good absorber should contain sufficient Mössbauer nuclei in the ground state to give an appreciable resonance absorption, but should avoid excessive loss in intensity of the transmitted gamma ray due to electronic absorption by thick absorbers. A finite thickness of the source or the absorber broadens the resonance line. The shape of the line is modified from a Lorentzian to a Gaussian function. Corrections for such broadening should therefore be made in careful structural analysis (25, 40, 54).

In order to obtain appreciable resonance absorption it is often necessary to cool either the source or the absorber or both. A number of cryostats have been described for this purpose (2, 32, 83).

IV. The Use of Mössbauer Spectroscopy in the Study of Hyperfine Structure

The possibility of using Mössbauer effect as a tool in chemical spectroscopy arose as a result of the finding that the chemical environment of the Mössbauer nucleus has a pronounced effect on the resonant absorption of the gamma ray (26, 27, 83). Extremely small changes are produced in the energy of the gamma ray as a result of the hyperfine interaction between the nucleus and its orbital electrons. These effects, which previously could not be observed by the usual methods, can be now clearly detected owing to the narrow linewidths of the resonance lines in the Mössbauer spectrum. The Mössbauer parameters of greatest interest to a chemist are isomer shift δ, quadrupole splitting ΔE_Q, and magnetic hyperfine splitting.

A. Isomer Shift

The isomer shift δ is defined as the displacement of the centroid of the Mössbauer spectrum from the zero velocity. It arises because the nucleus of an atom is not a point charge but has a finite volume. The electrostatic interaction between the charge distribution of the nucleus and those orbital electrons that have a finite probability of existing at the region of the nucleus, brings about a change in the energy levels of the nucleus. The electrostatic shift δE of a nuclear energy level is given by the relation (83)

$$\delta E = \frac{2\pi}{5} Ze^2 |\psi(0)|^2 R^2 \qquad (9)$$

where $|\psi(0)|^2$ represents the total electron density at the nucleus, R is the radius, and Z is the atomic number of the nucleus.

In practice, only the s electrons have a finite density at the nucleus, and hence $|\psi(0)|^2$ may be approximated with the electron density of the s electrons, $|\psi_s(0)|^2$. Relativistic corrections, however, introduce also a $p^{1/2}$ electron density in a heavy nucleus. Since a nuclear transition involves an excited and a ground state (see Fig. 7), the electrostatic change in the energy of a gamma transition is given by the expression

$$\delta E_{ex} - \delta E_{gd} = \frac{2\pi}{5} Ze^2 |\psi(0)|^2 (R_{ex}^2 - R_{gd}^2) \tag{10}$$

where R_{ex} and R_{gd} are the radii of the excited and ground states respectively. In a Mössbauer experiment, where a source–absorber pair is involved, we measure only a difference in the nuclear electrostatic energy changes of the source and the absorber. The isomer shift is therefore given by the relation

$$\delta = \frac{2\pi}{5} Ze^2 (R_{ex}^2 - R_{gd}^2)\{|\psi(0)|^2_{absorber} - |\psi(0)|^2_{source}\} \tag{11}$$

$$\delta = \frac{4\pi}{5} Ze^2 R^2 \frac{dR}{R} \{|\psi(0)|^2_{absorber} - |\psi(0)|^2_{source}\} \tag{12}$$

where $dR = R_{ex} - R_{gd}$. Here the electron density at the source and the absorber nuclei have been shown separately, since they have different values in

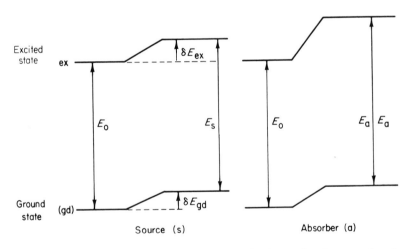

FIG. 7. The origin of isomer shift. The electrostatic interaction between the charge distribution in the nucleus and the extranuclear electrons shifts the nuclear energy levels in the source and the absorber.

dissimilar chemical environments. In practice, the values of the isomer shifts are measured and designated with reference to a standard source or a standard absorber, since Mössbauer experiments are always conducted with a source–absorber pair. The isomer shifts obtained with respect to any one source or the absorber can be converted with respect to any other source or the absorber by suitable correction factors, for which a number of tables have been prepared (*32, 64*). One such table (*37*), prepared by Herber for ^{57}Fe, is Table II. It is not

TABLE II

ISOMER SHIFTS (*37*) FOR SEVERAL ^{57}CO SOURCE LATTICES AND REFERENCE ABSORBERS AT ROOM TEMPERATURE

^{57}Co source or reference absorber	Observed isomer shifts of standard (mm/sec)	Correction factor for	
		Isomer shift relative to $Na_2[Fe(CN)_5NO]\cdot 2H_2O$ (mm/sec)	Isomer shift relative to 310 stainless steel (mm/sec)
Pt	−0.607	0.607	0.425
Cu	−0.483	0.483	0.301
Pd	−0.442	0.442	0.260
Fe	—	0.348	0.166
Fe	—	0.351	0.169
$K_4[Fe(CN)_6]\cdot 3H_2O$	—	0.215	0.033
310 stainless steel	−0.182	0.182	0
302 stainless steel	−0.175	0.175	−0.007
Cr	−0.075	0.075	−0.107
$Na_2[Fe(CN)_5NO]\cdot 2H_2O$	—	0	−0.182

unusual to find that two Mössbauer sources prepared by the same person and by the same method give slightly different values for the isomer shift. It is therefore advisable to find out the correction factor for a new source by actually determining with it the isomer shift for a standard absorber—e.g., $K_4[Fe(CN)_6]\cdot 3H_2O$, 310 stainless steel, $Na_2[Fe(CN)_5(NO)]\cdot 2H_2O$, etc.—with respect to which it is desired to express the isomer shift. The isomer shift depends on a nuclear term dR/R and a chemical term

$$\{|\psi(0)|^2_{\text{absorber}} - |\psi(0)|^2_{\text{source}}\}$$

that gives the difference in the total electron densities at the absorber and source nuclei. If the nucleus in the excited state has a smaller radius than that in the ground state, as happens in ^{57}Fe, dR/R is negative and the values of the isomer shift relative to a fixed standard decrease with an increase in the total electron density at the absorber nuclei. On the other hand, if dR/R is positive, as for ^{119}Sn (8), the values of the isomer shift would increase with an increase in $|\psi(0)|^2$. The importance of knowing the sign of dR/R in correlating the value of δ with $|\psi(0)|^2$ can best be illustrated by the two Mössbauer nuclei of iodine (72). dR/R is $+ve$ for ^{129}I and $-ve$ for ^{127}I. Hence the values of δ with respect to a ZnTe source for ^{127}I and ^{129}I in two chemical compounds of the same nominal composition, $Na_3H_2IO_6$, have been found to be $+1.19$ mm/sec and -3.35 mm/sec respectively. If the sign of dR/R is known, the electron density at the nucleus in a series of absorbers can be compared from the values of the isomer shifts relative to a fixed standard.

B. Quadrupole Splitting

The interaction of the quadrupole moment Q of a nucleus with the electric field gradient (EFG) q at the nucleus manifests itself as the splitting of the resonance spectrum. The nuclear quadrupole moment is a measure of the departure of a nucleus from spherical symmetry. A nucleus with spin I equal to 0 or $\frac{1}{2}$ is spherically symmetric and hence has $Q = 0$. A negative value of Q signifies that the nucleus is an oblate spheroid, while a positive Q shows that it is a prolate spheroid.

The Hamiltonian describing the quadrupole coupling has been described by Abragam, and we simply quote the result (1):

$$\mathcal{H} = \frac{e^2qQ}{4I(2I-1)}[3I_z^2 - I(I+1)] + \eta(I_+^2 - I_-^2) \tag{13}$$

where I_+ and I_- are raising and lowering operators and η is the asymmetry parameter related to the magnitude of the three components of the electric-field gradient, viz.,

$$\eta = \frac{\partial^2 V}{\partial x^2} - \frac{\partial^2 V}{\partial y^2} \bigg/ \frac{\partial^2 V}{\partial z^2} \tag{14}$$

which are usually chosen so that $0 \leqslant \eta \leqslant 1$. The eigenvalues of the above Hamiltonian are given by

$$E_{Q(m_I)} = \frac{e^2qQ}{4I(2I-1)}[3m_I^2 - I(I+1)](1 + \eta^2/3)^{1/2} \tag{15}$$

in which the magnetic quantum number m_I has the values $I, I-1, \ldots, -I$. If $I = \frac{3}{2}$, m_I can have the values $\frac{3}{2}$ and $\frac{1}{2}$. The quadrupole splitting E_Q, as shown in Fig. 8, is given by the difference in the eigenvalues of the above Hamiltonian for the two possible values of m_I—i.e.,

$$\Delta E_Q = E_{Q(3/2)} - E_{Q(1/2)}$$
$$= \tfrac{1}{2}e^2qQ(1 + \eta^2/3)^{1/2} \tag{16}$$

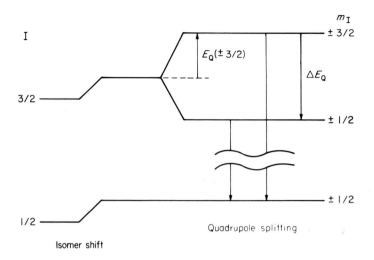

FIG. 8. The origin of quadrupole splitting. In an electric-field gradient the nuclear level with spin $I = \frac{3}{2}$ is split into two sublevels.

If the EFG has an axial symmetry, $\eta = 0$, hence

$$\Delta E_Q = \tfrac{1}{2}e^2qQ \qquad (17)$$

A typical quadrupole split spectrum of ^{57}Fe is shown in Fig. 9, in which the significance of δ and ΔE_Q is also indicated. The quadrupole splitting here gives the product of the nuclear quadrupole moment and the EFG and not its sign because the expression for the eigenvalue (Eq. 15) contains the square of

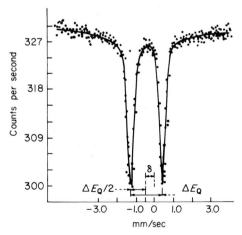

FIG. 9. A quadrupole-split spectrum obtained with a ^{57}Co source in Pd and a $Na_2[Fe(CN)_5NO]\cdot 2H_2O$ absorber at 25°C. The significance of δ and ΔE_Q is indicated.

m_I. Hence the states whose m_I differ only in sign are degenerate. If one of the nuclear levels involved in a Mössbauer transition has spin $> \frac{3}{2}$, there would be more than two lines in the spectrum, and in many cases it is possible to determine the sign of the EFG from an inspection of the asymmetric distribution of the resonance lines in the Mössbauer spectrum, as has been done for KIO_3 where the 27.75-keV Mössbauer transition in ^{129}I takes place from an excited state with spin $\frac{5}{2}$ to the ground state with spin $\frac{7}{2}$ (33). The sign of the quadrupole interaction for levels even with $I = \frac{3}{2}$ can be obtained from the nature of the Mössbauer spectra obtained using single crystals (17) or from the effect of intense magnetic fields on the quadrupole splitting in Mössbauer spectra of polycrystalline samples (12).

C. Magnetic Hyperfine Splitting

The Hamiltonian that describes the magnetic hyperfine interaction between a nucleus and its associated electrons in an atom can be written (50) as

$$\mathcal{H} = -\mu H \tag{18}$$

where μ is the nuclear magnetic dipole moment and H is the magnetic field that the electrons produce at the nucleus. The nuclear magnetic dipole moment is related to the nuclear spin I by

$$\mu = g\beta_n I \tag{19}$$

where g is the gyromagnetic ratio, often called the nuclear g factor, and β_n is the nuclear magneton. Equation (18) can therefore be written as

$$\mathcal{H} = -g_n \beta_n I \cdot H \tag{20}$$

For a nuclear level with $I > 0$, the magnetic hyperfine interaction splits the nuclear levels into $2I + 1$ sublevels whose eigenvalues are

$$E_{m_I} = -g\beta_n H \cdot m_I \tag{21}$$

where m_I, the nuclear magnetic quantum number, has as before the values $m_I = I, I - 1, \ldots, -I$. The above equations show that each nuclear level is split into $(2I + 1)$ equally spaced sublevels. Figure 10 illustrates the magnetic splitting of the nuclear levels in ^{57}Fe. The 14.4-keV gamma transition in ^{57}Fe is a magnetic dipole ($M1$) and subject to the selection rule $\Delta m = 0, \pm 1$. Hence in the presence of an internal magnetic field one can observe only six transitions (see also Fig. 18). The angular dependence of the individual transitions and the various selection rules have been discussed by Wertheim (83). For an axially symmetric EFG tensor with the symmetry axis at an

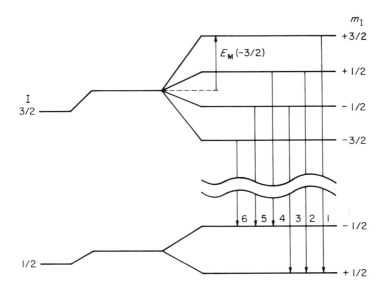

FIG. 10. The origin of magnetic hyperfine splitting of ^{57}Fe. The excited state with $I = \frac{3}{2}$ is split into four and the ground state with $I = \frac{1}{2}$ is split into two substates. The six allowed transitions ($\Delta m = 0, \pm 1$) and their angular dependence are shown. For lines 1 and 6 the angular dependence is $\frac{3}{2}(1 + \cos^2 \theta)$; for lines 3 and 4, $\frac{1}{2}(1 + \cos^2 \theta)$; for lines 2 and 5, $2 \sin^2 \theta$. (See also Fig. 18.)

angle θ with respect to the magnetic axis, the nuclear energy levels are given by the relation (83)

$$E = -g\beta_n Hm_I + (-1)^{|m_I| + 1/2} \frac{e^2 qQ}{4} \frac{(3 \cos^2 \theta - 1)}{2} \qquad (21)$$

if $eqQ/\mu H \ll 1$. The quadrupole splitting in such a case may be expressed as (9)

$$\Delta E_Q = \frac{e^2 qQ}{8I(2I - 1)} [3m_I^2 - I(I + 1)](3 \cos^2 \theta - 1 + \eta \sin^2 \theta \cos 2\phi) \qquad (22)$$

where θ and ϕ are the polar angles that define the direction of the magnetic field H with the crystal axes. If $I = \frac{3}{2}$ and $\eta = 0$, then

$$\Delta E_Q = \frac{e^2 qQ}{8} (3 \cos^2 \theta - 1) \qquad (23)$$

The Mössbauer parameters δ and ΔE_Q, and the effective magnetic field

H can be experimentally determined from the magnetic hyperfine spectrum by the following expressions (46, 65):

$$\delta = \frac{S_1 + S_2 + S_5 + S_6}{4} \tag{24}$$

$$\Delta E_Q = \frac{S_1 + S_6 - S_2 - S_5}{4} \tag{25}$$

and

$$H = -\frac{S_6 - S_1}{5.43}\frac{1}{\mu_0} \tag{26}$$

where $S_1, S_2, \ldots, S_6$ are the positions of the six resonance lines (as shown in Fig. 18) and μ_0 is the magnetic moment of the ground state.

V. Applications in Inorganic Chemistry

We have seen that the hyperfine interactions between the nuclear energy levels and the extranuclear electrons in an atom give rise to a number of parameters in the Mössbauer spectrum. These Mössbauer parameters can give, in favorable cases, a great deal of information on the electronic configuration of the atoms and their local symmetry. We shall now discuss how Mössbauer spectroscopy has been used to obtain information on the nature of bonding between a metal atom and its ligands, the structure of metal–ligand complexes, the identification of different oxidation states of the elements in their compounds, the amount of long- and short-range order in solids, and the magnetic properties of the materials.

A series of Mössbauer experiments with spin-free compounds of iron in different valency states shows a systematic behavior with respect to the isomer shift. Characteristic values of the isomer shift are obtained for different oxidation states, as illustrated in Fig. 11. In a similar manner the isomer shifts of the Sn^{2+}, Sn^{4+}, Sn^0, and the organotin compounds fall in characteristic

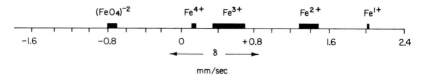

FIG. 11. The isomer shifts relative to 310 stainless steel for spin-free iron in different oxidation states.

regions (*14, 26, 27*). The first attempt to quantitatively relate the ^{57}Fe isomer shift with the details of chemical bonding was that of Walker et al. (*82*). They assumed that the value of the total s-electron density calculated by Watson (*80, 81*) for the free-ion $3d^n$ configuration may be taken to represent the value of $|\psi(0)|^2$ in an ionic salt. Purely ionic salts are, however, rarely found, and there is invariably a greater or smaller tendency towards covalency. It was therefore assumed that for each valency state the salt that gave the most positive isomer shift was completely ionic, and whenever isomer shift was found smaller than the maximum, it was attributed to the effect of the covalency which was represented by the configuration $3d^n 4s^x$. Walker et al. thus constructed for the various $3d^n$ configurations plots of the total s-electron density versus the $4s$-electron density and calibrated the isomer shift in terms of the total s-electron density, using the experimental value of the most positive isomer shift for ionic Fe^{2+} [$FeSO_4 \cdot 7H_2O$] and Fe^{3+} [$Fe_2(SO_4)_3 \cdot 6H_2O$] salts. They were thus able to account for the lower isomer shifts of weakly covalent spin-free compounds of a fixed oxidation state. They also showed that the increase in the total s-electron density (hence a decrease in the isomer shift) in going from the higher oxidation state to the lower oxidation state is due to a decrease in the shielding of the $3s$ electrons as a result of the removal of the d electrons. Danon (*18*) has recalibrated the isomer shift scale in the Walker et al. plot (see Fig. 12), as a result of which many of the discrepancies reported earlier on the application of the plot interpretation have been removed. Mössbauer spectroscopy thus provides a unique method of estimating s-electron covalency in transition-metal complexes, a piece of information valuable in molecular orbital calculations.

It is expected that in ionic salts, since the isomer shift depends upon the extent of $4s$-electron participation in bonding, a correlation should exist between the isomer shift and some bonding parameter such as electronegativity or ionicity. It has been shown that in the FeX compounds, where X is a Group-VIb dianion, the isomer shift increases with the electronegativity of the anion (*66*). In Fig. 13 the increasing isomer shifts of the quadrivalent tin compounds are clearly related to the increasing ionic character and the increasing electronegativity difference (*14*). Similar correlations (*7*) between the ionic character and the isomer shifts of the aurous, ferrous, and ferric compounds derived by Bhide et al. (*7*) are also shown in the same figure.

For spin-paired compounds of iron, matters are more complicated, because here the covalency cannot be described in terms of a single parameter such as the s-electron density. It is necessary to take into account the relative contributions of the s, p, and d wave functions in bonding as well as the delocalization of the electrons from the metal to the ligand orbitals. The isomer shifts of spin-paired compounds of iron are rather independent of the oxidation state and lie in the region of zero velocity relative to stainless steel. Isomer shifts

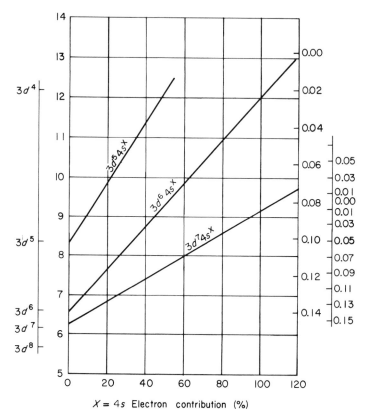

$X = 4s$ Electron contribution (%)

FIG. 12. The values of the isomer shift of ^{57}Fe with respect to 310 stainless steel as a function of $3d$ and $4s$ electron charge densities. The inner scale on the left-hand side gives $[2\sum_{n=1}^{3}|\psi_{ns}(0)|^{2} + x|\psi_{4s}(0)|^{2} - C]$ in units of a_{0}^{-3}. The scales on the right-hand side give δ (cm/sec) w.r.t. for stainless steel: the outer scale gives the original isomer-shift scale of Walker *et al.* (*82*). The inner scale gives the revised calibration proposed by Danon (adapted from Ref. *18*).

(*49*) relative to stainless steel for $K_{4}[Fe(CN)_{6}]\cdot 3H_{2}O$ and $K_{3}[Fe(CN)_{6}]$ are 0.083 ± 0.01 mm/sec and 0 ± 0.01 mm/sec respectively, indicating that the *s*-electron density on the iron atom in the two compounds is nearly the same. Molecular orbital calculations of Shulman and Sugano (*75*) show that as a result of the difference in the back-donation of the electrons from the metal ions in the ferro- and ferricynide complexes, the iron ions are approximately isoelectronic in these two complexes. Hence there is practically no difference in the isomer shifts of $K_{4}[Fe(CN)_{6}]\cdot 3H_{2}O$ and $K_{3}Fe(CN)_{6}$. The additional electron in the ferrocynide complex should therefore spend most of its time on the cynide radicals and not on the iron atom. The Mössbauer studies thus give information on the extent of the back-donation of the electrons from the metal to the

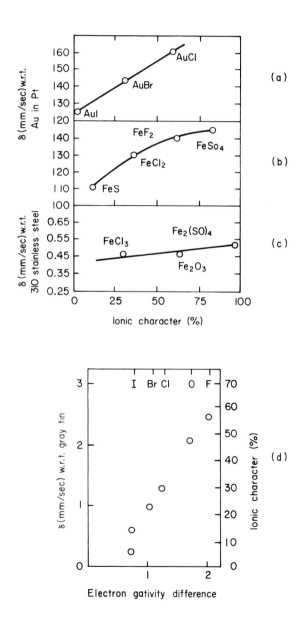

FIG. 13. Curves (a), (b), and (c) show the relationship of isomer shift to ionic character in Au^+, Fe^{2+}, and Fe^{3+} compounds (7). Curve (d) shows how the isomer shift of Sn^{4+} halides increases with ionic character and electronegativity difference (14).

ligand in the low-spin complexes. In a series of hexacyanoferrate derivatives $Fe(CN)_5L^{-n}$ in which a cyano group is replaced by some other ligand L (*17*, *49*), the value of the isomer shift will depend mainly on the π-bond strength of this particular ligand. The withdrawal of d_π electrons into the empty ligand orbitals will decrease the screening of the s electrons and hence decrease the isomer shift. Increasing values of the isomer shift observed for a series of different ligands L, shown below, indicate the decreasing π-bond strength of the ligands:

$$\delta: \quad NO^+ \quad CO \quad CN^- \quad SO_3^{2-} \quad Ph_3P \quad NO_2^- \quad NH_3$$

$$\xleftarrow{\hspace{5cm}}$$
$$\text{Increasing } \pi\text{–bond strength}$$

If the ligands around an atom are in a regular tetrahedral or octahedral arrangement, they have a cubic symmetry and the EFG $q = 0$. Hence the Mössbauer spectrum shows no quadrupole splitting. The presence of quadrupole splitting in a Mössbauer spectrum therefore indicates a significant departure from the cubic symmetry and a finite value of the EFG. For example, both $Na_4[Fe(CN)_6]$, in which Fe is surrounded by six CN^- radicals in a symmetrical octahedral arrangement, and $Sn(C_6H_5)_4$, in which the $C_2H_5^-$ radicals are arranged in the form of a symmetrical tetrahedron, show characteristic single-line Mössbauer spectra (*14*, *49*). Replacement of one of the CN^- radicals by NO^+ or one of the $C_6H_5^-$ radicals by Cl^- destroys the cubic symmetry and we get a well defined doublet in the Mössbauer spectra of $Na_2[Fe(CN)_5NO]\cdot2H_2O$ (Fig. 9) and $(C_6H_5)_3SnCl$ (*14*).

The value of the EFG in a noncubic symmetry may in general be expressed as

$$q = q_{\text{valence}}(1 - R) + q_{\text{lattice}}(1 - \gamma_\infty)$$

where q_{valence} is the EFG due to nonspherically symmetric distribution of the orbital electrons (incompletely filled electron shells) and q_{lattice} is the EFG due to noncubic symmetry of the distant charges on the ligand atoms in the crystal lattice. $(1 - R)$ and $(1 - \gamma_\infty)$ are the Sternheimer antishielding factors (*42*, *83*), which correct for the polarization of the completely filled electron shells in the presence of q_{valence} and q_{lattice} respectively. In compounds of Fe^{3+} ion, which has a spherically symmetric half-filled $3d$ shell, q_{valence} may be neglected; but for ionic compounds of Fe^{2+} ion the major contribution to the EFG would be from q_{valence}. The calculation of q therefore requires information on detailed crystal structure and charges on the ligands in the lattice as well as knowledge of the Sternheimer antishielding factors. This is not always possible. However, the nature of the quadrupole split spectra and a qualitative estimate of the EFG's in a number of related chemical compounds can often be used to advantage in the elucidation of molecular structure.

One of the earliest examples of the application of Mössbauer spectroscopy in the elucidation of molecular structure is that of triiron dodecacarbonyl, $Fe_3(CO)_{12}$ (*39*, *49*). A number of models had been proposed for this molecule. Some of these are shown in Fig. 14. The Mössbauer spectrum of $Fe_3(CO)_{12}$ also shown in the same figure consists of three lines of equal intensity, the central line being slightly broader than the outer two. If all three atoms of iron in $Fe_3(CO)_{12}$ were equivalent, as suggested in the triangular structures (c) and (d) and as originally proposed on the basis of x-ray studies, the Mössbauer spectrum should have either a single line or a doublet and not three lines. The observed Mössbauer spectrum is consistent with the presence of two iron atoms in equivalent positions with an appreciable EFG and one iron atom with a small EFG, which produces only a line broadening. The three-line spectrum is therefore in agreement with either of the two structures (a) and (b). When it was found that the Mössbauer spectrum of $Na^+[Fe_3(CO)_{11}H]^-$ (*22*) is virtually similar to that of $Fe_3(CO)_{12}$, these two structures had to be rejected, since it is not possible to replace one of the CO groups by H^- in any of the above two structures and still keep two iron atoms equivalent. A new structure, shown in (e), was therefore proposed. Here it is possible to replace one of the bridging CO groups by H^- without making the two iron atoms inequivalent. Subsequent x-ray studies have confirmed this structure (*16*). Similarly the Mössbauer spectrum of ferrous formate dihydrate contains four lines (*41*). The position, separation, and intensities of the four lines show that the spectrum may be interpreted as a combination of two doublets, which give nearly the same value of $\delta = 1.6$ mm/sec but different values of the quadrupole splitting, viz., $\Delta E_{Q_1} = 1.48$ mm/sec and $\Delta E_{Q_2} = 3.36$ mm/sec. This shows that Fe^{2+} ion is present at two inequivalent sites in the ferrous formate dihydrate. Crystal-structure studies on the monoclinic form of the transition-metal formate dihydrates confirm that Fe^{2+} ion is present at the center of distorted octahedra of two types. In one case the octahedron is formed by six O^{2-} ions and in the other case by two O^{2-} ions and four H_2O molecules (*41*).

Mössbauer effect studies have been used to obtain information on the cation oxidation state (*89*) and the site distribution (*3*, *90*) in mixed oxide systems, especially those containing cations of variable valency and capable of occupying more than one possible site. This can best be illustrated with the results of the studies reported on the mixed oxide systems with spinel structure (*56*, *59*, *90*). In an ideal spinel structure the anions form a cubic close packing in which the cations partly occupy the tetrahedral and partly the octahedral interstices. The spinel structure may be represented as $A[B_2]O_4$, where A represents the tetrahedral and B the octahedral sites. A series of Mössbauer spectroscopic studies on model cubic spinels so chosen that each contains either an Fe^{2+} or a Fe^{3+} ion at only one of the two possible cation sites show the following:

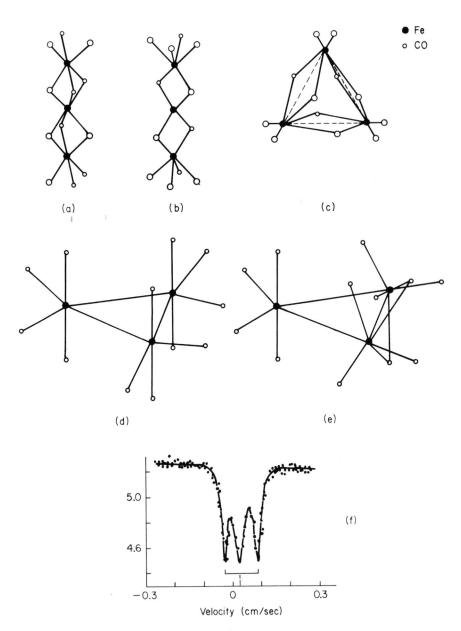

FIG. 14. The Mössbauer spectrum of $Fe_3(CO)_{12}$ at 78°K with a $^{57}Co(Cr)$ source (39), along with some of the proposed structures for $Fe_3(CO)_{12}$. In the structures, solid dots represent Fe, open circles represent CO.

(1) The Mössbauer spectrum of a normal cubic spinel such as $Fe^{2+}[Cr_2^{3+}]O_4$ shows a single resonance line because the Fe^{2+} ions are present at the tetrahedral (A) sites, which have cubic point symmetry. In a regular tetrahedral symmetry the sixth d electron outside the spherosymmetrical d^5 shell has equal probability of occupying any of the two degenerate $dx^2 - y^2$ and dz^2 states. The charge distribution in such a case has a cubic symmetry. The EFG at the nucleus is therefore zero.

(2) Mössbauer spectra of cubic spinels such as $Cd^{2+}[Fe^{3+}]O_4$ and $Ge^{4+}[Fe^{2+}]O_4$ containing either an Fe^{3+} or Fe^{2+} ion at the octahedral (B) sites only show distinct quadrupole splitting. This has been attributed to the presence of a net EFG at the B sites in the spinel structure, on account of the assymmetry of the next nearest neighbors, even though the first six neighbors form a regular octahedron.

(3) Inverse cubic spinels such as $Fe^{3+}[Cr^{3+}Ni^{2+}]O_4$ containing Fe^{3+} ions at the A sites only also give quadrupole-split Mössbauer spectra. This arises as a result of the asymmetric charge distribution on the next nearest twelve B neighbors of each A site, in spite of the fact that the first four O^{2-} neighbors form a regular tetrahedron.

The presence of the quadrupole splitting in the Mössbauer spectra of cubic spinels containing either an Fe^{2+} or an Fe^{3+} ion shows that the absence of long-range order can be detected in an Mössbauer spectrum.

The Mössbauer spectrum of a partially inverse spinel

$$Fe^{2+}_{1-x}Al^{3+}_x[Fe^{2+}_xAl^{3+}_{2-x}]O_4$$

should therefore be a resultant of two quadrupole-split spectra on account of reasons (2) and (3) given above. Here x is defined as the degree of inversion and has a value $0 < x < 1$. The Mössbauer spectrum of this spinel, shown in Fig. 15, can indeed be resolved into two quadrupole-split spectra shown by the continuous and the dotted lines. The outer pair of lines with $\delta = 1.52$ mm/sec and $\Delta E_Q = 2.76$ mm/sec have been attributed to the Fe^{2+} ions at the octahedral sites, and the inner doublet with $\delta = 1.10$ mm/sec and $\Delta E_Q = 1.39$ mm/sec to the Fe^{2+} ions at the tetrahedral sites. If the Debye–Waller factors for the Fe^{2+} ions at the A and B sites are assumed to be equal, the ratio of the areas under the two resolved Mössbauer spectra will give the ratio of Fe^{2+} ions on the two sites. The value of x has thus been found to be 23 % and is in good agreement with the value of 25 % determined by the x-ray diffraction method (90). Mössbauer spectra of the naturally occurring glasses, tektites, indicate they contain iron predominantly as Fe^{2+} and that there is a considerable degree of short-range order in these glassy materials (55). In a similar manner the Mössbauer spectra of the meteorites such as "Plainview," "Holbrook," and "Johnstown" along with those of

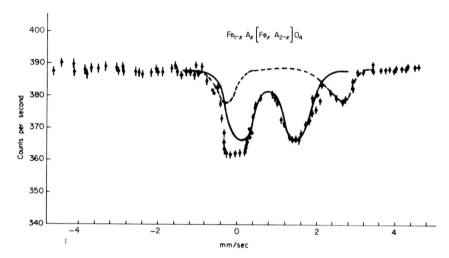

FIG. 15. Mössbauer spectrum of the partially inverse spinel $Fe_{1-x}Al_x[Fe_xAl_{2-x}]O_4$ at 25°C, taken with a $^{57}Co(Cu)$ source. The spectrum has been analyzed into two quadrupole-split Mössbauer spectra. The outer pair of lines (dashed curve) is attributed to the Fe^{2+} ions at the octahedral sites. The inner pair of lines is due to the Fe^{2+} ions at the tetrahedral sites. The experimental points have been shown with error bars (90).

the naturally occurring silicate minerals olivine and orthopyroxene have been used to determine the ratio of reduced iron to oxidized iron and have thus helped in the classification of these meteorites (78).

One of the most striking applications of Mössbauer spectroscopy is the evidence it has provided for the hitherto unknown $XeCl_4$ and its structural relationship to the square planar XeF_4 and ICl_4^- (68). Earlier work (69) had shown that it was possible to prepare by β decay ($^{127}I \rightarrow {}^{129}Xe$) xenon compounds with the same structure as that of the parent iodine compound. Using a source of K $^{129}ICl_4 \cdot H_2O$ and an absorber of 26.4% ^{129}Xe as hydroquinone clathrate, both at 4.2°K, Perlow and Perlow (68) obtained a Mössbauer spectrum of $XeCl_4$ as shown in Fig. 16. The Mössbauer spectrum of an absorber of square planar XeF_4 obtained with an unsplit K $^{129}IO_4$ source is also shown for comparison. These studies show that $XeCl_4$ is stable for a period longer than the lifetime of the 40-keV excited state in ^{129}Xe. Studies on aligned crystals and randomly oriented microcrystalline samples established that in the formation of $XeCl_4$ (a) the square planar structure of ICl_4^- and the orientation of its plane are preserved, (b) there is no appreciable yield of either $XeCl_2$ or Xe, (c) quadrupole moment of ^{129}Xe is negative, and (d) the asymmetry in the intensities of the two wings of the spectra arises as a result of accidental orientation of the needlelike crystals in the K $^{129}ICl_4H_2O$ source.

Mössbauer spectra have also been used to give information on the chemical

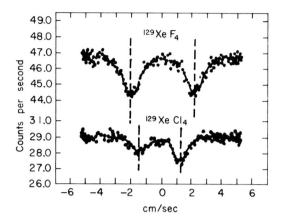

FIG. 16. Mössbauer spectrum of XeCl₄ obtained by taking a source of K^{129}ICl₄·H₂O and a single-line absorber of 26.2% ^{129}Xe as hydroquinone clathrate, both at 4.2°K (68). The Mössbauer spectrum of square planar XeF₄ is also shown for comparison.

state of an atom following the radioactive decay of its nucleus. It is known that in nuclear transitions, especially those by electron capture or isomeric transition, highly oxidized states (76) of an atom are produced as a result of the secondary Auger processes (5). These highly charged chemical states relax to their stable valency state in a very short time ($< 10^{-12}$ sec) by the capture of the electrons from the neighboring atoms, as indeed happens in metallic lattices. Soon after the discovery of the Mössbauer effect it was realized that it may be possible to detect by Mössbauer spectroscopy these highly charged metastable states if the radioactive parent is allowed to decay in a dielectric lattice in which their lifetimes may become comparable to the Mössbauer transition time. Theoretical calculations by Pollak (70) indicated that in the decay of $^{57}Co^{2+}$, iron atoms with valencies up to $+7$ are likely to be produced as the result of the Auger effect. Wertheim (84) was the first to report the existence of Fe^{3+} in addition to the expected Fe^{2+} in the electron-capture decay of $^{57}Co^{2+}$ in CoO. This original observation was followed by a number of other Mössbauer-effect measurements (6, 85), which produced evidence for the formation of Fe^{3+} and Fe^{2+} and possibly Fe^{4+} in one case (43). Clear evidence for the presence of Fe^{4+} in addition to Fe^{3+} and Fe^{2+} is provided by a recent paper (44). Figure 17 shows the Mössbauer spectrum taken with a source of $^{57}Co[Cr_2]O_4$ matched with a single-line $K_4[Fe(CN)_6]\cdot3H_2O$ absorber. The spectrum can be analyzed to consist of three unsplit resonance spectra with isomeric shifts characteristic of Fe^{4+}, Fe^{3+}, and Fe^{2+} in the spin-free state (44). The decisive coincidence Mössbauer experiments by Triftshauser and Craig (79) have established that these highly charged states are produced rapidly (10^{-15}–10^{-9} sec) and stabilized in the source

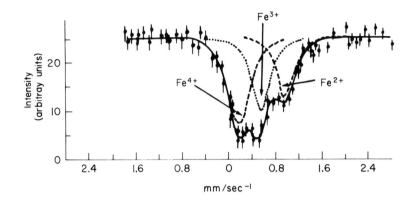

FIG. 17. Mössbauer spectrum obtained with a $^{57}Co[Cr_2]O_4$ source at $-80°C$, matched against a $K_4Fe(CN)_6\cdot 3H_2O$ absorber at 25°C. A synthetic curve composed of three single-line spectra with isomer shifts characteristic of spin-free Fe^{4+}, Fe^{3+}, and Fe^{2+} is denoted by a continuous line superposed over the experimental points, shown with error bars (44).

lattice (probably because of nonstoichiometry) for periods that are long as compared to the arbitrary time scale of the mean life (10^{-7} sec) of the 14.4-keV Mössbauer transition in ^{57}Fe.

If the highly ionized atom produced as a result of the Auger cascade is part of a molecule, it should tend to distribute its excess charge to other atoms of the molecule, since the time of Auger cascade is much smaller than the period of vibration of the atoms in a molecule. The distribution of positive charges throughout the molecule will then produce a strong Coulombic repulsion between different atoms of the molecule, leading to its virtual decomposition (86, 87). It has recently been reported that the Mössbauer spectrum of $^{57}Co(III)(1-10\ phenanthroline)_3(ClO_4)_3\cdot 2H_2O$, matched against a $K_4[Fe(CN)_6]\cdot 3H_2O$ absorber, can be resolved into two quadrupole-split spectra characteristic of low-spin Fe^{III}† and high-spin Fe^{2+}. The presence of high-spin Fe^{2+}, one oxidation state lower than the expected low-spin Fe^{III}, has been taken as an evidence for the disintegration of the $Co(III)(1-10\ phenanthroline)_3(ClO_4)_3\cdot 2H_2O$ molecule (45). The decomposition of the complex is likely to produce a number of species, including Fe^{3+}, which can easily trap an electron and be reduced to Fe^{2+}, since a large number of electrons will be produced in the Auger cascade and the subsequent fragmentation of the molecule.

Mössbauer spectra can also be used to investigate the magnetic state of a material. In the absence of an external magnetic field a randomly oriented sample of ferromagnetic, antiferromagnetic, or ferrimagnetic material

† Spin-paired iron is written as Fe^{II} and Fe^{III} and spin free iron as Fe^{2+} and Fe^{3+}.

containing iron shows a magnetic hyperfine splitting. In a thin absorber the intensities of the six lines as defined in Fig. 10 have the ratio $3:2:1:1:2:3$ (25, 83). If an external magnetic field is applied parallel to the direction of propagation of gamma rays in a ferromagnetic material, the atomic spins align parallel to the external magnetic field, and as a result of the angular dependence (83) of the individual lines (see Fig. 10), the ratio of the intensities of the six lines at magnetic saturation becomes $3:0:1:1:0:3$. For the same reasons, the ratio of the intensities changes to $3:4:1:1:4:3$ when the external magnetic field is adjusted perpendicular to the direction of propagation of the gamma rays (35, 71). In an antiferromagnetic material there is no net magnetic moment per unit cell to interact with the external magnetic field. Hence in a polycrystalline antiferromagnetic material the external magnetic field brings about only a general broadening of the six resonance lines, without changing the center of gravity of the individual lines or their relative intensities (31). The situation in the case of a ferrimagnetic material is more complicated. The presence of a net magnetic moment per unit cell and the presence of iron at two inequivalent sublattices is expected, in an external magnetic field, to give rise to two sets of Mössbauer spectra, each consisting of six lines. The atomic spins of the sublattice with the larger magnetic moment align parallel, while those of the other sublattice align antiparallel to the external magnetic field. The external magnetic field has the same effect on the relative intensities of the lines in the two sets of spectra as in the case of the ferromagnetic material, discussed earlier. For example, the Mössbauer spectrum of ferrimagnetic $Ga_{0.8}Fe_{1.2}O_3$ is a six-line spectrum with line shapes that indicate the presence of at least two hyperfine spectra (24). On the application of an external magnetic field the spectrum changes dramatically. As the magnitude of the external magnetic field is increased, the intensities of the lines 2 and 5 corresponding to $\Delta m = 0$ (see Fig. 10) decrease and the outer lines corresponding to $\Delta m = \pm 1$ split into two doublets of unequal intensities. The Mössbauer spectrum of ferrimagnetic $Ga_{0.8}Fe_{1.2}O_3$ was thus shown to consist of two distinguishable hyperfine split spectra corresponding to two inequivalent sublattices containing ^{57}Fe.

Mössbauer spectroscopy can also be used in determining not only the magnitude (3, 57) [Eq. (26)] but also the direction of the effective magnetic field as well as the sign of the EFG. The Mössbauer spectrum of the ferrimagnetic inverse spinel $Fe^{3+}[Cr^{3+}Ni^{2+}]O_4$ above its Curie temperature of 325°C has a symmetrical doublet, indicating the existence of quadrupole splitting and hence a finite EFG at the A sites. This is expected, as explained earlier, because of the asymmetric charge distribution on the twelve B sites that surround each A site in the spinel structure. If, however, the Mössbauer spectrum is taken in the ferrimagnetic state, a symmetrical hyperfine split spectrum is obtained with no apparent existence of quadrupole interaction (see Fig. 18). If we recall that the quadrupole splitting is given by Eq. (23) in the presence of an axially symmetric EFG tensor with its symmetry axis inclined at angle θ to the magnetic axis, the quadrupole splitting can be absent in a magnetic hyperfine split spectrum in spite of the presence of a finite EFG if $(3 \cos^2 \theta - 1) = 0$—i.e., if the magnetic field H is inclined at an angle $\theta = \cos^{-1} \sqrt{\frac{1}{3}}$ to the symmetry axis of the EFG (56). This appears to be the general property of the spinel lattice (56).

Mössbauer-effect measurements also provide a simple method of determining the magnetic transition temperature (15, 28, 29, 47). An appropriate Mössbauer source is so chosen that maximum resonance absorption in the paramagnetic state occurs at or near zero Doppler velocity. A plot of the intensity of the gamma ray transmitted through or scattered by the absorber at zero Doppler velocity as a function of temperature will show a break in the curve at the magnetic transition temperature. Figure 19 shows that the magnetic transition temperature of $328 \pm 3°C$ determined for the spinel $Fe^{3+}[Ni^{2+}Cr^{3+}]O_4$ from Mössbauer experiments is in good agreement with

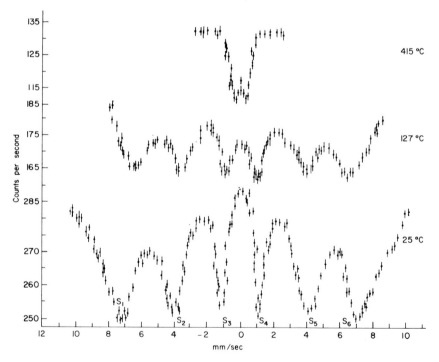

FIG. 18. Mössbauer spectra of a ferrimagnetic $Fe^{3+}[Cr^{3+}Ni^{2+}]O_4$ absorber at temperatures above and below the Curie point (325°C). $^{57}Co/Cu$ source at 25°C (56).

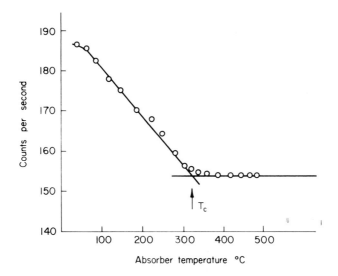

FIG. 19. A plot of the count rate at zero Doppler velocity versus the temperature of $Fe^{3+}[Cr^{3+}Ni^{2+}]O_4$ absorber; $^{57}Co(Cu)$ source at 25°C. The paramagnetic-to-ferrimagnetic transition is indicated by a break in the curve at 328 ± 3°C.

the reported Curie temperature of 325°C (56). This method has an advantage over the conventional techniques, since no external magnetic field is required in the measurement and the usual problem of extrapolation to zero magnetic field is avoided.

The magnetic hyperfine splitting of a Mössbauer spectrum, however, does not always require a macroscopic magnetization or an ordered magnetic state. If the atomic-spin relaxation time in a paramagnetic material is longer than the reciprocal of the Larmor precession frequency of the nucleus, each nucleus sees an internal magnetic field of its own electrons and hence a complete magnetic hyperfine splitting is observed. If, on the other hand, the atomic-spin relaxation rate is larger than the Larmor frequency of the nuclear spin, the nucleus sees only a fluctuating internal field that averages to zero. In such a situation the Mössbauer spectrum shows no magnetic hyperfine splitting. The atomic-spin relaxation time depends upon the spin–spin and the spin–lattice interactions. The spin–spin relaxation time can be increased by diluting a paramagnetic ion in a diamagnetic lattice, and the spin–lattice relaxation time can be increased by cooling the material. In addition, spherically symmetric ions such as Fe^{3+} are likely to have particularly long spin–lattice relaxation times because of their weak spin–lattice coupling. Wertheim was the first to report on the observation of the hyperfine splitting of ^{57}Fe in the paramagnetic state from his Mössbauer spectroscopic studies of a dilute solution of Fe_2O_3 in Al_2O_3 (*83*). Thus Mössbauer spectroscopic studies can be used to estimate the spin relaxation times without an external magnetic field. The effects of long spin relaxation times are particularly noticeable at low temperatures in biologically important macromolecules containing iron. In these large molecules the Fe–Fe distances are very large, so that the spin–spin interactions are almost absent and the spin–lattice interactions are reduced to a minimum at low temperatures. The spin relaxation time in Ferrichrome A has thus been found to have a value of 0.12 ± 0.02 μ sec at 2.5°K.

VI. Concluding Remarks

We have attempted to illustrate the applications of Mössbauer spectroscopy in inorganic chemistry by means of a few selected examples from the recent literature on Mössbauer spectroscopic studies on the nature of chemical bonding, molecular structure and symmetry, chemical effects of nuclear transformation, and the magnetic properties. In a chapter of this size it is difficult to deal with several other interesting applications of Mössbauer spectroscopy, such as the effect of pressure on chemical bonding, the kinetics and mechanism of solid-state reactions, structure of adsorbed layers and nature of adsorption processes, occurrence of low-spin ⇌ high-spin equilibria, and several other properties. Many of these have been discussed in a recent book by Goldanskii and Herber (*27*).

In the ten years since the original report of Mössbauer on the recoiless emission and absorption of gamma rays in solids, Mössbauer spectroscopy has rapidly advanced from a technique-oriented to a problem-oriented discipline. It is now possible by means of Mössbauer spectroscopy not only to corroborate physical evidence obtained by other standard physical methods, such as x-ray diffraction, NMR, and infrared, but also to obtain new kinds of structural information. For example, Mössbauer spectroscopy cannot solve a crystal structure as can x-ray diffraction, but can give information on the local symmetry of an atomic site in a crystal lattice or detect early stages of a phase transformation that cannot be detected by x-ray diffraction—by x-ray diffraction we can examine only the effects of long-range order. Mössbauer studies also provide a sensitive scale (10^{-7} sec for ^{57}Fe) for measurement of reaction rates not normally available in other techniques. Although one can determine the magnitude of the internal magnetic field by NMR with a precision greater than that obtainable in Mössbauer spectroscopy, yet it is difficult to detect the resonance effect by NMR without a prior knowledge of the order of magnitude of the internal magnetic field. In many cases the magnetic resonance effect is too broad to be observed by NMR. On the other hand, Mössbauer spectroscopy offers a very simple and foolproof method of measuring the magnitude of the internal magnetic field. Although the Mössbauer effect has been actually observed in the nuclides of a number of elements (Fig. 3), yet applications to inorganic chemistry are essentially limited to iron and tin. This is because Mössbauer spectra of these elements are easily obtained at room temperature, while for other elements, such as iodine, gold, and the rare-earth elements, one invariably needs liquid-nitrogen or liquid-helium temperatures. This difficulty can, however, be overcome to a large extent by developing new methods of preparation of Mössbauer sources so that one may obtain a reasonable resonance absorption at room temperature for other nuclides as well. For example, until 1965 no suitable source matrix for ^{119m}Sn was known; the best available Mössbauer source up to that time was in the form of $^{119m}SnO_2$, which gave a rather large linewidth of 1.4 mm/sec. Now, with a Ba $^{119}SnO_3$ source, we can obtain 50% more resonance absorption and yet obtain a linewidth of 0.8 mm/sec, close to the theoretical value of 0.69 mm/sec. The availability of reliable commercial Mössbauer spectrometers in the last few years, particularly from the U.S.A. and Israel, has been of great help to chemists desirous of using Mössbauer spectroscopy but not having facilities for building a reliable Mössbauer spectrometer.

References

1. Abragam, A., and Boutron, F., *Compt. Rend.* **252**, 2404 (1961).
2. Barrett, P. H., and Grodzins, L., *Rev. Sci. Instr.* **36**, 1607 (1965).
3. Bauminger, B., Cohen, S. G., Marinov, A., and Ofer, S., *Phys. Rev.* **122**, 318 (1961).
4. Bearden, A. J., (ed.), Proc. Intern. Conf. Mössbauer Effect, 3rd, *Rev. Mod. Phys.* **36**, 333 (1964).
5. Bergstrom, I., *in* "Beta and Gamma Ray Spectroscopy" (K. Siegbahn, ed.), p. 624. North Holland Publ., Amsterdam, 1955.
6. Bhide, V. G., and Shenoy, G. K., *Phys. Rev.* **143**, 309 (1966).
7. Bhide, V. G., Shenoy, G. K., and Multani, M. S., *Solid State Commun.* **2**, 221 (1964).
8. Bocquet, J. P., Chu, Y. Y., Kistner, O. C., Perlman, M. L., and Emery, G. T., *Phys. Rev. Letters* **17**, 809 (1966).
9. Boyle, A. J. F., and Hall, H. E., *Rept. Progr. Phys.* **25**, 441 (1962).
10. Boyle, A. J. F., Bunbury, D. St., and Edwards, C. E., *Proc. Phys. Soc.* (*London*) **77**, 1062 (1961b).
11. Breit, G., and Wigner, E., *Phys. Rev.* **49**, 519 (1936).
12. Collins, R. L., *J. Chem. Phys.* **42**, 1072 (1965).
13. Compton, D. M. J., and Schoen, A. H., (eds.), *Proc. Intern. Conf. Mössbauer Effect, 2nd,* Wiley, New York, 1962.
14. Cordey-Hayes, M., Report of a Panel on Application of the Mössbauer Effect in Chemistry and Solid State Physics, p. 156. Tech. Rept. Ser., No. 50, International Atomic Energy Agency, Vienna, 1966.

15. Craig, P. P., and Steyert, W. A., *Phys. Rev. Letters* **13**, 802 (1964).
16. Dahl, L. F., and Blount, J. F., *Inorg. Chem.* **4**, 1373 (1965).
17. Danon, J., *J. Chem. Phys.* **41**, 3378 (1964).
18. Danon, J., Report of a Panel on Application of the Mössbauer Effect in Chemistry and Solid State Physics, p. 89. Tech. Rept. Ser. No. 50, International Atomic Energy Agency, Vienna (1966).
19. DeBenedetti, S., Fernando, DeS. B., and Hoy, G. R., *Ann. Rev. Nucl. Sci.* **16**, 31 (1965).
20. Delyagin, N. N., Sphinel, V. S., Bryukhanov, V. A., and Zvenglinskii, G., *Soviet Phys. JETP (English Transl.)* **12**, 619 (1961).
21. Duncan, J. F., and Golding, R. M., *Quart. Rev.* **19**, 36 (1965).
22. Erickson, N. E., and Fairhall, A. W., *Inorg. Chem.* **4**, 1320 (1965).
23. Flinn, P. A., *Rev. Sci. Instr.* **34**, 1422 (1963).
24. Frankel, R. B., Blum, N. A., Foner, S., Freeman, A. J., and Schieber, M., *Phys. Rev. Letters* **15**, 958 (1965).
25. Frauenfelder, H., "The Mössbauer Effect." Benjamin Press, New York, 1962.
26. Goldanskii, V. I., "The Mössbauer Effect and its Applications in Chemistry," p. 5. Consultants Bureau, New York, 1964.
27. Goldanskii, V. I., and Herber, R. H. (eds.), "Chemical Applications of the Mössbauer Effect." Academic Press, New York, 1968.
28. Gonser, U., Meechan, C. J., Muir, A. H. Jr., and Wiedersich, H., *J. Appl. Phys.* **34**, 2373 (1963).
29. Grant, R. W., Wiedersich, H., and Gonser, U., *Bull. Am. Phys. Soc.* **10**, 708 (1965).
30. Grant, R. W., Kaplan, M., Keller, D. A., and Shirley, D. A., *Phys. Rev.* **A133**, 1062 (1964).
31. Grant, R. W., *in* "The Mössbauer Effect and its Application in Chemistry" (R. F. Gould, ed.), p. 34. American Chemical Society, New York, 1967.
32. Gruverman, I. J., (ed.), "Mössbauer Effect Methodology," Vol. I. Plenum Press, New York, 1965.
33. Hafemeister, D. W., DePasquali, G., and deWaard, H., *Phys. Rev.* **B135**, 1089 (1964).
34. Hafemeister, D. W., and Brooks-Shera, E., *Phys. Rev. Letters* **14**, 593 (1965).
35. Hanna, S. S., Heberle, J., Perlow, G. J., Preston, R. S., and Vincent, D. H., *Phys. Rev. Letters* **4**, 513 (1960).
36. Hay, H. J., *in* The Mössbauer Effect, *Proc. Intern. Conf. Mössbauer Effect, 2nd, Saclay, France*, September 13–15, 1961. D. M. J. Shoen and A. H. Compton, (eds). Wiley, New York, 1962.
37. Herber, R. H., *in* "Mössbauer Effect Methodology," (I. J. Gruverman, ed.), Vol. I, p. 7. Plenum Press, New York, 1965.
38. Herber, R. H., and Spijkerman, J. J., *J. Chem. Phys.* **42**, 4312 (1965).
39. Herber, R. H., Kingston, W. R., and Wertheim, G. K., *Inorg. Chem.* **2**, 153 (1963).
40. Housley, R. M., Erickson, N. E., and Dash, J. G., *Nucl. Instr. Methods.* **27**, 29 (1964).
41. Hoy, G. R., and Barros, F. de S., *Phys. Rev.* **139**, A929 (1965).
42. Ingalls, R., *Phys. Rev.* **128**, 1155 (1962).
43. Ingalls, R., and DePasquali, G., *Phys. Letters* **15**, 262 (1965).
44. Jagannathan, R., and Mathur, H. B., *J. Inorg. Nucl. Chem.* **31**, 3363, (1969).
45. Jagannathan, R., and Mathur, H. B., *Inorg. Nucl. Chem. Letters* **5**, 89 (1969).
46. Johnson, C. E., Ridout, M. S., Cranshaw, T. E., and Madsen, P. E., *Phys. Rev. Letters* **6**, 450 (1961).
47. Johnson, C. E., *Proc. Phys. Soc.* **88**, 943 (1966).
48. Kankeleit, E., *in* "Mössbauer Effect Methodology" (I. J. Gruverman, ed.), Vol. I. Plenum Press, New York, 1965.
49. Kerler, W., Neuwirth, W., Fluck, E., Kuhn, P., and Zimmermann, B., *Z. Physik* **173**, 321 (1963).

50. Kopfermann, H., "Nuclear Moments." Academic Press, New York, 1958.
51. Kuhn, W., *Phil. Mag.* **8**, 625 (1929).
52. Lee, Y. K., Keaton, Jr. P. W., Ritter, E. T., and Walker, J. C., *Phys. Rev. Letters* **14**, 957 (1965).
53. Major, K. K., *in* "Proc. Intern. Conf. Mössbauer Effect, 2nd," (D. M. J. Compton and A. H. Schoen, eds.), p. 242. Wiley, New York, 1962.
54. Margulies, S., and Ehrman, J. R., *Nucl. Instr. Methods* **12**, 131 (1961).
55. Marzolf, J. G., Dehn, J. T., and Salmon, J. F., *in* "The Mössbauer Effect and its Application in Chemistry" (R. F. Gould, ed.), p. 61. American Chemical Society, New York, 1967.
56. Mathur, H. B., Sinha, A. P. B., and Yagnik, C. M., *Indian J. Pure Appl. Phys.* **5**, 155 (1967).
57. Meyer-Schutzmeister, L., Preston, R. S., and Hanna, S. S., *Phys. Rev.* **122**, 1717 (1961).
58. Miller, C. A., and Broadhurst, J. H., *Nucl. Instr. Methods* **36**, 283 (1965).
59. Mizoguchi, T., and Tanaka, M., *J. Phys. Soc. Japan* **18**, 1301 (1963).
60. Morrison, R. J., Atac, M., Debrunner, P., and Frauenfelder, H., *Phys. Letters* **12**, 35 (1964).
61. Mössbauer, R. L., *Z. Physik* **152**, 124 (1958).
62. Mössbauer, R. L., *Z. Naturforsch.* **14a**, 211 (1959).
63. Mössbauer, R. L., *Ann. Rev. Nucl. Sci.* **12**, 123 (1962).
64. Muir, A. H., Jr., Andoo, K. J., and Coogan, H. M., "Mössbauer Effect Data Index 1958–65." Wiley (Interscience), New York, 1966.
65. Nicol, M., and Jura, G., *Science* **141**, 1035 (1963).
66. Ono, K., Ito, A., and Hirahara, E., *J. Phys. Soc. Japan* **17**, 1615 (1962).
67. Pasternak, M., *Proc. Symp. Faraday Soc. Mössbauer Effect* p. 119. The Faraday Society, London, 1967.
68. Perlow, G. J., and Perlow, M. R., *J. Chem. Phys.* **41**, 1157 (1964).
69. Perlow, G. J., and Perlow, M. R., *Rev. Mod. Phys.* **36**, 353 (1964).
70. Pollak, H., *Phys. Status Solidi* **2**, 720 (1962).
71. Preston, R. S., Hanna, S. S., and Heberle, J., *Phys. Rev.* **128**, 2207 (1962).
72. Reddy, K. R., Barros, F. de S., and DeBenedetti, S., *Phys. Letters* **20**, 297 (1966).
73. Ruby, S. L., and Holland, R. E., *Phys. Letters* **14**, 591 (1965).
74. Seyboth, D., Obenshain, F. E., and Czjzek, G., *Phys. Rev. Letters* **14**, 954 (1965).
75. Shulman, R. G., and Sugano, S., *J. Chem. Phys.* **42**, 39 (1965).
76. Snell, A. H., Pleasonton, F., and Need, J. L., *Phys. Rev.* **116**, 1548 (1959).
77. Spijkerman, J. J., Ruegg, F. C., and DeVoe, J. R., *in* "Application of the Mössbauer Effect in Chemistry and Solid State Physics," Tech. Rept. Ser. No. 50, p. 53. International Atomic Energy Agency, Vienna, 1966.
78. Sprenkel-Segal, E. L., and Hanna, S. S., *Geochim. Cosmochim. Acta* **28**, 1913 (1964).
79. Triftshauser, W., and Craig, P. P., *Phys. Rev.* **162**, 274 (1967).
80. Watson, R. E., MIT Tech. Rept. No. 12 (1959).
81. Watson, R. E., *Phys. Rev.* **119**, 1934 (1960).
82. Walker, L. R., Wertheim, G. K., and Jaccarino, V., *Phys. Rev. Letters* **6**, 98 (1961).
83. Wertheim, G. K., "Mössbauer Effect: Principles and Applications." Academic Press, New York, 1964.
84. Wertheim, G. K., *Phys. Rev.* **124**, 764 (1961).
85. Wertheim, G. K., and Guggenheim, H. J., *J. Chem. Phys.* **42**, 3873 (1965).
86. Wexler, S., and Anderson, G. R., *J. Chem. Phys.* **33**, 850 (1960).
87. Wexler, S., *Science* **156**, 901 (1967).
88. Wood, R. W., "Physical Optics." Macmillan, New York, 1954.
89. Yagnik, C. M., and Mathur, H. B., *Indian J. Pure Appl. Phys.* **6**, 211 (1968).
90. Yagnik, C. M., and Mathur, H. B., *J. Phys.* (C) **1**, 469 (1968).

AUTHOR INDEX

Numbers in parentheses are reference numbers and indicate that an author's work is referred to although his name is not cited in the text.
Numbers in italics show the page on which the complete reference is listed.

A

Aberg, T., 223(1, 75), *244, 246*
Abouaf-Marguin, L., 93(81), *103*
Abragam, A., 250(1), *296*, 360, *379*
Abramowitz, S., 78(1), 79(2), *100*
Acquista, N., 78(1), 79(2), 89(3), *100*
Adams, A. C., 278(2), *296*
Adamson, A. W., 43, 44, 45, 54, *55, 56*
Adrian, F. J., 98, *100*
Affsprung, H. E., 153(60), *166*
Agarwala, U., 134(103), *167*
Agazzi, E. J., 233, *246*
Agenäs, L-B., 199(63), *207*
Ahearn, 174(1), *205*
Ahrland, S., *164*
Airey, P. L., 51, *55*
Akamatu, H., 138(142), *168*
Alderdice, D. S., 28
Alei, M., 264(3), *296*
Alexander, R. E., 134(2), *164*
Alford, D. O., 153(3), *164*
Allavena, M., 78(5, 30), 79(5), *100, 101*
Allen, L. C., 285(78), *298*
Allerhand, A., 150(4), 156(4), *164*
Allred, A. L., 160, *166*, 322(1), *344*
Amis, E. S., 263(58), *297*
Andersen, C. A., 242, *244*
Anderson, A. W., 53, *56*

Anderson, G. R., 374(86), *381*
Anderson, J. S., 78(6), *100*
Anderson, W. A., 294, *297*
Andoo, K, J., 354(64), 356(64), 359(64), *381*
Andrew, E. R., 315(2), 321(2), *344*
Andrews, L., 78(10), 87, 88, 89(8, 9, 32), 97 (8), *100*
Andrews, L. J., 109(5), 119(5), 123(5), 124 (5), 131(5), 132(5), 134(5), 136(5), 137(5), *164*
Andrews, W. L. S., 86(7), 89(7), *100*
Angell, C. L., 93(119), 94(119), *105*
Appleton, A., 221, *244*
Arai, H., 138(6), *164*
Araki, M., 269(95), *298*
Ard, W., 313(59), *346*
Arefev, I. M., 153, *164*
Arkell, A., 89(11, 109), *101, 105*
Arnett, E. M., 124(8), *164*
Arnold, J., 153(9, 10), *164*
Arrhenius, G., 223, *244*
Ashbough, A. L., 135(112), *167*
Asprey, L. B., 260(88a), *298*
Aston, F. W., 174, *205*
Atac, M., 355(60), *381*
Aten, A. C., 139(11), *164*
Atkins, P. W., 259(4), 260(4), *296*
Aulinger, F., 196(3), *205*
Azumi, T., 14(42), *27*

383

SUBJECT INDEX

A

Acetylacetonates, 199, 273, 287
Acidity scales, 158
Adducts
  of acids, 153, 154
  of Lewis acids, 124, 134
  of Lewis bases, 124, 134, 154
Ag^+ complexes, 123, 134, 135, 163
$AlCl_3$ complexes, 125, 134, 135
$AlCl_nI_{4-n}^-$, 278
Allowed transitions, 8
Ammonia, 151
Ammonium ion, 154
Anthracene anion, 41
Antiferromagnetism, 375
Applications
  matrix isolation spectroscopy
    correlation of vibrational properties
      for asymmetric MX_2 species, 87
      for symmetric MX_2 species, 84–87
    determination of thermodynamic
      properties, 91
    of vibrational and geometric struc-
      ture by IR, 77–81
    electronic structure, 92–96
    future, 98–100
    geometry and vibrations of poly-
      atomic solids, 82–84
    study of ions, 90, 91
  NQR
    Group III halides, 320–322

Group IV halides, 322, 323
  hexahalides, 327–331
  minerals, 334–339
  miscellaneous halides, 331, 332
  ^{14}N resonance, 332–334
  transition metal, 323–327
  quantitative analyses, 339–344
X-ray
  analysis of plasmas, 241
  determination of bonding, 241
    of coordination, 241
  distribution of solar spectra, 241
    of stellar spectra, 241
  intensity of solar spectra, 241
    of stellar spectra, 241
  qualitative analysis, light elements,
    241
  quantitative analysis, light elements,
    241
  FeO, 241
  O, 242
  Si, 242, 244
  silicon oxynitride, 242
Aqueous solutions, 149, 264
Auger process, 373

B

^{11}B chemical shifts, 292
Basicity scales, 158
^9Be resonance, 255
Benzene complexes with acceptors, 120

401